The Immersion Method for Microbially Induced Calcite Precipitation

The Immersion Method for Microbially Induced Calcite Precipitation: Applications for Sustainability comprehensively explores the synergy between microbiology and engineering, focusing on MICP and its applications in material improvement and environmental remediation. Highlighting the immersion method, the book elucidates MICP's ability to enhance mechanical properties uniformly throughout treated materials, offering a sustainable alternative for various engineering challenges. Emphasizing its applications in civil engineering and environmental management, the book showcases MICP's efficacy in ground improvement, crack repair, and soil stabilization, alongside its potential for mitigating environmental contamination.

- Examines the potential to create environmentally sustainable and resilient construction materials.
- Discusses various applications of MICP across critical areas of civil engineering and environmental management, including ground improvement, crack repair, and remediation of contaminated environments.
- Provides a detailed examination of MICP-treated materials' long-term durability and performance under various environmental conditions.

The Immersion Method for Microbially Induced Calcite Precipitation

Applications for Sustainability

Lin Li, Shihui Liu, Kejun Wen, and Xinyu Lu

CRC Press
Taylor & Francis Group
Boca Raton London New York

CRC Press is an imprint of the
Taylor & Francis Group, an **informa** business

Designed cover image: Shutterstock

First edition published 2025
by CRC Press
2385 NW Executive Center Drive, Suite 320, Boca Raton FL 33431

and by CRC Press
4 Park Square, Milton Park, Abingdon, Oxon, OX14 4RN

CRC Press is an imprint of Taylor & Francis Group, LLC

ISBN: 978-1-032-85739-8 (hbk)
ISBN: 978-1-032-85740-4 (pbk)
ISBN: 978-1-003-51963-8 (ebk)

DOI: 10.1201/9781003519638

Typeset in Times
by Apex CoVantage, LLC

Contents

Preface

The critical role of microbially induced calcite precipitation (MICP) in sustainable engineering practices is widely recognized across diverse fields, including construction, environmental remediation, and soil stabilization. MICP technology leverages the natural mineralization processes of microorganisms to address pressing challenges, such as ground improvement, erosion resistance, and material durability. As the global demand for sustainable solutions intensifies, MICP has emerged as a promising approach to reduce environmental impact while enhancing material performance and engineering efficiency.

This book focuses on the development and application of the immersion method in MICP, which offers a controlled and efficient way to achieve uniform calcite precipitation across treated materials. The immersion method has been widely explored for its potential to produce bio-bricks, bio-beams, and other innovative construction materials with enhanced mechanical properties. The book presents a comprehensive overview of MICP technology, discussing the preparation of bacterial and cementation media, the design of molds, and the experimental methodologies employed in various applications. It also addresses critical factors affecting the MICP process, including bacterial and urease concentration, reaction time, and environmental conditions, providing valuable insights into optimizing treatment procedures.

In addition to fundamental research, this book highlights the application of MICP in fields such as erosion control, slope stabilization, and the bioremediation of industrial by-products. Through detailed experimental studies and case analyses, it explores the potential of MICP to transform traditional engineering practices into sustainable, eco-friendly solutions. While this book primarily emphasizes the material performance and engineering properties of MICP-treated samples, it does not delve into broader geotechnical stability considerations, which have been thoroughly addressed by the authors in other publications.

This book originates from several major research projects conducted by the authors (Lin Li, Kejun Wen, Shihui Liu) for the US National Science Foundation and US Department of Transportation – University Transportation Center beginning in 2011 at Jackson State University and at Tennessee State University. The findings and methodologies shared here reflect the authors' commitment to advancing the understanding and application of MICP technology. It is important to note that the views and conclusions contained in this book are those of the authors and do not necessarily represent the official policies of any affiliated organizations. The mention of specific products or trade names is for informational purposes only and does not constitute an endorsement or recommendation for their use.

Thanks are also extended to our collaborators, researchers, and students, including Dr. Farshard Amini and Dr. Yadong Li of Jackson State University; Dr. Junke Zhang and Dr. Chi Li of Inner Mongolia Institute of Technology; Dr. Qian Zhao and Dr. Mingdong Li of East China University of Technology; Stacy Holton of the US Army Engineer Research and Development Center; Charles McKenzie of the US Army Corps of Engineer's Coastal Hydraulics Laboratory; Dr. Changming Bu and Dr. Wei Huang of Chongqing University of Science and Technology; Dr. Rong Zhang, Dr. Ubani Ogbonnaya, and Dr. Junliang Tao at Arizona State University; and Yongfei Li, Yanhai Liu, Dr. Catherine Armwood, Karayan Cole, Katie Lagmago, Kang Du, Dr. Peidong Su, Alesha Jackson, Dr. Yang Li, Benjamin Douglas, Ayanna Hardy-Fuller. Our sincere appreciation is extended to all involved in helping us develop this book.

Lin Li, PhD, PE, Fellow of ASCE
Tennessee State University
Nashville, TN, USA

About the Authors

Dr. Lin Li is Professor in the Department of Civil and Architectural Engineering at Tennessee State University (TSU). He is serving as Dean of the College of Engineering at TSU. Dr. Li received his PhD in civil engineering from the University of Wisconsin–Madison. He is a PE in three states and a fellow of ASCE. He teaches geotechnical engineering courses, including foundation engineering, unsaturated soil mechanics, geoenvironmental engineering, advanced soil mechanics, and soil dynamics. His expertise is in innovative levee testing and protection, bio-mediated ground improvement, sustainable infrastructure, and geoenvironmental area. He is recognized as a leading expert in the research field. He chairs and speaks at international conferences on those subjects and has published five books and book chapters and over 150 heavily cited refereed journal papers. He has served as the principal investigator or co-principal investigator on over 30 sponsored research projects representing well over $11.5 million from the National Science Foundation, Department of Homeland Security, US Army Corps of Engineer, US Department of Transportation, US Department of Housing and Urban Development, Tennessee DOT, Mississippi DOT, Recycled Materials Resource Center under the Federal Highway Administration, and Institute for Multimodal Transportation under the Federal Transit Administration.

Dr. Shihui Liu is Assistant Professor in the Department of Civil and Architectural Engineering at Tennessee State University. Dr. Liu holds a PhD in civil engineering from Jackson State University and completed her postdoctoral research at Tennessee State University. Her research interests are ground improvement, sandy soil remediation, and bio-inspired and smart construction materials and development. Dr. Shihui Liu is an accomplished educator and researcher. With a passion for advancing knowledge and a commitment to academic excellence, Dr. Liu has made significant contributions to the field of innovative building and construction materials. Dr. Liu won three research grants during her first academic year at Tennessee State University.

Dr. Kejun Wen, serving as Assistant Professor within the Department of Civil and Environmental Engineering at JSU, is a dedicated educator and accomplished researcher. Holding a PhD in civil engineering from Jackson State University, Dr. Wen also holds a professional engineering license in the state of Texas. In her role, Dr. Wen imparts knowledge across a spectrum of civil engineering courses, catering to both undergraduate and graduate students. Her expertise spans topics ranging from foundational engineering mechanics to specialized areas, such as strength of materials, introduction to geotechnical engineering, and introduction to transportation engineering. With a comprehensive curriculum that includes water resource engineering, foundation engineering, advanced foundation engineering, and traffic engineering, Dr. Wen boasts over a decade of research experience, particularly in the development of innovative, bio-inspired construction materials and nature-based solutions for ground improvement.

Xinyu Lu is Research Associate in the Department of Civil and Architectural Engineering at Tennessee State University (TSU). She received her graduate degree in civil engineering at TSU. She previously earned a master of science in resource and environmental engineering and a bachelor of science degree in engineering management from Chongqing University of Science and Technology (CQUST), China. Her research interests include sustainable construction materials, bio-mediated ground improvement, and the recycling of agricultural and industrial waste. She is particularly focused on developing sustainable innovative materials, such as hempcrete and microbially induced carbonate precipitation (MICP) bio-bricks. Her expertise lies in blending biotechnology with civil engineering to promote sustainable construction practices, reduce carbon emissions, and enhance soil stability. Xinyu Lu is the author or co-author of eight peer-reviewed published articles, with total citation of 148, and five articles are under review. She has published extensively on topics such as rubberized concrete, recycled aggregate concrete, and bio-mediated soil improvement. She has also contributed to practical advancements, holding six invention patents related to sustainable construction technologies. Her academic excellence is reflected in her numerous awards, including the prestigious China National Scholarship (top 0.2%) for both undergraduate and graduate studies and the Academic First-Class Scholarship (awarded four times, top 1%). She also graduated with the highest honors from CQUST. Beyond her academic achievements, Xinyu actively contributes to the scholarly community as a peer reviewer for renowned journals, including the *Environmental Science and Pollution Research and Geomicrobiology Journal.* She has participated in several high-impact workshops and symposia, fostering collaborations and advancing research in sustainable civil engineering.

1 Introduction

1.1 BACKGROUND

With the growing environmental problems and energy crisis, sustainability has become a global consensus. Consequently, there has been a marked increase in research aimed at discovering or developing sustainable materials and processes. Microbially induced carbonate precipitation (MICP) technology, which is energy-saving, environmentally friendly, green, and sustainable, has attracted an extensive attention from researchers at home and abroad. MICP is a widespread form of biologically induced mineralization with applications in disciplines such as biotechnology, geotechnology, paleontology, civil engineering, and other fields.

1.1.1 MICP Mineralization Mechanism

MICP can be categorized based on various bacterial mineralization mechanisms, such as photosynthesis, sulfate reduction, denitrification, ammonification, and ureolysis. These mechanisms primarily involve the utilization of inorganic carbon, specifically CO_2, either for the generation of organic matter or for energy acquisition through organic matter decomposition. Subsequently, in the presence of environmental calcium ions (Ca^{2+}), calcium carbonate is formed.

Initially, the utilization of microorganisms was recognized as a novel approach to augment oil recovery from petroleum reservoirs (Bu et al. 2022; Fe Rris et al. 1996; Li et al. 2023). Subsequently, the technology found broader applications, including the repair of limestone buildings and porous media surfaces (Métayer-Levrel et al. 1999; Stocks Fischer et al. 1999). A subsequent study by Mitchell et al. posited the considerable potential for microbially induced techniques in geotechnical soil engineering (Mitchell and Santamarina 2005). In 2004, Whiffin (2004) proposed that MICP technology can be effectively deployed to bind loose sand particles, which can improve the strength and stiffness of sand. In 2006, bio-soils were identified as an important research topic by the US National Research Council (NRC) in the 21st century (Millennium et al. 2006). In 2007, the first international workshop on biogeotechnical engineering facilitated interdisciplinary discussions on emerging research topics within this field (DeJong et al. 2007), fueling a surge in MICP-related research both nationally and internationally.

In the process of MICP, according to the different ways of assimilation of bacterial metabolism, it can be divided into autotrophic bacteria, such as nitrifying bacteria (Yang et al. 2022) and iron bacteria, and heterotrophic bacteria, such as *Bacillus subtilis* (Chiet et al. 2016; Fazelikia et al. 2023). Autotrophic bacteria use inorganic carbon (CO_2) as carbon source to produce organics through methanogenesis or photosynthesis. When there are calcium ions in the medium, calcium carbonate crystals can be formed. Heterotrophic bacteria obtain energy by decomposing organic matter (such as urea), and heterotrophic bacteria induce calcium carbonate precipitation,

DOI: 10.1201/9781003519638-1

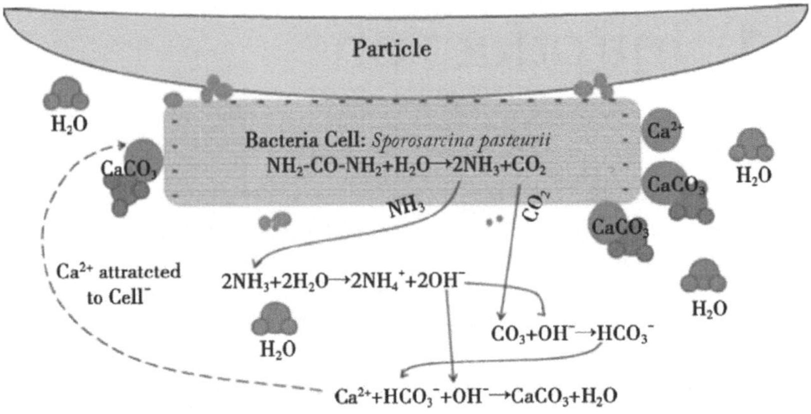

FIGURE 1.1 Calcium carbonate precipitation induced by ureolytic bacteria.

Source: DeJong et al. (2010).

including sulfur cycle and nitrogen cycle. In the sulfur cycle, sulfate reduction occurs mainly, and the nitrogen cycle includes the ammonification of amino acids, denitrification, and ureolysis (Braissant et al. 2003; Castanier et al. 1999, 2000; Jain 2024; Riding et al. 2000; Sridhar et al. 2021). Among them, ureolysis is the most effective way of MICP. It is the most widely used because of its fast cementation rate, easy control of the reaction, and high mineralization efficiency (Dhami et al. 2014). In this way, usually, microorganisms use urea as energy source to produce urease through their own metabolic activities. Urea releases NH_3 and CO_2 under the catalysis of urease, which increase the pH of the environment; microorganisms adsorb Ca^{2+} in the solution continuously, thereby causing calcium carbonate deposition. The reaction equation is shown in formulas (1.1)–(1.4), and the schematic diagram is shown in Figure 1.1.

$$CO(NH_2)_2 + H_2O \xrightarrow{\text{urease}} 2NH_3 + CO_2 \tag{1.1}$$

$$NH_3 + H_2O \leftrightarrow NH_4^+ + OH^- \tag{1.2}$$

$$CO_2 + OH^- \rightarrow HCO_3^- \tag{1.3}$$

$$Cell - Ca^{2+} + HCO_3^- + OH^- \rightarrow Cell - CaCO_3 \downarrow + H_2O \tag{1.4}$$

1.1.2 MICP Treatment Methods

The process of microbially induced calcite precipitation (MICP) requires the injection of bacterial solution and cementation solution into the sand, and MICP treatment methods greatly affect the yield and efficiency of calcium carbonate precipitation. Currently, common MICP treatment methods include surface infiltration, injection, pre-mixing, and immersion.

Surface infiltration involves spraying a bacterial solution and cementation solution onto the soil surface, allowing the solution to infiltrate through gravity and capillary action (Cheng and Cord-Ruwisch 2012). This method is simple, cost-effective, and suitable for large-scale applications, particularly for highly permeable porous materials, such as coarse sand and gravel (Cheng and Cord-Ruwisch 2014; Mahawish et al. 2017). Sun et al. (2022) used the MICP method to improve the erosion resistance of surface loess slopes. Kong et al. (2019) have demonstrated enhanced wind erosion resistance of desert aeolian sand using surface spraying. Similarly, Xie et al. (2019) improved the water stability and erosion resistance of clay by surface-spraying bacterial and cementation solutions. Liu et al. (2016) applied spraying method to reinforce the surface of dams, achieving an unconfined compressive strength of 9 MPa and a reduction in permeability coefficient by three orders of magnitude.

Injection involves injecting a bacterial solution and cementation solution into the soil to form samples. Stocks Fischer et al. (1999) injected a mixture of bacterial solution and cementation solution directly into sand, leading to rapid pore blockage near the injection points (Rahim et al. 2015; Stocks Fischer et al. 1999). Whiffin et al. (2007) introduced a two-phase injection method, first injecting bacterial solution, followed by cementation solution, to prevent blockage and ensure more uniform distribution of calcium carbonate. Harkes et al. also achieved satisfactory MICP results using a similar method (Harkes et al. 2010; Martinez et al. 2013). Sun et al. (2022) also achieved satisfactory MICP results using a similar injection method by adjusting factors like temperature, bacterial concentration, reaction time, and injection rate. This optimization improved bacterium retention and carbonate precipitation, leading to stronger bio-treated soil. Nagy and Kustermann (2023) studied the use of MICP injections to rehabilitate porous building materials, like recycled aggregate mortar and deteriorated masonry. They found that multiple pressure injections reduced porosity, increased density and weight, and lowered capillary water absorption by 48.3%. Strength tests showed a 45.16% increase in flexural strength and a 35.64% increase in compressive strength.

Pre-mixing involves pre-mixing a bacterial solution with soil for treatment, which is generally more effective for fine-grained soils. Soon et al. (2014) pre-mixed bacterial solution with soil samples and compacted them, followed by the injection of cementation solution using air pressure, resulting in a 69% increase in shear strength and a 90% decrease in hydraulic conductivity. Oliveira et al. (2016) mixed bacteria, cementation solution, and powdered soil at optimum moisture content, then layered and compacted them into PVC molds, resulting in a 40–106% increase in strength. Fu et al. (2021) uniformly mixed powdered soil with bacterial solution and cementation solution, finding that the unconfined compressive strength was 1.5 times that of untreated samples when the cementation solution concentration was 1.5 mol/L.

Immersion involves completely immersing reinforced samples in a bacterial solution, allowing calcium carbonate crystals to form through liquid penetration. This method produces uniformly distributed samples with high strength as bacterial solution and cementation solution fully contact the samples. Bu et al. (2018b) found that precipitation of $CaCO_3$ was relatively uniform using the immersion method. Zhao et al. used immersion to solidify sand using a fully contacting flexible mold, resulting in a more uniform distribution of calcium carbonate (Zhao et al. 2014a). Wen et al. (2018b) improved the engineering properties of sand through immersion, with unconfined compressive strength reaching 6.4 MPa. Li et al. (2020) used immersion to cement sand into bio-bricks, increasing compressive strength by 50–70%. Liu et al. (2019) found that multiple immersion treatments improved the durability of MICP-treated samples. Bu et al. (2018a) used the immersion method and four types of molds to treat beam-shaped samples of different sizes with MICP technology and found that bio-beams made with rigid, fully contacting molds and geotextiles exhibited the best mechanical performance. Liu and Wen found that multiple MICP immersion significantly enhanced the resistance of bio-bricks to accelerated erosion (Liu et al. 2021; Wen et al. 2018a).

1.1.3 MICP Application

Through MICP technology, the strength, stiffness, erosion resistance, permeability, and other properties of the soil can be improved. Loose sand particles can be integrated into a whole through biological cementation, and environmentally friendly cementitious materials can be generated. MICP technology also has been used in many fields widely, such as in ground improvement, crack repair, construction materials, soil stabilization, environmental remediation, etc.

1.1.3.1 Ground Improvement

In nature, there exist some poor-quality ground foundations whose bearing capacity is insufficient, posing threats to human life and property safety. Therefore, researchers have attempted to find an environmentally friendly and low-emission method to reinforce these foundations. Harkes et al. (2010) proposed a method for optimizing the fixation of bacteria in porous media, which is suitable for foundation reinforcement. Gomez et al. (2015) conducted field experiments based on MICP technology, ultimately forming a 2.5 cm thick hard cemented layer. The research indicated that MICP treatment of the field test plots effectively increased the surface strength of the test plots, rendering them more erosion-resistant. van Paassen et al. (2010b) reinforced a 100 m^3 sand base model with grouting, and the stiffness of the biogrout significantly increased after one day of treatment. Whiffin et al. (2007) conducted MICP reinforcement treatment on a 5 m long sand column, demonstrating that calcium carbonate was generated throughout the length of the sand column, resulting in significant improvements in strength and stiffness. This method can enhance the soil's bearing capacity, making it an efficient and effective reinforcement technique compared to other underground soil treatment methods. Han et al. (2024) also achieved satisfactory liquefaction resistance results using a similar MICP injection method in their centrifuge shake table tests on untreated and MICP-cemented sand models. Li et al. (2024) developed a reactive transport model to simulate soil improvement via MICP in shallow submerged conditions, incorporating bacterial attachment and fluid density. Mwandira et al. (2024) explored using fruit and vegetable (FV) waste as a low-cost bacterial growth medium for bio-cementation, an eco-friendly method to enhance geomaterial stability. Their study demonstrated effective bacterial growth, with maximum urease activity and compressive strength increases in clay and locomotive ash, indicating FV waste's potential for bio-cementation applications.

1.1.3.2 Crack Repair

Cracks are a major factor leading to the collapse and damage of buildings, with many buildings developing cracks annually, resulting in property losses and casualties. An application involving MICP, which induces the deposition of calcium carbonate by microorganisms, is arousing significant interest among scholars. This application can be utilized to repair natural stone structures. Initially observed in petroleum extraction, it was discovered that rock fractures surrounding oil reservoirs could be repaired by microorganisms. Interestingly, this microbially induced mineral bonding continues to be effective even after cell death.

Currently, there are numerous ancient buildings worldwide that have undergone extensive weathering and erosion over time, resulting in numerous cracks and localized damage. To address this issue, some researchers have employed MICP technology to protect ancient buildings. Adolphe et al. (1990) were among the first to utilize MICP technology for stone preservation as early as 1990. Métayer-Levrel et al. (1999) investigated the surface protection coatings (i.e., calcium carbonate) produced through the MICP mineralization process on limestone buildings, monuments, and statues. Utilizing MICP technology for repairing and protecting the surfaces of stone or cultural artifacts offers significant advantages. Tiano et al. (2006), based on bioremediation techniques, conducted MICP repair treatments on commemorative calcium stone sculptures, demonstrating that the proposed method had a certain reinforcing effect on calcium stone materials. Xu et al. (2024) investigated using cellulose-immobilized bacteria for self-healing concrete via microbially induced calcium carbonate precipitation (MICP). They developed microcapsules with enhanced mechanical strength that effectively immobilized bacteria, allowing for significant crack repair under water curing conditions. After 28 days, cracks up to 0.3 mm were repaired, achieving over 90% repair for smaller cracks. Anand et al. (2024) developed a silica fume–based bacterial agent for bio-cementitious grout to remediate cracks in concrete structures using microbially induced calcite precipitation (MICP). This ready-to-use grout enables field-scale applications, effectively sealing cracks through enhanced bacterial activity. Their study evaluated mechanical and watertightness properties, revealing improved pore densification and reduced water permeability in repaired concrete.

In addition to repairing ancient buildings, MICP technology can also be applied to modern building crack repairs. Wang et al. (2018b) used different repair materials to repair concrete with cracks. The research results showed that using MICP repair solution for repairs could achieve 70% of the mechanical properties of benchmark concrete. Additionally, it was found to be feasible for repairing cracks in solid concrete. Chen et al. (2019) employed two microbial repair methods, injection and immersion, to repair minor cracks in concrete. The results indicated that the reduction in permeability after injection repair was more pronounced, and the strength of the repaired specimens exceeded 90% of that of specimens without cracks. Furthermore, Wiktor et al. (2011) used porous lightweight aggregates loaded with bacteria and lactate calcium mixture. By creating cracks and immersing them in water for maintenance, it was found that the effect of microbial loading on inducing mineralization on the crack surface was more efficient. The maximum crack repair width reached 470 μm, whereas the width of cracks repaired without bacteria was only 210 μm. Luo et al. (2021)

prepared artificial functional aggregates using raw materials such as fly ash, silica fume, kaolin, limestone powder, and sodium carbonate powder to improve the self-repair efficiency of concrete. Moreover, Wang et al. (2014) encapsulated microorganisms in microcapsules to protect them from the high-alkaline environment of concrete, achieving better results in repairing concrete cracks. Jonkers (2011) found that spores and mineralization essentials of bacteria loaded in ceramic granules could prolong the effective repair action of bacteria to over six months. Xu et al. (2018) conducted crack self-repair tests using ceramic granules as bacterial carriers. The results showed that within 28 days, the maximum crack repair width reached 273 μm, the crack repair rate was 86%, the compressive strength increased by 24%, and the water absorption coefficient decreased by 27%.

1.1.3.3 Construction Materials

Due to extensive global energy usage and greenhouse gas emissions, climate change resulting from massive construction activities necessitates the development of sustainable alternative building materials. The application of novel MICP technology to cement sand into bio-bricks holds potential value, representing an innovative approach in the field of construction materials by offering a new, more sustainable, and energy-efficient solution in the form of "bio-bricks" and "bio-beams."

Bernardi et al. (2014) proposed a novel technique for manufacturing bio-bricks using a bio-induced natural cementation process, introducing bacteria and nutrients into silica-rich masonry sand. Through microbially induced carbonate precipitation, the particles are bonded together, ultimately forming sandstone materials. Results showed that the compressive strength of bio-bricks could reach 2 MPa, and longitudinal wave velocity measurements demonstrated relatively uniform stiffness in bio-bricks. Bio-bricks were found to be comparable to bricks prepared using more traditional cement and lime additives. Bu et al. (2018b) utilized Ottawa silica sand as the treatment material and produced bio-bricks via immersion, comparing them with bricks modified with lime and cement. Experimental results indicated that the flexural strength of MICP-treated bio-bricks was 950 kPa, comparable to specimens treated with 20–25% cement but significantly higher than those treated with 30% lime. The compressive strength of MICP-treated bio-bricks reached 500 kPa, equivalent to specimens treated with 30% lime. Unconfined compressive strength (UCS) tests revealed that the UCS of MICP-treated specimens was approximately 1,300 kPa, higher than that of specimens treated with 10% cement (900 kPa) and significantly higher than that of lime-treated specimens (approximately 140 kPa). Li et al. (2020) enhanced the compressive strength of bio-bricks using discrete random distribution fibers, coarse sand, clay additives, and various immersion treatments. Results showed that the compressive strength of all bio-bricks increased with the number of treatments, and the addition of synthetic and natural fibers increased the compressive strength by 50–70%. Furthermore, FTIR, XRD, and SEM results confirmed the formation of calcite and aragonite during MICP. Lambert and Randall (2019) used human urine instead of urea to manufacture bio-bricks using MICP technology, achieving a maximum compressive strength of 2.7 MPa, comparable to traditional bricks. Farajnia et al. (2022) introduced bacterial strains with high urease activity found in historical adobe structures in Iran and deserts for bio-brick production, confirming the potential of local bacteria as an economical and sustainable alternative for bio-brick manufacturing. Bio-bricks prepared from silica sand via immersion exhibited an average compressive strength and water absorption rate of 3,000 kPa and 8.5%, respectively. The bio-bricks also demonstrated good durability during wet–dry cycles, with a peak bending stress–strain curve reaching 1,300 kPa. Moreover, X-ray diffraction and SEM confirmed the formation of aragonite crystals during the microbial cementation process at the microscopic level.

Wen et al. (2018a) investigated the flexural performance of bio-beams by adding randomly distributed fibers and bamboo materials into Ottawa silica sand (containing 99.7% quartz) and subjecting them to multiple immersion-based MICP treatments. The results indicated that the addition of fibers could enhance the ductility of bio-beams, with the optimal fiber content being approximately 0.3%. Additionally, the inclusion of bamboo materials increased the flexural strength of bio-beams from 890 kPa to 1,750 kPa. Multiple MICP immersion treatments significantly improved the flexural strength of bio-beams, reaching up to 3 MPa at maximum. However, it is noteworthy that after multiple treatments, the ductility of fiber-reinforced bio-beams decreased. Bu et al. (2018c) developed an improved permeable rigid mold for preparing bio-beam specimens via immersion method and compared the effects of different MICP immersion treatments on bio-beams. The permeable mold provided support for bacterial growth by ensuring sufficient contact with the cementitious medium, inducing uniform precipitation of calcium carbonate within the sand matrix to form bio-beams. The results showed that the flexural strength of bio-beams subjected to two and three MICP treatments increased by 63% and 100%, respectively, compared to single MICP treatment. Bu et al. (2019) investigated four different types of molds, including uncovered plastic beam rigid molds, fully contacting flexible molds, geotextile rigid fully contacting molds, and rigid fully contacting molds with filter paper, for MICP treatment of beam-shaped specimens using the immersion method. The results demonstrated that the samples prepared using geotextile rigid fully contacting molds formed the optimal beam shape and the most uniform calcite precipitation, exhibiting the best mechanical performance.

1.1.3.4 Soil Stabilization

MICP technology finds its widest application in the field of soil stabilization, particularly in the reinforcement of sandy soils. Presently, due to the loose nature and good permeability of sandy soils, microbes exhibit high compatibility with sand particles, leading to successful outcomes of MICP technology in sandy soil applications.

DeJong et al. (2010) demonstrated that through biologically mediated improvement processes, a range of soil properties could be significantly altered. Experimental results indicated a decrease in permeability by 10^{-3}, a reduction in compressibility by 10^{-2}, a stiffness increase by 10^{2}, a shear strength enhancement by 10^{2}, and a transition in volume response from shrinkage to expansion. Cheng et al. (2017) investigated the effects of factors such as temperature, enzyme activity, rainwater scouring, and freeze–thaw cycles on MICP-mediated soils. Their study revealed that rainwater scouring had an inhibitory effect on bio-cementation, while MICP-treated soil exhibited good durability against freeze–thaw erosion. Salifu et al. (2016) conducted relevant research on stabilizing slopes under tidal flow cycles and mitigating slope erosion using MICP technology. Results showed that MICP-treated slopes exhibited significantly enhanced durability after 30 tidal flow cycles compared to untreated slopes, with improved stability and erosion resistance.

Liu et al. (2022) studied the influence of rainfall-induced erosion on the stability of sandy slopes treated with MICP technology. After 24 h of rainfall erosion, the slope surface treated with 0.5 M Ca MICP appeared smoother than the non-eroded surface. Li et al. (2017) investigated the bioremediation of aeolian sand using MICP, finding that the formation of biocrusts could effectively prevent desert erosion. Li et al. (2019) examined the corrosion resistance and stability of biogeomaterials in sulfate environments and the induced corrosion mechanism. The average strength of corroded samples remained at 35.7% relative to non-corroded samples after soaking in sulfate acid solution for up to 7 days and reached 57.3% after 28 days. Liu et al. (2019) investigated the long-term engineering performance of MICP-treated sandy soils under critical environmental parameters, such as dry–wet cycles, freeze–thaw cycles, and acid rain conditions, through indoor experiments. After dry–wet and freeze–thaw cycles, the UCS of multiple MICP-treated samples did not significantly decrease. For samples treated with four times MICP, the UCS remained at 51.9% after soaking in acid rain solution with a pH of 3.5 for 15 days.

1.1.3.5 Environmental Remediation

Due to rapid industrialization, heavy metal and radioactive nuclide pollutants generated by industrial activities pose significant threats to the environment due to their toxicity, non-biodegradability, and persistent accumulation. Heavy metal ion pollutants are a serious concern in the environment, and the removal of heavy metal ion pollutants from contaminated soil and wastewater is crucial. In recent years, researchers have discovered that the use of MICP technology is an effective, economical, and environmentally friendly approach with great potential in the remediation of metal-contaminated solid waste and the mechanical enhancement of materials. Its advantages include, but are not limited to, mild reaction conditions, low energy consumption, simple process, minimal environmental impact, and applicability to various residues and contaminated soils.

Warren et al. (2001) studied the microbial degradation of strontium (Sr) groundwater pollutants based on urea hydrolysis, showing that microbially induced calcium carbonate deposition, as a form of bioremediation, is effective in addressing Sr pollution issues. Arsenic is a highly toxic metal substance with significant environmental hazards. As(III) is highly mobile in soil and easily leaches into groundwater. Achal et al. (2012b) found that using MICP technology for bioremediation significantly reduced the mobility of arsenic in the soil's exchangeable fraction, while the concentration of arsenic in the carbonate fraction significantly increased. Various biomineralization products, such as calcite, vaterite, aragonite, and vanadinite, were present in the soil samples after bioremediation. Zhang et al. proposed modifying industrial by-product steel slag using MICP technology to reduce natural aggregate consumption and promote resource recycling (Zhang et al. 2021a, 2023) and introduced MICP to immobilize chromium (Cr) in stainless steel pickling sludge (SSPS) (Zhang et al. 2022). Fly ash is one of the major by-products of coal combustion, with nearly half of it being unable to be reused or recycled. Zhang et al.

used urease-producing bacteria to modify fly ash and assessed its impact on metal leaching, finding that bacterial organic products induced bioleaching and potential competition for carbonate promoted the leaching of Ca, Mg, Mn, and V, which would facilitate metal recovery (Zhang et al. 2021b). Fly ash and sand mixtures treated by bioprocessing were used to prepare building materials (Zhang et al. 2020).

1.2 CONTENTS OF THIS BOOK

Microbial mineralization represents a prevalent natural phenomenon wherein certain microorganisms utilize their metabolic activities to catalyze the formation of diverse mineral crystals. This intrinsic capability has spurred a global research trend among scholars towards the application of microbially induced calcite precipitation (MICP) technology in engineering. Given the novelty of MICP, there currently exist no standardized protocols to constrain scholarly investigations, resulting in a diverse array of MICP treatment methods, including the injection method, surface infiltration method, and pre-mixing method. The choice of MICP treatment method significantly influences its effectiveness. This book primarily focuses on elucidating the immersion method within MICP and its applications in engineering.

Each chapter delineates specific aspects of MICP, ranging from an overview of the technology to its varied applications in construction materials, erosion control, and environmental remediation. Additionally, the durability of MICP-treated materials is thoroughly investigated, with a focus on understanding the influence of key environmental parameters on long-term engineering properties. This book aims to contribute to the academic discourse surrounding MICP technology and its potential implications for sustainable engineering practices.

Chapter 1 serves as the introduction. Chapter 2 provides an overview of MICP. Chapter 3 offers a detailed examination of the immersion method in MICP, along with suitable sample preparation techniques. Chapter 4 elucidates the application of MICP in construction materials, highlighting its potential for environmentally friendly technologies, such as bio-bricks and bio-beams. Chapter 5 explores the role of MICP in erosion control, addressing its efficacy in mitigating rainfall erosion, desert sand deposition, surface erosion, and surface acid corrosion. Chapter 6 discusses the application of MICP in environmental remediation, focusing on its use in preparing construction materials from recycled industrial by-products to reduce the consumption of natural aggregates and promote resource recycling. Furthermore, it examines the utilization of MICP in treating coal fly ash and recycled steel slag for beneficial purposes, thereby mitigating environmental impacts and enhancing material recycling. Finally, Chapter 7 provides a conclusion summarizing the key findings and insights gleaned from the preceding chapters, thereby contributing to the academic discourse on MICP technology and its potential for fostering sustainable engineering practices.

2 MICP Overview

MICP has emerged as a significant technology in the domains of energy conservation, environmental protection, and sustainable ecology, witnessing rapid advancements in recent years. To comprehensively and systematically understand the knowledge structure and development context of this field, this study collected and examined a substantial amount of data. A total of 1,935 articles were retrieved from the Web of Science (WOS) Core Collection, spanning the years 2012 to 2021. Using the bibliometric visualization software CiteSpace, the collaboration networks among authors, institutions, and countries and the co-occurrence network of keywords, as well as the co-citation network of journals and literature in this field, were analyzed. The data indicates that the field of MICP is still in its emerging stage. The systematic analysis of the knowledge structure in this field aims to identify the frontier hot spots and challenges.

2.1 PUBLICATION YEARS AND JOURNALS ABOUT MICP

Figure 2.1 shows the annual publication involving MICP. It can be seen that from 2012 to 2014, there were about 100 papers published on MICP; from 2015 to 2018, the annual number of publications increased slowly. In 2019, there was an obvious increase in trend. From 2019 to 2021, MICP research developed rapidly, with 995 papers published, accounting for 51.4%.

These publications on MICP have been published in 500 journals, which means the field of research on MICP is extensive. *Construction and Building Materials* ranked the top in the number of published articles (80), followed by the *Geomicrobiology Journal* (65) and the journal *Palaeogeography, Palaeoclimatology, Palaeoecology* (49). According to published journals, MICP technology is mainly used in civil engineering, architecture, geology, and environmental fields. Table 2.1 lists the ten most fruitful journals; these journals are worth referring to by researchers.

2.2 COOPERATIVE NETWORK ANALYSIS

Cooperative network analysis includes macro co-country, medium co-institution, and micro co-author. In the cooperation network, the node size indicates the number of published articles; the thickness of the connection line indicates the density of cooperation between authors, institutions, or countries; and the change of color from dark to bright symbolizes the publication time of the article from far to near. Centrality quantifies the significance of nodes, which act as a bridge in the research field. In all the figures in this section, these pivotal nodes are depicted as purple circles.

2.2.1 Co-Country

Geographic distribution analysis of MICP studies was conducted using CiteSpace, with the elimination of duplicate entries. The findings reveal that over the past

DOI: 10.1201/9781003519638-2

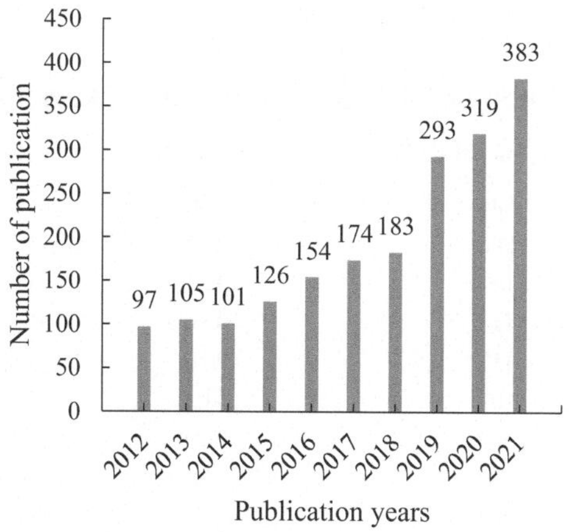

FIGURE 2.1 Annual publication trends in MICP research within the WOS database.

TABLE 2.1
Top 10 Productive Journals in MICP Research

Number	Name of Journal	Number of Published Papers
1	*Construction and Building Materials*	80
2	*Geomicrobiology Journal*	65
3	*Palaeogeography, Palaeoclimatology, Palaeoecology*	49
4	*Journal of Geotechnical and Geoenvironmental Engineering*	43
5	*Sedimentology*	42
6	*Sedimentary Geology*	37
7	*Geobiology*	34
8	*Scientific Reports*	34
9	*Frontiers in Microbiology*	28
10	*Journal of Materials in Civil Engineering*	28

decade, MICP research has been undertaken in 91 countries or regions worldwide. China occupies a leading position in this domain, contributing 587 publications, which account for 30.0% of the total scholarly output. The United States follows closely, with 506 publications, representing 26.1% of the total. Other significant contributors include Australia, Germany, England, India, France, and others. The top 15 countries or regions, along with their respective publication volume, are detailed in Table 2.2.

Figure 2.2 presents a visual representation of the national cooperation network as analyzed through CiteSpace. Over the past decade, the development of MICP research in several countries with significant influence has exhibited relative stability, and their publication output has steadily increased. In the figure, countries such as China, USA, Australia, Germany, France, and Japan have high centrality,

TABLE 2.2
Geographic Distribution of MICP Research (Top 15 Regions/Countries)

Number	Number of Published Papers	Centrality	Country	Percentage
1	587	0.30	China	30.3%
2	506	0.32	USA	26.1%
3	141	0.14	Australia	7.3%
4	118	0.16	Germany	6.1%
5	112	0.10	England	5.8%
6	102	0.03	India	5.3%
7	100	0.29	France	5.2%
8	89	0.04	Canada	4.6%
9	82	0.16	Japan	4.2%
10	77	0.01	South Korea	4.0%
11	74	0.07	Italy	3.8%
12	73	0.09	Spain	3.8%
13	65	0.01	Iran	3.4%
14	53	0.04	Singapore	2.7%
15	50	0.02	Switzerland	2.6%

Note: Percentage refers to the percentage of the top 15 countries in the total number of literatures (1,935) over the past decade.

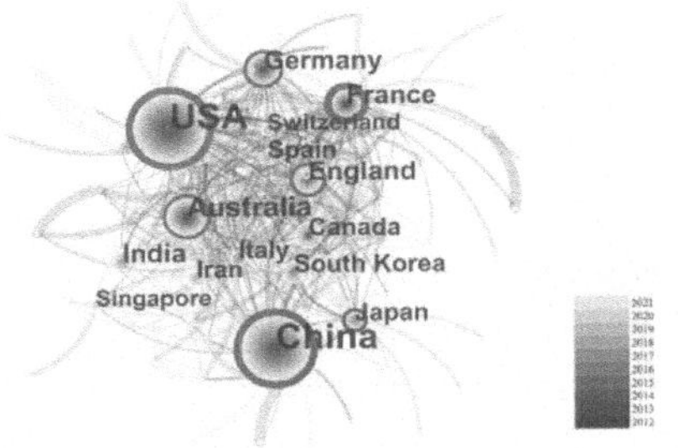

FIGURE 2.2 Map of national cooperation network.

indicating frequent collaborative efforts among research institutions within these countries. Notably, France's centrality is particularly noteworthy. Although it ranks seventh globally in terms of publication volume, its high centrality suggests that its research findings are of substantial value and attract the attention of scholars worldwide. Meanwhile, although China leads in the number of published papers, its centrality ranking of second (0.30) indicates that there is still room for further expansion of China's influence in this evolving research field.

2.2.2 Co-Institution

Figure 2.3 presents the institutional cooperation network map, which has 368 nodes and 610 links, with a threshold of 20. The visualization results derived from an analysis of 1,935 publications provide insights into the principal research institutions active in this field over the past decade. The interconnections between nodes offer a clear representation of the extent of collaborative relationships among these research organizations.

Prominent research institutions in the MICP field include the Chinese Academy of Sciences, Southeast University, and Nanyang Technological University. The Chinese Academy of Sciences, with 121 publications and a centrality score of 0.46, leads the research landscape. Their research focuses on bioremediation, MICP soil reinforcement, the effects of environmental changes on MICP (Achal et al. 2012a, 2012c; Chen et al. 2014b; Kumari et al. 2014, 2016; Wang et al. 2018a), etc. Southeast University focuses on contaminant immobilization (Jiang et al. 2019), bio-cement (Qian et al. 2018; Sun et al. 2018; Yu et al. 2016), self-healing of cement-based materials (Ling et al. 2017), etc. Nanyang Technological University mainly studies biocementation (Cheng et al. 2018a; Choi et al. 2016a, 2017a, 2017b), soil improvement (Chu et al. 2013), MICP improvement of recycled aggregate (Qiu et al. 2014), etc. It is noteworthy that the Chinese Academy of Sciences maintains relatively loose connections with both Southeast University and Nanyang Technological University. However, each of these institutions has established robust collaborative networks with other top-tier research institutions.

The analysis of research institutions reveals that universities play a central role in MICP research. While the Chinese Academy of Sciences leads in terms of publications, it is noteworthy that 9 out of the top 10 institutions are universities, contributing to 73.9% of the top 10 institutions' publications. The color distribution in

FIGURE 2.3 Institutional collaboration network mapping in MICP research.

Figure 2.3 indicates increasing global attention to MICP research, particularly in recent years. In the figure, large nodes cluster together, while small nodes are more scattered, indicating that the cooperation between various countries and institutions needs to be strengthened.

2.2.3 Co-Author

Author's cooperation analysis can identify the core authors in a certain field, and the core authors are the backbone to enhance the academic influence of the journal. Figure 2.4 is the map of the author's cooperation network, which has 437 nodes and 632 links, with a threshold set at 10. Nodes highlighted with red rings signify instances of citation bursts. According to the statistics of CiteSpace software, there are 22 core authors with ten or more published papers, and a total of 374 papers have been published, accounting for 19.3% of the total number of papers. Within this network, four significantly larger nodes stand out prominently, representing the scholars Jian Chu, Satoru Kawasaki, Varenyam Achal, and Kazunori Nakashima, and there is a certain cooperative relationship between them. Notably, Satoru Kawasaki and Kazunori Nakashima wield substantial academic influence, as evidenced by their prominently featured nodes with citation bursts, signifying their high impact and underscoring their significance in the field. Additionally, the dense connection between these two nodes indicates a strong collaborative relationship between these two authors. From their published papers, it is found that their research directions are also very similar. Specifically, Satoru Kawasaki's publications span from 2015 to 2021, while Kazunori Nakashima's work covers the period from 2017 to 2021.

According to CiteSpace analysis, among the top 10 most cited articles published by these two authors, they are co-authors of 5 articles (Dilrukshi et al. 2018; Fujita et al. 2017; Gowthaman et al. 2019b, 2020; Mwandira et al. 2017). They are committed to the study of using MICP to improve the strength and other properties

FIGURE 2.4 Author collaboration network mapping in MICP research.

of sand or soil (Dilrukshi et al. 2018; Gowthaman et al. 2019b, 2020; Khan et al. 2015; Li et al. 2018b; Mwandira et al. 2017). They also explored the feasibility of using different microbial species, like *Pararhodobacter* sp. (Mwandira et al. 2017); *Bacillus megaterium* (Li et al. 2015d); *Pararhodobacter* sp. SW, a ureolytic bacteria (Dilrukshi et al. 2018); locally isolated bacteria, in Japan (Gowthaman et al. 2019b); and the novel indigenous ureolytic bacteria isolated from the cold subarctic region (Gowthaman et al. 2019a), to reinforce sand or soil with MICP. Jian Chu is devoted to the research of bio-cementation for sand or soil (Cheng et al. 2018a; Choi et al. 2016a, 2016b, 2017a, 2017b; Xiao et al. 2019), while Varenyam Achal mainly studies the bioremediation of soil contaminated by heavy metals (Achal et al. 2012b, 2012c, 2013, 2014; Kumari et al. 2016). Some small groups are not represented on this map, such as L. C. Miao, X. H. Sun, and H. X. Wang forming a group of authors. In addition, there are also some small groups with only one connection, which is limited to two people's cooperation. The network has not been launched, and the cooperation force was weak.

2.3 CO-OCCURRENCE ANALYSIS OF KEYWORDS

In the keyword co-occurrence network, the node size indicates the occurrence frequency of keyword. The node connection indicates that these keywords appear in the same article or multiple articles. The more times they appear at the same time, the thicker the connection, indicating the strength of the relationship between these keywords. The color shift from dark to bright signifies the chronological order of keyword appearances, with darker colors representing earlier appearances and brighter colors indicating more recent ones. Red circles represent keywords with citation bursts, indicating that these specific keywords garnered significant attention within a relatively short timeframe.

2.3.1 Research Topics

The analysis of keywords provides a rapid overview of research topics, encapsulating the fundamental content of a research paper (Zhu et al. 2017). It aids researchers in promptly identifying the prevailing themes within a particular field. Utilizing CiteSpace for keyword co-occurrence analysis, and after merging similar keywords, the keyword co-occurrence network map shown in Figure 2.5 was developed. To enhance clarity, the network is pruned with the "Pathfinder" method, and only keywords with a frequency greater than 50 are displayed in Figure 2.5. The most frequently occurring keywords in the network are "carbonate precipitation" and "MICP," which are also included in the search criteria for this study and, therefore, won't be discussed further. After their exclusion, we identified the top 10 keywords by frequency (Table 2.3) and the top 20 keywords with centrality scores exceeding 0.1 (Table 2.4) based on keyword frequency and centrality values.

In terms of high-frequency keywords, researchers are keenly interested in utilizing MICP technology to improve properties such as the strength of soil or sand. Notably, *Sporosarcina pasteurii* emerges as the most extensively studied bacterium. However, it is essential to recognize that not all high-frequency keywords exhibit

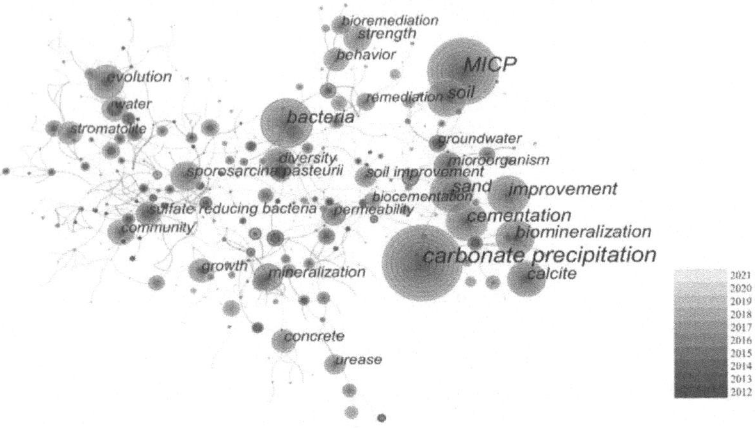

FIGURE 2.5 Keyword co-occurrence network visualization in MICP research.

TABLE 2.3
High-Frequency Keywords in the MICP Field

Number	Count	Centrality	First Occurrence	Keyword
1	195	0.11	2012	Bacteria
2	186	0.05	2013	Cementation
3	168	0.00	2013	Improvement
4	154	0.02	2014	Soil
5	149	0.02	2012	Sand
6	139	0.03	2012	Biomineralization
7	112	0.00	2012	Calcite
8	85	0.01	2016	Strength
9	84	0.12	2012	Evolution
10	80	0.01	2013	*Sporosarcina pasteurii*

high centrality. Therefore, relying solely on high-frequency keywords may not suffice to comprehensively discuss research hotspots in this field.

In CiteSpace, keywords with centrality values ≥ 0.1 are indicative of their significant influence. Noteworthy keywords, such as "biofilm," "chemistry," and "calcification," hold particular importance, indicating that researchers not only pay attention to improving the properties of materials through MICP but also are equally concerned with its underlying mechanisms and the mineralization processes involved. Moreover, keywords like "cementitious material" and "nanoparticle" have emerged as research hotspots in the past five years. This suggests that researchers are exploring practical applications that integrate novel materials with MICP technology. Consequently, it appears that utilizing MICP to enhance the performance of novel materials may represent a future research trend.

TABLE 2.4
Keywords with High Centrality in the MICP Field (Top 20)

Number	Centrality	Count	First Occurrence	Keyword
1	0.43	18	2014	Biofilm
2	0.36	13	2012	Chemistry
3	0.34	29	2013	Calcification
4	0.31	16	2016	Cementitious material
5	0.26	3	2021	Corrosion
6	0.24	6	2020	Nanoparticle
7	0.23	8	2012	Fractionation
8	0.22	18	2012	Morphology
9	0.21	38	2012	Climate change
10	0.21	19	2012	Bacterial
11	0.20	33	2013	Ureolysis
12	0.20	18	2012	Oxidation
13	0.20	5	2020	Stable isotope
14	0.19	3	2020	Variability
15	0.18	10	2013	Microbial mat
16	0.17	51	2012	Groundwater
17	0.17	23	2012	Reef
18	0.16	74	2012	Mineralization
19	0.16	24	2016	Compressive strength
20	0.16	18	2012	Adsorption

2.3.2 Research Frontiers

The time zone map of the co-occurrence of keywords reflects the research progress of the topic, offering valuable insights into the evolving research landscape of MICP. It allows for an understanding of the evolving research hotspots in the MICP field from a temporal perspective, thereby indicating the frontiers of research in this domain. To simplify the analysis, the "top 30" keywords with the highest occurrence frequency in each time slice were extracted, and the network was pruned with the "Pathfinder" method. The chronological progression of keywords within the MICP field is illustrated in Figure 2.6, with each node's position corresponding to the year of its first appearance. It is worth noting that once the keyword appears, it is fixed to the year of its initial appearance. Although the keyword may continue to appear in subsequent papers, it will no longer be displayed in the figure. The line of nodes indicates that two keywords appear together in one or more articles.

From 2012 to 2014, in addition to the keywords retrieved in this study, other frequently occurring keywords included "biomineralization," "sulfate reducing bacteria," "groundwater," "soil," "sand," etc., which indicated that the research hotspots were not only about biomineralization mechanisms but also research on groundwater, soil, and sand, which has been the hot topic of research so far. In 2015–2016,

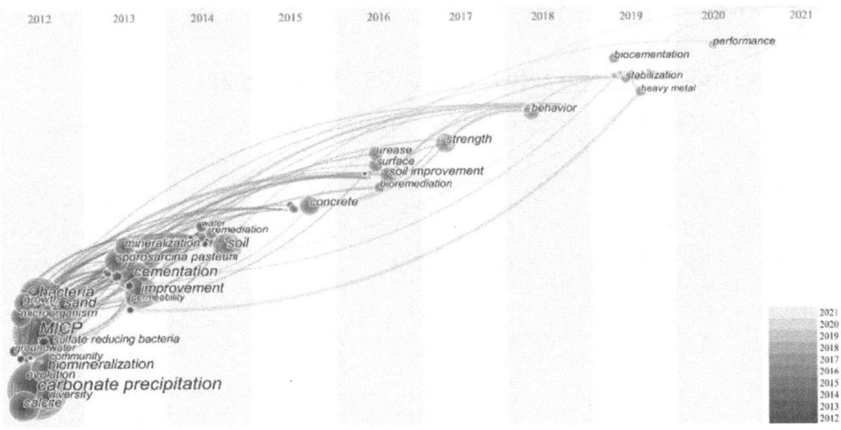

FIGURE 2.6 Temporal analysis of keywords in MICP research.

the keywords "concrete" and "bioremediation" emerged, indicating that MICP was considered a promising and eco-friendly alternative for addressing environmental challenges across various disciplines, that it can be used to solve traditional and current remediation technologies (Anbu et al. 2016). It implies that MICP was poised to extend its applications into diverse fields. In 2019, "heavy metal" became closely associated with "contaminated soil" in 2013, highlighting the enduring research focus on MICP technology's application in the bioremediation of heavy metal–contaminated soil. Future research in this domain is expected to focus increasingly on the microscale phenomena and characteristics of MICP treatment. The application of MICP technology is anticipated to expand into a broader range of fields, which is consistent with the conclusions in Section 2.3.1.

2.4 CO-CITATION ANALYSIS

Co-citation analysis involves the co-appearance of two works in the reference list of a third work, thus forming a co-citation relationship (Li and Chen 2016). In the co-citation network, the node size indicates the citation frequency of the journal or literature, with larger nodes indicating higher citation frequencies and greater influence. The node connection line represents the co-citation relationship of the journal or literature, and its thickness indicates the co-citation strength. The color shift from dark to light symbolizes the time of the journal or literature from far to near.

2.4.1 Co-Citation Analysis of Journals

Journal co-citations reflect the correlation between various journals and disciplines. The network is pruned with the "Pathfinder" method, resulting in a co-citation network comprising 109 nodes and 504 connections, as shown in Figure 2.7. The purple circle represents centrality in the figure. The thicker the circle, the higher the centrality of the node, and the greater the importance of the node in cooperation.

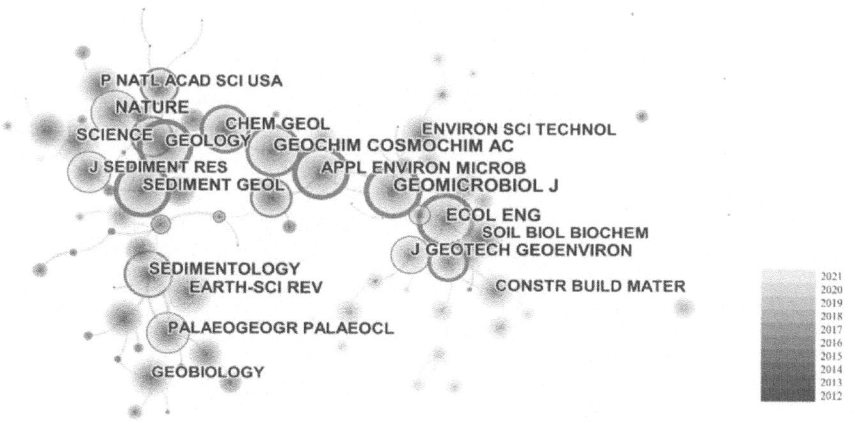

FIGURE 2.7 Visualization of journal co-citation network in MICP research.

Geomicrobiology Journal and *Ecological Engineering* are the most frequently cited journals, with citation counts exceeding 800 times. However, it is worth noting that although *Applied and Environmental Microbiology* ranks sixth in citation frequency, it ranks first in centrality. This suggests that the journal possesses high-quality and substantial influence in the field.

The publication of MICP research is primarily associated with journals related to terms like "ecological," "geology," and "environment." This observation underscores the environmentally friendly and sustainable nature of MICP technology, highlighting its potential applications across a wide range of fields and disciplines.

2.4.2 Co-Citation Analysis of Literatures

Each paper cites some references generally, and these references can be visualized as nodes in a co-citation network, where the connections between nodes represent the frequency of citations within the same paper (Chen et al. 2013). This approach allows for a clear understanding of the research focus and direction within a specific research area (Chen et al. 2014a; Small 1973). CiteSpace was used to divide the co-citation network into multiple co-citation reference clusters. In this study, the LLR algorithm in CiteSpace was used to analyze the collected 1,935 articles, and the cluster graph formed was shown in Figure 2.8.

The co-citation network results in references being tightly connected within the same cluster but loosely connected between different clusters. Each cluster is labeled by title, keyword, and abstract terms of the papers cited with that cluster Chen et al. (2013). For example, the central rose-red area is labeled as "#3 microbial diversity," indicating that papers on the topic "microbial diversity" cited cluster #3, with the largest node representing the paper topic about "bio-cemented sand" as a prominent citation within this cluster. The modularity value (Q) was 0.6568, and the mean silhouette value (S) was 0.8793 of this cluster. It is generally believed that $Q > 0.3$ means significant clustering structure; $S > 0.5$ means reasonable clustering.

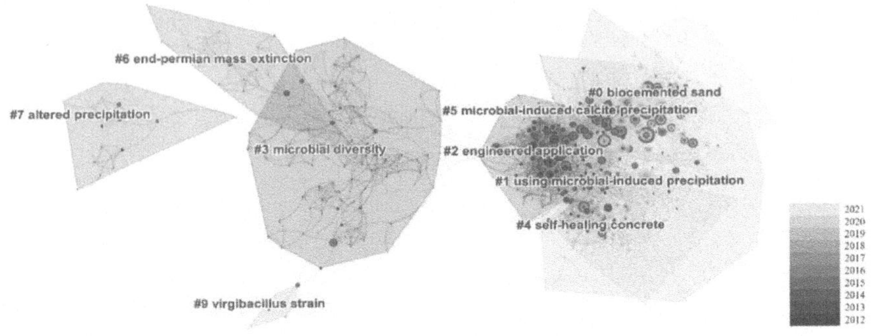

FIGURE 2.8 Visualization of literature co-citation clusters in MICP research.

TABLE 2.5
Temporal Properties of Major Clusters

Cluster ID	Size	Silhouette	From	To	Duration	Median	Theme
0	128	0.834	2014	2020	6	2017	Bio-cemented sand
1	105	0.781	2012	2020	8	2016	Using microbially induced precipitation
2	98	0.897	2007	2015	8	2010	Engineered application
3	93	0.947	2007	2014	7	2009	Microbial diversity
4	90	0.923	2011	2020	9	2015	Self-healing concrete
5	69	0.856	2009	2020	11	2014	Microbially induced calcite precipitation

These values indicate a remarkable clustering structure and reasonable clustering in this study. Different colors in various regions represent the time when the co-citation links within those regions first appeared in the figure. The nodes within the red tree-like structure indicate the period of citation bursts, signifying that these articles experienced widespread citations in a short timeframe. The cluster number is automatically marked by CiteSpace in the form of increasing from 0. The smaller the number, the more articles contained in the corresponding cluster, indicating that it is more important in MICP research.

Table 2.5 shows the temporal characteristics of the first six clusters, which reflect the start year, end year, and median year of each cluster; the "size" represents the number of cited references contained in each cluster. The silhouette of each cluster is greater than 0.75, which indicates that the six clusters have a high degree of homogeneity. Except for cluster #1 and cluster #5, which share similarities with the theme words studied in this book, the four largest clusters will be analyzed in detail in what follows.

Cluster #0, labeled as "bio-cemented sand," is the largest cluster, with 128 papers. The span is from 2014 to 2020, with a median publication year of 2017. Cluster #0 is the most recently formed cluster; the ten most cited references and the ten citing articles in this cluster (Table 2.6) were selected for this study. The three most cited

TABLE 2.6
Cited References and Citing Articles of Cluster #0 on Bio-Cemented Sand

Cited References			Citing Articles		
Cites	**Author (Year)**	**Journal**	**Coverage (%)**	**Author (Year)**	**Title**
73	K. Feng (2016)	*J Geotech Geoenviron* (Feng and Montoya 2016)	56	S. C. Chuo (2020)	"Insights into the current trends in the utilization of bacteria for microbially induced calcium carbonate precipitation" (Chuo et al. 2020)
71	L. Cheng (2017)	*J Geotech Geoenviron* (Cheng et al. 2017)	46	Y. Xiao (2021)	"Homogeneity and mechanical behaviors of sands improved by a temperature-controlled one-phase MICP method" (Xiao et al. 2021b)
64	E. Salifu (2016)	*Eng Geol* (Salifu et al. 2016)	39	Y. Xiao (2021)	"Kinetic biomineralization through microfluidic chip tests" (Xiao et al. 2021a)
56	N. J. Jiang (2017)	*Geotechnique* (Jiang and Soga 2017a)	35	Y. Xiao (2021)	"Liquefaction modeling for bio-cemented calcareous sand" (Xiao et al. 2021c)
55	M. J. Cui (2017)	*Acta Geotech* (Cui et al. 2017)	34	M. Sharma (2021)	"State of the art review of emerging and biogeotechnical methods for liquefaction mitigation in sands" (Sharma et al. 2021)
54	N. J. Jiang (2017)	*J Geotech Geoenviron* (Jiang et al. 2017b)	34	D. Mori (2021)	"A review on qualitative interaction among the parameters affecting ureolytic microbially induced calcite precipitation" (Mori and Uday 2021)
51	H. Lin (2016)	*J Geotech Geoenviron* (Lin et al. 2016)	33	D. Terzis (2019)	"A decade of progress and turning points in the understanding of bio-improved soils: a review" (Terzis and Laloui 2019)
50	M. D. Li (2016)	*J Mater Civil Eng* (Li et al. 2016b)	32	A. I. Omoregie (2021)	"Bioprecipitation of calcium carbonate mediated by ureolysis: a review" (Omoregie et al. 2021)
48	M. G. Gomez (2017)	*J Geotech Geoenviron* (Gomez et al. 2017)	31	M. M. Rahman (2020)	"State-of-the-art review of microbially induced calcite precipitation and its sustainability in engineering applications" (Rahman et al. 2020)
48	P. Xiao (2018)	*Soil Dyn Earthq Eng* (Xiao et al. 2018)	30	M. Naveed (2020)	"Application of microbially induced calcium carbonate precipitation with urea hydrolysis to improve the mechanical properties of soil" (Naveed et al. 2020)

Note: Coverage indicates the proportion of references in cluster #2 cited in this paper.

references include Feng, Cheng, and Salifu (Cheng et al. 2017; Feng and Montoya 2016; Salifu et al. 2016). The most cited reference in this cluster, Feng and Montoya (2016), mainly discussed the influence of cementation level and confining pressure on MICP. The results show that the increase of peak friction angle, residual friction angle, and initial elastic modulus is related to the degree of cementation. The second most cited reference, Cheng et al. (2017), investigated the effects of temperature, enzyme activity, rain erosion, and freeze–thaw cycles on soils mediated by MICP. Their study revealed that rain erosion had an inhibitory effect on biological cementation, while MICP-treated soil exhibited durability against freeze–thaw erosion. The third most cited reference, Salifu et al. (2016), focused on stabilizing slopes under tidal cycles and mitigating slope erosion through MICP. Their findings indicated significantly improved stability and erosion resistance in MICP-treated slopes, particularly after 30 tidal flow cycles. Other cited references in this cluster also focus on improving various mechanical properties of soil through MICP. Cluster #0 consists of 128 co-cited references, and the ten selected citing articles, published in 2020 and 2021, basically, cite at least 30% of the references within this cluster. The most cited review article was published by Chuo et al. (2020), who mainly discussed the MICP responses caused by various bacteria and the influencing factors of MICP, with a focus on the frequently used bacterium *Sporosarcina pasteurii*.

Cluster #2, labeled as "engineered application," includes two highly cited references that experienced citation bursts. De Muynck et al. (2010a) mainly reviewed the application of bio-cementation in many fields and prospected its application prospects. DeJong et al. (2010) focused on various engineering properties of bio-mediated soil improvement and pointed out the advantages and prospects. Table 2.7 shows some important citing articles in cluster #2. These articles have high citation counts, which reflects not only that they are the most representative and influential citing articles in cluster 2 but also that the citing articles in Table 2.7 are closely related to cluster #2. Among the six highly cited articles in Table 2.7, Phillips has published two of them, underscoring his importance as a scholar in this field. From a research perspective, it's apparent that scholars are not only interested in enhancing various mechanical properties of soil using MICP but are also keen on applying MICP to various engineering practices. The application of MICP technology in engineering is a major challenge and focus within this field.

Cluster #3, labeled as "microbial diversity," focuses on the mineralization response of different microorganisms. Microbial carbonate mineralization is widespread in nature and in microorganisms, holding significant ecological and geological implications (Arp et al. 2012; Shiraishi 2012). However, there are a variety of bacteria that can induce calcium carbonate deposition. Coman et al. (2015) focused on the mineralogy, macroscopic phenomena, microstructure, and microbial diversity of the microbiome. Their findings revealed that calcium carbonate predominantly precipitated in the form of calcite. Kamennaya et al. (2012) demonstrated that cyanobacteria can serve as biocatalysts for carbonate mineralization, while Couradeau et al. (2012) highlighted the significant role of cyanobacteria in calcium carbonate formation, which expands the diversity of organisms capable of forming amorphous calcium carbonate. Gallagher et al. (2012) showed that sulfate-reducing bacteria possess the capacity to induce calcium carbonate deposition and play a crucial role in organic

TABLE 2.7
Citing Articles of Cluster #2 on Engineered Application

Coverage (%)	GCS	Bibliography
46	182	Phillips et al. (2013a), "Engineered applications of ureolytic biomineralization: a review." *Biofouling*, DOI 10.1080/08927014.2013.796550.
36	298	Dhami et al. (2013), "Biomineralization of calcium carbonates and their engineered applications: a review." *Frontiers in Microbiology*, DOI 10.3389/fmicb.2013.00314.
22	241	Zhu and Dittrich (2016), "Carbonate precipitation through microbial activities in natural environment, and their potential in biotechnology: a review." *Frontiers in Bioengineering and Biotechnology*, DOI 10.3389/fbioe.2016.00004.
20	237	Anbu et al. (2016), "Formations of calcium carbonate minerals by bacteria and its multiple applications." *Springerplus*, DOI 10.1186/s40064-016-1869-2.
20	108	Phillips et al. (2013b), "Potential CO_2 leakage reduction through biofilm-induced calcium carbonate precipitation." *Environmental Science and Technology*, DOI 10.1021/es301294q.
17	112	Achal et al. (2015), "A review of microbial precipitation for sustainable construction." *Construction and Building Materials*, DOI 10.1016/j.conbuildmat.2015.04.051.

Note: Coverage indicates the proportion of references in cluster #2 cited in this paper; GCS represents the frequency of citations in Web of Science.

mineralization processes. Microbial species represent a pivotal factor in the MICP process, and investigating microbial diversity in the field of MICP contributes to the advancement.

Cluster #4, labeled as "self-healing concrete," focuses on repairing cracks in concrete through MICP, with a particular emphasis on self-healing methods. Notably, the two most frequently cited references are both review articles by Zhu and Dittrich (2016) and Anbu et al. (2016), respectively. Zhu and Dittrich (2016) mainly reviewed the different metabolic modes of MICP, MICP in natural environments, and the practical application. Anbu et al. (2016) delved into environmental factors affecting urease production and carbonate precipitation, alongside practical applications, such as heavy metal removal and quality enhancement of building materials. At least 15% of the references in cluster #4 were cited, mainly in 2020 and 2021, indicating that self-healing concrete is still a hot research topic. Concrete is a very common and vital construction material with high compressive strength but susceptibility to cracking under tensile stress. These cracks can significantly impact concrete's durability and service life (Qian et al. 2021). Therefore, the development of eco-friendly and sustainable technologies for repairing concrete cracks is crucial. Although various microorganisms can induce calcium carbonate precipitation, which can be used to repair concrete cracks, the high alkalinity of concrete often poses a challenge to microbial survival. Therefore, it is very important to choose suitable materials to encapsulate

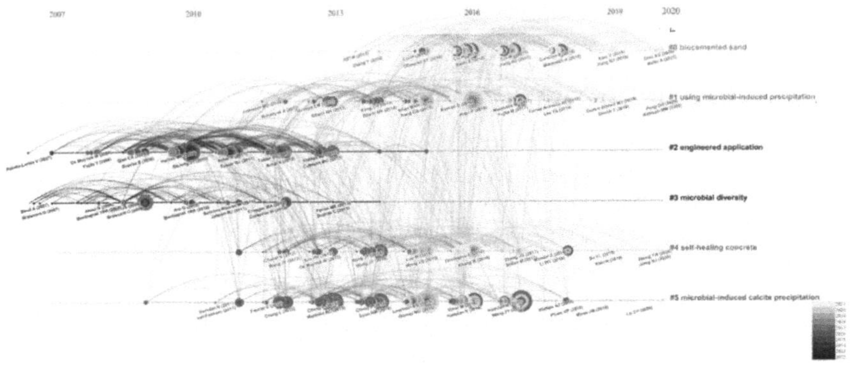

FIGURE 2.9 Temporal evolution of co-citation clusters in MICP research.

TABLE 2.8
Top 10 References with the Strongest Citation Bursts

References	Title	Strength	Duration	Range (2012–2021)
De Muynck et al. (2010a)	"Microbial carbonate precipitation in construction materials: a review"	33.81	2012–2015	
DeJong et al. (2010)	"Bio-mediated soil improvement"	31.70	2012–2015	
Dupraz et al. (2009)	"Processes of carbonate precipitation in modern microbial mats"	27.07	2012–2014	
Harkes et al. (2010)	"Fixation and distribution of bacterial activity in sand to induce carbonate precipitation for ground reinforcement"	14.99	2012–2015	
Al Qabany et al. (2012)	"Factors affecting efficiency of microbially induced calcite precipitation"	14.30	2013–2017	
DeJong et al. (2013)	"Biogeochemical processes and geotechnical applications: progress, opportunities and challenges"	23.43	2015–2018	
Phillips et al. (2013a)	"Engineered applications of ureolytic biomineralization: a review"	15.25	2015–2018	
Martinez et al. (2013)	"Experimental optimization of microbially induced carbonate precipitation for soil improvement"	15.75	2016–2018	
Al Qabany et al. (2013)	"Effect of chemical treatment used in MICP on engineering properties of cemented soils"	15.00	2016–2018	
Montoya et al. (2015)	"Stress–strain behavior of sands cemented by microbially induced calcite precipitation"	17.72	2017–2021	

bacteria so that it can withstand the high alkalinity of concrete and improve the durability of concrete (Su et al. 2021; Zheng et al. 2020a, 2020b). Algaifi et al. (2020) also evaluated the effect of microbial activity on the compressive strength of concrete. In the future, efforts are also needed to extend the durability of concrete structures in the field.

Figure 2.9 shows the timeline view of the top 6 largest clusters. In this visualization, literature from the same cluster is aligned on the same horizontal line, providing a clear overview of the time span for each cluster's publications and the growth and decline trends of each research area (Li and Chen 2016). The largest cluster is displayed at the top, and colored curves represent co-cited connections added in corresponding years. Articles marked with red circles in the graph are noteworthy as they have experienced citation bursts, signifying their importance and influence in the field. Below each timeline, the two most cited references each year are displayed, with the bottom one being the most cited, followed by the upper one. Some clusters have lifetimes exceeding ten years; others have a relatively shorter lifetime. Except for clusters #2 and #3, the other four clusters have been active so far. The connections between each cluster are very dense in the figure, indicating that each cluster has a close relationship and MICP is widely used.

Literatures with strong citation bursts are characterized by sudden and significant increases in citations, indicating major milestones in the development of a research field (Chen 2017). Table 2.8 lists the ten references with the highest burst strength in the entire dataset between 2012 and 2021. The reference with the highest value in the "Strength" column represents a crucial milestone in MICP research. Among these ten papers, five were from cluster #5; four were from cluster #2. This suggests that a significant portion of the literature with strong citation bursts focuses on the study of MICP-reinforced sand or soil in this field. The two most influential papers are De Muynck et al. (2010a) and DeJong et al. (2010), which deserve close attention from researchers.

3 Immersion Method and MICP Sample Preparation

3.1 MICP SAMPLE PREPARATION

3.1.1 Preparation of Bacteria Solution

3.1.1.1 Growth Medium (ATCC 1376 *Bacillus pasteurii* NH4-YE Medium)

The components of the growth medium per liter of deionized water include 20 g of yeast extract, 10 g of ammonium sulfate $(NH_4)_2SO_4$, and 0.13 M Tris buffer (adjusted to a pH of 9.0). First, autoclave the ingredients separately to sterilize them. Then, dissolve the yeast extract, ammonium sulfate, and Tris buffer in 1 L of deionized water. After sterilization, mix the components thoroughly to ensure homogeneity.

3.1.1.2 *Sporosarcina pasteurii* (ATCC 11859)

Begin by opening the vial containing the bacteria. Using a pipette, withdraw 0.5 to 1 mL of ATCC 1376 growth medium (5–6 mL) to rehydrate the entire pellet. Transfer the rehydrated pellet back into the same growth medium tube and mix well. From this suspension, take 1 mL and add it to a second tube containing 10 mL of growth medium. Use several drops from the second tube to inoculate additional tubes (1:10 inoculation rate). This step ensures the growth is not hindered by the cryoprotectant present in the initial medium.

Incubate all inoculated tubes in a shaker at 200 rpm and 30°C for 48 to 72 h. If growth is insufficient, transfer the bacteria from the incubated medium into fresh tubes to stimulate bacterial activity. Centrifuge the tubes at 4000 g for 20 minutes, discard the supernatant, and replace it with fresh growth medium. Repeat the incubation and centrifugation process as needed. The bacterial concentration for MICP experiments should be adjusted by measuring the optical density at 600 nm wavelength using a spectrophotometer (Genesys 20, Thermo Scientific).

3.1.2 Preparation of Cementation Media

Cementation media is used to provide chemical compositions for ureolysis, including urea, $CaCl_2{\cdot}2H_2O$, NH_4Cl, $NaHCO_3$, and nutrient broth (Zhao et al. 2014a). A summary of the components and concentrations in the cementation media is shown in Table 3.5. The urea–Ca^{2+} molar ratio was fixed at 1:1. A cementation media concentration ranging from 0.125 M Ca to 0.75 M Ca was used to initiate MICP treatment in this study. The cementation media concentration was selected

DOI: 10.1201/9781003519638-3

TABLE 3.1
Chemical Components for Cementation Media

	Chemical Concentration (g/L)					
Chemical	**0.125 M Ca**	**0.25 M Ca**	**0.50 M Ca**	**0.75 M Ca**	**1 M Ca**	**1.5 M Ca**
NH_4Cl	10	10	10	10	10.0	10.0
Nutrient broth	3	3	3	3	3.0	3.0
$NaHCO_3$	2.12	2.12	2.12	2.12	2.12	2.12
Urea	7.5	15	30	45	60.0	90.0
$CaCl_2 \cdot 2H_2O$	18.4	36.8	73.5	110.3	147.0	220.5

based on the previous study, which indicated that whole cell *S. pasteurii* urease was tolerant to concentrations of up to 3 M urea and 2 M Ca, and allowed the controlled precipitation of up to 1.5 M Ca within one treatment (Whiffin 2004). Zhao et al. (2014b) also found that urea and calcium ions in cementation solution had not been fully utilized when its concentration was higher than 1 M Ca. However, when cementation media concentration was as low as 0.125 M Ca, the MICP-treated sample with immersing method failed, so that the sand was not able to bond together.

3.1.3 Detection of Bacteria and Urease Activity

3.1.3.1 Bacteria Biomass

Bacterial cell concentration was quantified by measuring absorbance (optical density) of the suspension using a spectrophotometer (UV-1700 UV-VISIBLE Spectrophotometer, Shimadzu) at 600 nm wavelength (OD_{600}).

3.1.3.2 Urease Activity

According to equation (3.1), ion concentration increases as urea hydrolysis is catalyzed by urease. As a result, the electrical conductivity of the solution increases, which is proportional to the concentration of active urease. The increase rate of electrical conductivity reflected the hydrolysis rate of urea, which was used to describe the urease activity of bacteria solution or urease solution (Whiffin 2004).

In this study, conductivity was monitored for 5 minutes using a conductivity meter (OAKTON, con 6 Acorn series) after 1.5 mL of bacteria solution was mixed with 13.5 mL 1.67 M urea solution. The *actual conductivity variation rate* (mS/min) is the measured conductivity variation rate multiplied by the dilution factor. The *dilution factor* is defined as the ratio of original bacteria concentration to the bacteria concentration after mixing with urea solution. The actual conductivity variation rate can be converted to urea hydrolysis rate (mM urea hydrolyzed/min), based on a correlation that 1 mS/min corresponds to a hydrolysis activity of 11 mM urea/min in the measured range of activities (van Paassen 2009). This urea hydrolysis rate is the urease activity in equation (4.1) (Whiffin 2004). Specific urease activity (mM hydrolyzed urea/min/OD) can be calculated by dividing urease activity by bacteria biomass (OD_{600}), which reflects the urease catalytic ability of urea hydrolysis.

$$\begin{array}{l}\text{Specific Urease Activity}\\ \text{(mMureahydeolysed min/OD)}\end{array} = \frac{\text{Urease Activity (mMureahydrolysed min)}}{\text{Biomass } (OD_{600})} \quad (3.1)$$

3.1.4 Preparation of Molds

3.1.4.1 Full-Contact Flexible Mold (FCFM)

The FCFM, made from geotextile fabric, offers flexibility and full contact with the sample. The geotextile properties include a grab tensile strength of 1,689 N, a tear strength of 667 N, and a water flow rate of 34 mm/s. The geotextile is cut into three parts: an annular section, a bottom, and a cover. These components are then sewn together to form the mold, which can be used for preparing various sample sizes, such as for unconfined compression or direct shear tests (Figure 3.1).

3.1.4.2 Rigid Full-Contact Mold (RFCM)

The RFCM includes a flexible geotextile layer and a rigid holder made from polypropylene perforated sheets with staggered 6.35 mm diameter holes. The clearance between holes is 9.53 mm. The mold comprises three chambers, each measuring 177.8 mm in length, 76.2 mm in width, and 38.1 mm in height. To assemble the mold, the polypropylene sheets are cut and fastened with plastic screws. Geotextile layers are then cut and placed as a bottom and cover over the sand sample (Figure 3.2).

3.1.4.3 Cored Brick Mold

This mold includes a rigid holder, flexible layer, and cardboard tubes. The cardboard tubes are 60 mm in diameter and 140 mm in length. Drill 61 mm holes in the rigid holder for the tubes, ensuring their alignment within each chamber. Place the flexible

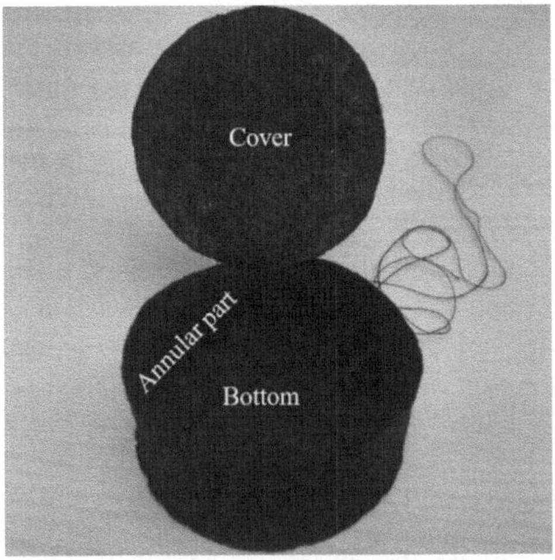

FIGURE 3.1 Full-contact flexible mold for direct shear tests.

FIGURE 3.2 Sample preparation of bio-bricks: (a) assembled mold for brick, (b) sand added into the mold, (c) flexible cover added on the top of sand sample.

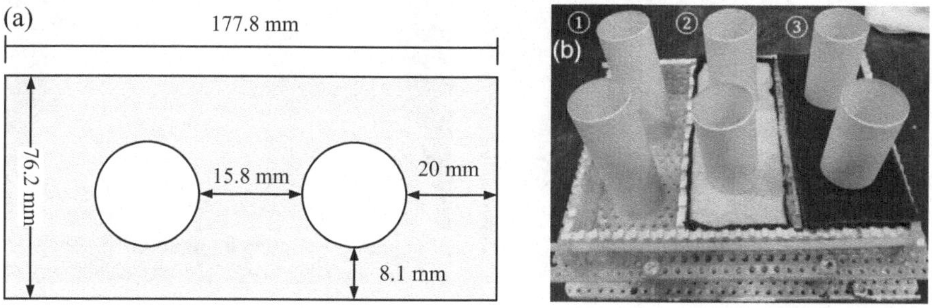

FIGURE 3.3 (a) Holes distribution on one chamber of mold. (b) Sample preparation of bio-cored bricks: (1) assembled mold for cored brick, (2) sand added into the mold, (3) flexible cover added on the top of sand sample.

layer inside the holder with corresponding holes. Once filled with sand, the mold is sealed with the flexible cover on top (Figure 3.3).

3.1.5 Preparation of Batch Reactor

The batch reactor setup, illustrated in Figure 3.4, includes a plastic container for holding soil samples, cementation media, a support shelf for samples, and air pumps to provide oxygen to the bacteria. To set up the reactor, connect the air pump to an air supply using plastic tubing and place it inside the plastic box. Then, pour the cementation media into the container, ensuring the soil samples are fully immersed for effective media diffusion.

3.1.6 Preparation of Soil Samples

3.1.6.1 MICP-Treated Soil Sample

To prepare MICP-treated samples, fill the molds (FCFM, RFCM, or cored brick mold) with sand, achieving a median density condition with a dry density between 1.58 and 1.64 g/cm³. After placing the mold cover, saturate the sample with the prepared bacteria solution. Position the samples on the support shelf and immerse them

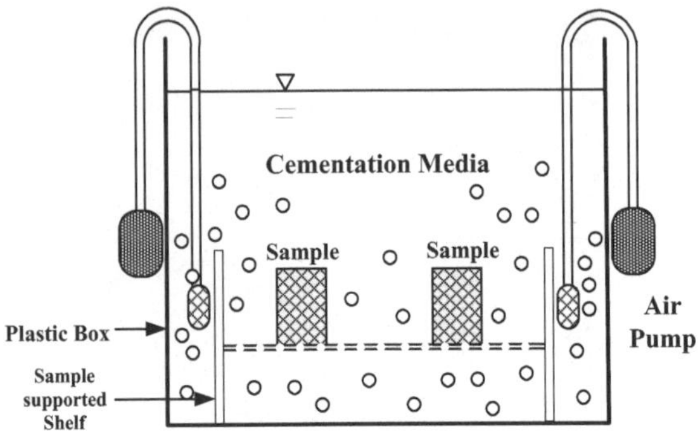

FIGURE 3.4 Sketch of batch reactor for MICP.

FIGURE 3.5 Samples placed on supported shelf.

in the batch reactor filled with cementation media. Turn on the air pump and allow the MICP reaction to proceed for seven days. Once complete, remove the samples, cut open the flexible mold or demold the rigid holder, and dry the samples in an oven at 105°C for 48 h until their weights stabilize (Figure 3.5).

3.1.6.2 Cement-Treated Bricks with Bio-Surface Treatment

Portland cement (type I/II) is used as the binder in cement-treated bricks. Mix 900 g of sand, 90 g of cement, and 200 mL of water to form a uniform mixture. Place the mixture into a rigid mold measuring 177.8 mm by 76.2 mm by 38.1 mm. Cure the samples at 25°C and 100% humidity for seven days. After curing, dry the bricks in an oven at 105°C for 48 h, until their weights stabilize. Additional testing or treatment can follow the drying process.

3.2 DEVELOPMENT OF DIFFERENT MOLD FOR PREPARING SAMPLES THROUGH MICROBIALLY INDUCED CALCITE PRECIPITATION

3.2.1 Experimental Materials

Ottawa silica sand (99.7% quartz) was used in this study. Figure 3.6 shows the particle size distribution. The Ottawa sand is uniformly graded, with a median particle

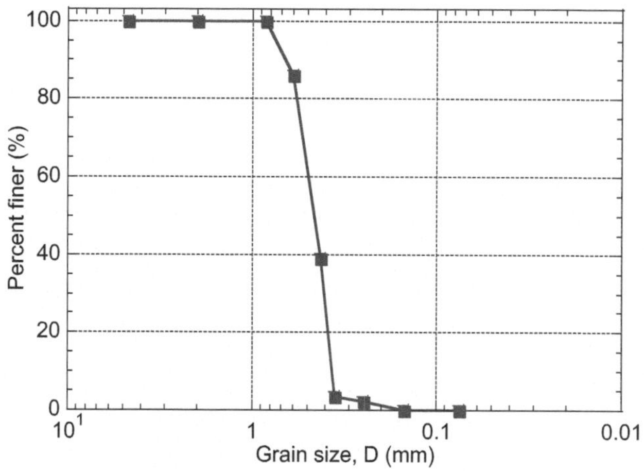

FIGURE 3.6 Particle size distribution of standard sand.

size of 0.47 mm and no fines included. According to the USCS classification, the Ottawa sand is poorly graded sand. The bacteria and growth medium and cementation media followed are those in Section 3.1.1 and Section 3.1.2.

3.2.2 Development of a Rigid Full-Contact Mold for MICP-Treated Sample

As a new environmentally friendly and sustainable technique for soil improvement, MICP has been studied widely. A suitable sample preparation mold is the first step to prepare the MICP treatment. The purpose of this study is to develop a rigid full-contact mold to prepare larger-size bio-beam specimens for MICP treatments with immersing method. Four different molds were developed, including a plastic beam rigid mold without cover, a full-contact flexible mold, a rigid full-contact mold with geotextile, and a rigid full-contact mold with filter paper. Beam-shaped specimens with different dimensions were treated with the four molds in the MICP process. Four-point bending test and split tensile strength test were conducted to evaluate the flexure behavior of MICP-treated specimens from the four sample preparation molds. The result indicated that the MICP-treated sample prepared by rigid full-contact mold with geotextile can form the best beam shape and the most uniform calcite precipitation inside the sample. Meanwhile, the bio-beam made by rigid full-contact mold with geotextile achieved the best mechanical properties. The flexure strength of the bio-beam made by rigid full-contact mold was 1125 kPa at the failure flexure strain of 1.5%, and the split tensile strength was 368 kPa, with a failure tensile strain of 1.8%.

3.2.2.1 Sample Mold and Preparation

3.2.2.1.1 Impermeable Rigid Mold

The impermeable rigid mold was made by a plastic beam box with a top opening. The size of the mold is 177.8 mm × 76.2 mm × 38.1 mm (L × W × H). The open top area in this mold was calculated as 135.5 cm^2, which is around 29% of total surface area (464.5 cm^2).

3.2.2.1.2 Full-Contact Flexible Mold

The FCFM was developed by Zhao et al. (2014a). It is made with geotextile, which is a polypropylene, staple-fiber, needle-punched nonwoven material with grab tensile strength of 1700 kN, grab elongation of 50%, trapezoidal tear of 670 kN, apparent opening size of 0.15 mm, water flow rate of 34 mm/s, thickness of 1.5 mm, and unit mass of 200 g/m^2. The mold's size is 177.8 mm × 76.2 mm × 38.1 mm (L × W × H).

3.2.2.1.3 Rigid Full-Contact Mold

The rigid full-contact mold (RFCM) includes a flexible layer and a rigid holder. The rigid holder was made with polypropylene (PP) perforated sheet with 6.35 mm thickness. This PP sheet has 6.35 mm diameter staggered holes, and the distance between holes is 9.53 mm. The rigid holder was assembled with a different piece of PP sheet by a long screw rod and a nut, and waterproof tape was wrapped around the screw members to prevent corrosion. The holder size can be varied to meet different needs. Figure 3.7 shows an example of a holder in the dimension of 355.6 mm × 101.6 mm × 38.1 mm (L × W × H). To prepare identical three specimens in one MICP treatment, the rigid holder was revised to include three internal short beams, as shown in Figure 3.8a. The size of each divided cell is 177.8 mm × 76.2 mm × 38.1 mm (L × W × H).

Two different flexible layers were used to cover the bottom and side walls of the rigid holder. One is geotextile (same as FCFM), and another is filter paper. The filter paper used in this study has particle retention of 11 μm and thickness of 180 μm. The geotextile was able to hold the sand. After the bacteria solution is added into the sand, a layer of geotextile or filter paper can be added on the top of the holder (Figure 3.5a).

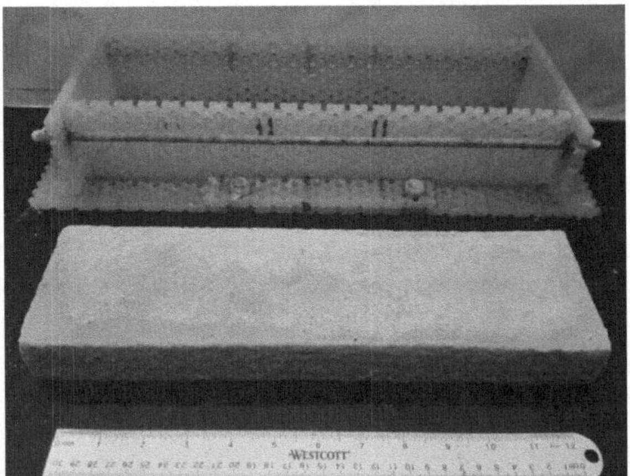

FIGURE 3.7 Larger-size rigid full-contact mold with MICP-treated specimen. The mold size is 355.6 mm × 101.6 mm × 38.1 mm (L × W× H).

3.2.2.1.4 Sample Preparation

For bio-beam specimen, 900 g sand was uniformly mixed with 250 mL bacteria solution and then air-pluviated into the mold to reach a median dense condition (relative density *Dr* in the range of 40~55%, and dry density of sand ranging from 1.58 to 1.64 g/cm^3). For RFCM, the flexible layer (either geotextile or filter paper) was placed inside the rigid mold. Ottawa sand with a controlled weight of 900 g was added into each internal mold and mixed uniformly with bacteria, and a cover of geotextile or filter paper was added, as shown in Figure 3.8d. For full-contact flexible mold, 900 g sand with bacteria was added into the geotextile mold, then it was closed with geotextile on top. All soil samples were immersed into the batch reactor with 0.5 M Ca cementation media and were allowed to react for 7 days, following the method of Zhao et al. (2014b). During the 7 days of reaction, only air was continuously pumped into the reactor. No bacteria, growth medium, or cementation medium was added to or pumped out of the reactor. After 7 days of reaction, all samples were removed from the solution, and the sample was taken out from each mold as shown in Figure 3.9. In Figure 3.9a, c, and d, the beam-shaped specimens had a length of 177.8 mm and a width of 76.2 mm and a height of 38.1 mm. In Figure 3.9b, the soil specimen is nearly beam-shaped due to the soft geotextile mold. All specimens were oven-dried at 50°C for at least 2 days before any testing.

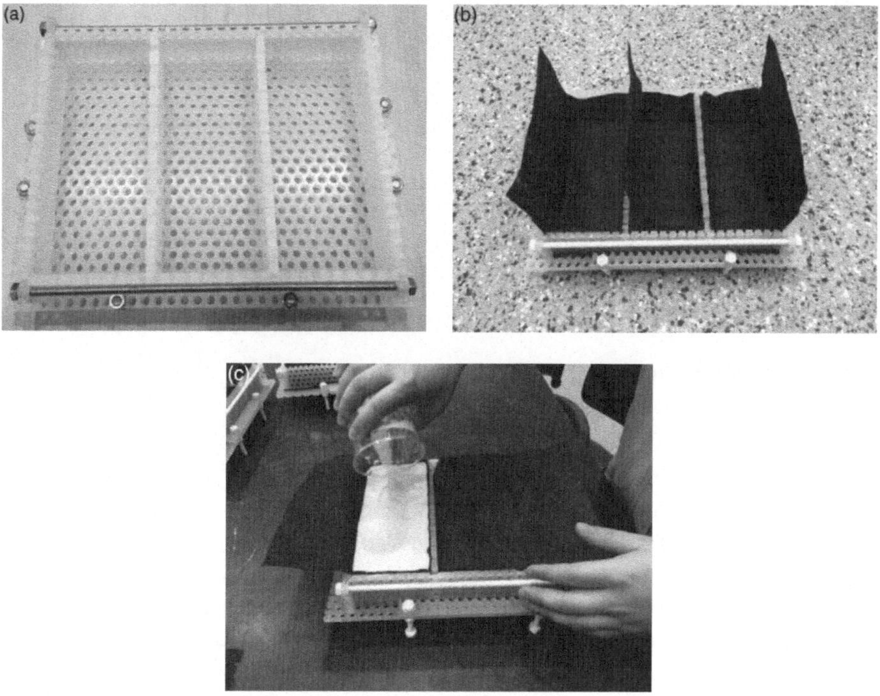

FIGURE 3.8 Rigid full-contact mold with geotextile layer.

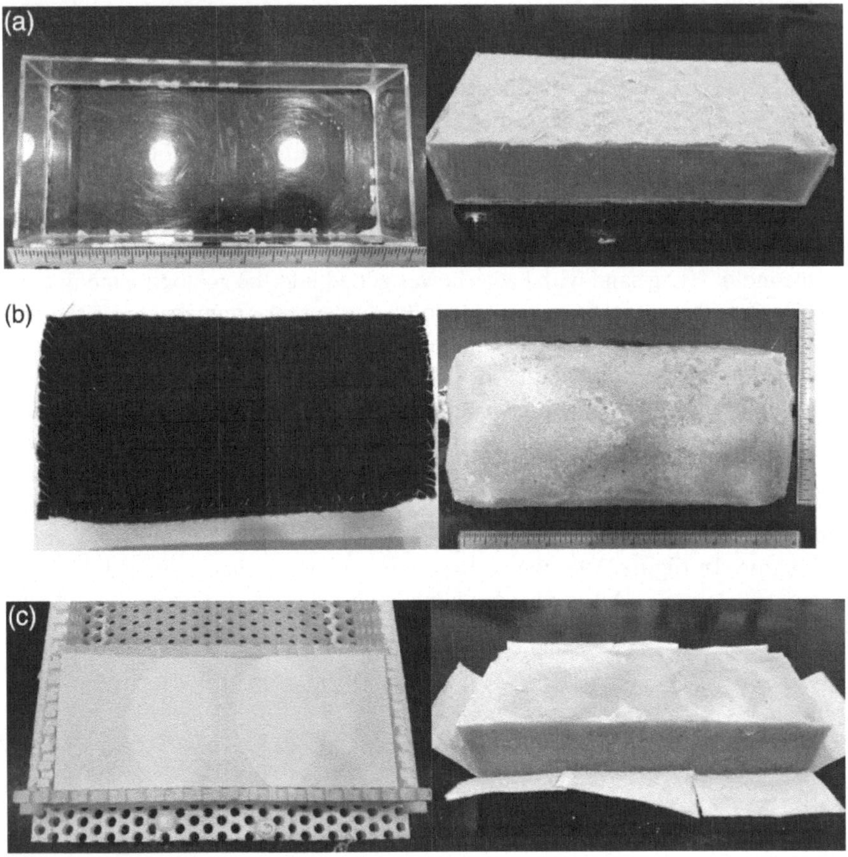

FIGURE 3.9 MICP-treated bio-beam with: (a) impermeable rigid mold, (b) full-contact flexible mold, (c) rigid full-contact mold with geotextile, and (d) rigid full-contact mold with filter paper. The left column is four molds; right column is the MICP-treated specimen (177.8 mm × 76.2 mm × 38.1 mm (L × W× H)) with four molds.

3.2.2.2 Experimental Tests

3.2.2.2.1 Four-Point Bending Test

Four-point bend tests were conducted on the bio-beam specimens after 7 days' treatment to explore the flexure behavior of MICP-treated soil. For four-point bending test, the specimen can be flipped to have 76.2 mm as height. The support system is shown in Figure 3.9a. As shown in Figure 3.10b, the specimens were set on two supports that were 152.4 mm apart from each other, and stress was applied on the top two points in the middle of the specimen with 50.8 mm distance. Following the testing method of ASTM D 6272, vertical load was conducted under strain-controlled conditions at a uniform loading rate of 1.5%/min until the bio-beam failed (Figure 3.10c). All samples were conducted at least in triplicate.

FIGURE 3.10 Four-point bending test.

3.2.2.2.2 Split Tensile Strength Test

Split tensile strength test was conducted to investigate the shear resistance of the bio-beam. The dimension of the test specimen was 76.2 mm × 38.1 mm × 38.1 mm (L × W × H), cut from the original size of the bio-beam specimen. Two thin plywood-bearing strips were used so that the load can be applied uniformly along the central height of the sample, as shown in Figure 3.11. Following the testing method of ASTM C 496, vertical load was conducted under strain-controlled conditions at a uniform loading rate of 1.5%/min until the bio-beam specimen failed. All samples were conducted at least in triplicate.

3.2.2.2.3 $CaCO_3$ Content

The $CaCO_3$ precipitated in the soil specimens was determined by acid-washing method, that is, by crushing the specimens, and soil in a different location was collected. All the collected samples were washed with an HCl solution (0.1 M) to dissolve the precipitated carbonates, rinsing, draining, and oven-drying. The difference in weight between the original specimen and the treated specimen was the weight of the carbonates that were precipitated in the specimen.

FIGURE 3.11 Split tensile strength test: (a) image of test setup, and (b) failure specimens after splitting tensile strength test.

3.2.2.3 Sample Observations of Bio-Beam Prepared with Four Molds

The right side of Figure 3.9 shows the sample after MICP treatment. The impermeable rigid mold was constructed by plastic plate materials. Only the opening top could allow the cementation media to flow though the specimens. After MICP treatment, the specimens prepared by impermeable rigid mold were removed. The top layer of the specimen showed an obviously denser $CaCO_3$ precipitation. The void space was reduced immediately when the calcite precipitated in the top opening layer, which inhabited the cementation media flow into the central and bottom layer of the specimens.

In the full-contact flexible mold, the high permeability of geotextile materials promotes the penetration of cementation media into the soil pores and provides a full-contact area during the MICP process. However, when 900 g sand was compacted into the full-contact flexible mold, the mold could not maintain the beam shape due to the low stiffness of the geotextile materials. After 7 days of treatment, the sand was bonded by calcite precipitations, but the shape of the sample could not maintain the beam shape. As shown in Figure 3.12, the shape of the treated specimen was taller in the middle section than in the sides.

For the MICP-treated samples prepared by RFCM, the soil specimen was supported by PP perforated sheet and was enclosed by flexible and permeable materials, including filter paper and geotextile. The opening size of the filter paper was around 10 μm. Once $CaCO_3$ precipitation occurred in the MICP, it was easy to block the pores on the filter paper, which could hinder further MICP reaction. When replacing the filter paper with geotextile materials, the entire MICP-treated sample was relatively uniform and could be easily removed from the mold. The advantages of geotextile were that it increased the uniformity of the precipitation and made the sample strong. Meanwhile, the PP perforated sheet helped make the treated specimen into a beam shape.

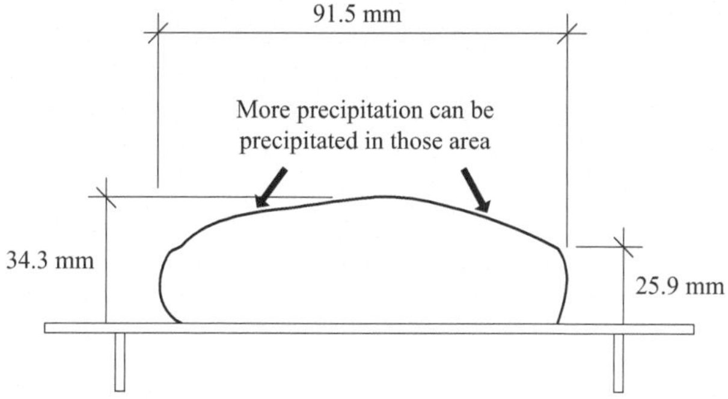

FIGURE 3.12 Cross section of bio-beam made by FCFM.

TABLE 3.2
Summary of Properties and Performance in MICP Treatment of Different Molds

Type of Mold	Properties	Advantage	Disadvantage
Impermeable rigid mold	Plastic, hard	Simple to assembly	Non-uniform $CaCO_3$ precipitation, lower strength
Full-contact flexible mold	Fabric, soft	Fair uniform	Cannot maintain beam shape; too soft to build large sample; sample hard to be tested
Rigid full-contact mold with geotextile	Plastic with soft fabric	Good uniform; maintain beam shape; higher strength performance	Complex to assemble
Rigid full-contact mold with filter paper	Plastic with soft paper	Fair uniform; maintain beam shape	Complex to assembly; filter paper hard to be removed after reaction

The advantages and disadvantages of each mold are summarized in Table 3.1. The rigid mold is easy to assemble, but the uniformity and the mechanical properties of the treated specimens are the lowest. The full-contact flexible mold improved the uniformity of the specimen, but it is too soft to build a lager sample or beam-shaped specimen. For the rigid full-contact mold, the rigid holder helped improve the stiffness of the mold to develop a better beam shape and large specimens. The geotextile as a flexible film increased the contact area between the specimen and cementation media, which can promote $CaCO_3$ distributed in the bio-beam uniformly. The use of filter paper inside the rigid full-contact mold reduced the contact area due to its limited pores. Moreover, the filter paper always sticks with the specimen after reaction.

3.2.2.4 Flexure Behavior of Bio-Beams Prepared with Four Molds

Figure 3.13a shows the force and deflection relationship of bio-beam developed by the four different sample preparation molds. As shown in this figure, the sample treated with FCFM achieved the highest vertical load as 2,194 N, and displacement at mid-span of the bio-beam was 0.86 mm. The sample prepared by the rigid mold with 29% open area had the lowest vertical load at 1,415 N, and displacement at mid-span of bio-beam was 0.7 mm. The rigid molds only allow the cementation media to flow though the sand from the open top. The bottom and side are blocked by the impermeable plastic sheet. In this case, most $CaCO_3$ precipitated in the open top layer. The sample treated with right full-contact mold with filter paper got the lowest vertical bending load (1,270 N) because the filter paper with limited size of pores could hinder further reaction. The sample prepared by RFCM with geotextile had a vertical load around 1,630 N, and the displacement at mid-span of the bio-beam was around 0.76 mm.

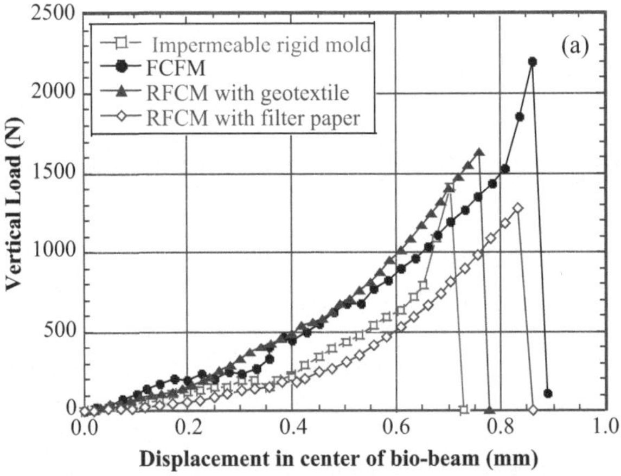

FIGURE 3.13 Flexure behavior of bio-beam prepared with impermeable rigid mold, full-contact flexible mold, rigid full-contact mold with geotextile, rigid full-contact mold with filter paper: (a) Vertical load as a function of displacement in the center of bio-beams. (b) Flexure stress–strain curve.

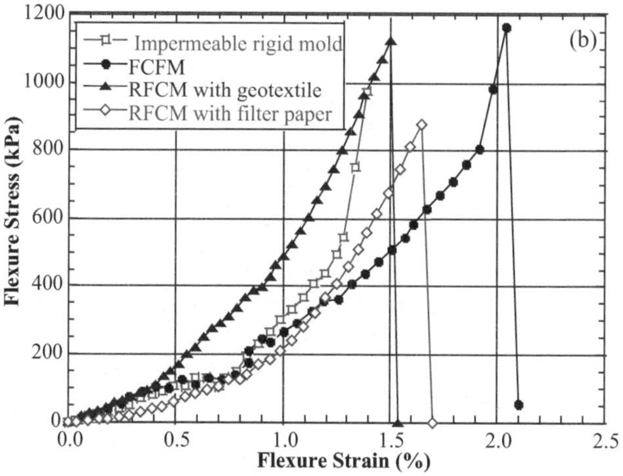

FIGURE 3.13 (Continued)

According to ASTM D6272, the flexure stress and flexure strain were also calculated shown in Figure 3.13b. For the beam-shaped specimens from the rigid mold and the rigid full-contact molds, the height of specimens was 76.2 mm, and the width of specimens was 38.1 mm. Due to the irregular bean shape of the MICP-treated specimen in FCFM, the height was expanded to 91.5 mm, and the width was reduced to 34.3 mm (Figure 3.12). The flexure strength of the treated specimen in RFCM with geotextile was 1,125 kPa at the flexure strain of 1.5%, while the flexure strength of treated specimen in FCFM was 1,160 kPa at the flexure strain of 2%.

3.2.2.5 Split Tensile Strength of Bio-Beams Prepared with Four Molds

Figure 3.14 shows the split tensile strength of the bio-beam made by four different molds, and the testing sample was cut from the regular size of the bio-beam. As shown in the figure, the sample prepared by RFCM with geotextile gained the highest split tensile strength at 368 kPa and failure tensile strain at 1.8%. For the sample prepared with FCFM, the split tensile strength was 206 kPa at the failure tensile strain of 1.55%. For the specimen made by RFCM with filter paper, the split tensile strength decreased to 180 kPa. The specimen prepared by impermeable rigid mold achieved the lowest split tensile strength (157 kPa). The open top area with a large amount of $CaCO_3$ did not help too much in this test, because the vertical load was applied perpendicular to the open top area. The specimen failed from the weaker bottom, where the $CaCO_3$ content was only 5.8%.

3.2.2.6 Effect of Molds on $CaCO_3$ Distribution

To evaluate the uniformity of the MICP process in the bio-beam specimen, the $CaCO_3$ content in five areas of the unconfined compression test samples was measured. As indicated in Figure 3.15, the sample was divided into different areas, which were both ends (I) (20 mm thick), two sides (II) (20 mm thick), and the middle part (III). Part III was, in turn, divided into three areas (III-1, III-2, III-3). The $CaCO_3$

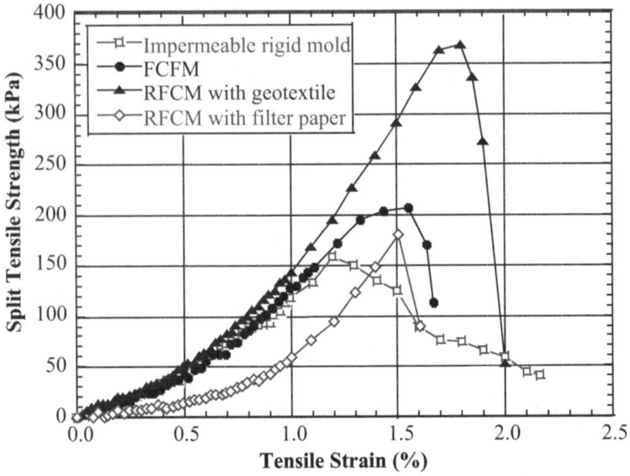

FIGURE 3.14 Split tensile strength of bio-beams as a function of tensile strain.

TABLE 3.3
$CaCO_3$ Content (in % by Weight of Dry Sand) of the Samples for Different Molds

Parts No. / Molds Type	I	II	III		
			III-1	III-2	III-3
Impermeable rigid mold	–	–	14.9 ± 0.4	5.6 ± 0.7	5.8 ± 0.3
Full-contact flexible mold	7.9 ± 0.1	7.6 ± 0.8	4.8 ± 0.3	3.0 ± 0.7	5.8 ± 0.3
Rigid full-contact mold with geotextile	9.4 ± 1.2	8.4 ± 1.5	8.5 ± 1.2	6.3 ± 0.5	7.9 ± 1.7
Rigid full-contact mold with filter paper	9.2 ± 0.2	10.9 ± 0.1	7.4 ± 0.3	3.7 ± 0.9	7.7 ± 0.6

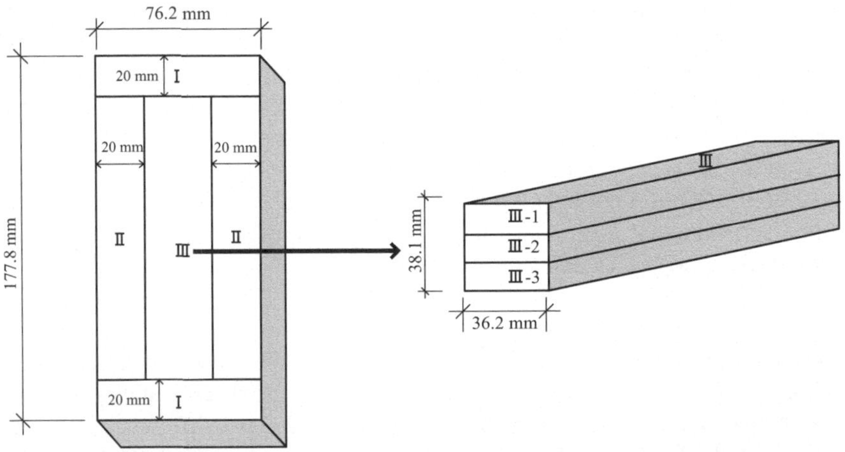

(a) Plan view of sample **(b) Three parts divided from part III**

FIGURE 3.15 Five different areas of bio-beam sample for sample uniformity evaluation.

content is indicated in Table 3.3. As shown in this table, the $CaCO_3$ content of the sample prepared by RFCM with geotextile ranged from 6.3% to 9.4%, which means $CaCO_3$ was distributed uniformly in the soil sample. For full-contact flexible mold sample, the irregular shape of the top surface caused non-uniform precipitation. The $CaCO_3$ content of top area III-1 was around 4.8%, and the central part (III-2) had the lowest $CaCO_3$ content, around 3.0%. For impermeable rigid mold sample, $CaCO_3$ precipitation was distributed non-uniformly within the entire sample, especially on the top part of the specimen. The highest $CaCO_3$ content of the rigid mold at the top layer (III-1) was 14.9%, while for the middle layer and the bottom layer, it reduced to 5.6% and 5.8%, respectively. For RFCM with filter paper sample, the outside layer part I and part II achieved the highest $CaCO_3$, which was around 9.2% and 10.9%, respectively. But the $CaCO_3$ precipitation in the inner part (III-1 and III-2) of the sample reduced to around 7.4%, and the central part (III-3) only had 3.7% $CaCO_3$.

Four different molds were developed, including a plastic beam rigid mold without cover, a full-contact flexible mold, a rigid full-contact mold with geotextile, and a rigid full-contact mold with filter paper, to prepare larger-size bio-beam specimens for MICP treatments with immersing method. Beam-shaped specimens with different dimensions were treated with the four molds in the MICP process. Four-point bending test and a split tensile strength test were conducted to evaluate the flexure behavior of MICP-treated specimens from the four sample preparation molds. The result indicated that the MICP-treated sample prepared by rigid full-contact mold with geotextile can form the best beam shape and the most uniform calcite precipitation inside the sample. The geotextile with fibrous structure increased the contact area of the MICP-treated sample, which effectively increased the penetration of cementation media. Furthermore, the large amounts of pores allowed more precipitation to occur inside the mold to improve the strength of samples. The $CaCO_3$ content test has also shown that the $CaCO_3$ content of the sample ranged from 6.3 ± 0.5% to 9.4 ± 1.2% by weight. This indicates that the sample prepared by RFCM with geotextile enhances sample uniformity. Four-point bending test and split tensile strength test both indicate that the bio-beam prepared by rigid full-contact mold with geotextile achieved the best performance on flexure behavior and shear resistance among the four sample preparation molds. The flexure strength of the bio-beam made by rigid full-contact mold was 1,125 kPa at the failure flexure strain of 1.5%, and the split tensile strength was 368 kPa, with a failure tensile strain of 1.8%.

3.2.3 Development of Nonwoven Geotextiles for Full-Contact Flexible Mold for MICP-Treated Sample

3.2.3.1 Sample Mold and Preparation

3.2.3.1.1 Geotextile

Geotextiles have been used in many geotechnical engineering applications, including filtration, separation, and drainage functions, in the recent decades because of their low cost and long durability. In this study, geotextiles acted as soil containers during the MICP process. Five different types of geotextiles were selected and labeled as G1, G2, G3, G4, and G5. The properties of these five different geotextiles are shown in Table 3.3, which includes mechanical properties and physical properties. G1, G3,

G4, and G5 are polypropylene, staple fiber, needle-punched nonwoven geotextiles. Geotextiles which are made of polypropylene generally have the advantages of high chemical resistance, low moisture absorption, low cost, and high mass per unit weight compared with other types of geotextiles. From Table 3.3, G1 was the thinnest and lightest geotextile, which has a nominal thickness of 1.07 mm and a unit mass of 162.5 g/m^2, but it had the largest apparent opening size of 0.21 mm. G5 was the thickest and densest geotextile, which had a nominal thickness of 2.91 mm, the smallest opening size of 0.15 mm, and the highest mass per unit area of 562.5 g/m^2. All opening sizes of these five geotextiles were smaller than the median particle size of sand used in this study. No leak happened in the immersing process.

G2 is also polypropylene, staple fiber, needle-punched nonwoven geotextile, but one side is heat-sintered, which acts as a sintered side. The schematic drawing and SEM image of G2 are shown in Figure 3.16a, respectively. From Table 3.3, G2, G4, and G5 had similar grab tensile strength, trapezoidal tear strength, and CBR puncture resistance. G4 and G5 had a water flow rate of 20.37 mm/s, which was much less than G2 of 33.95 mm/s. G2 had a unit mass of 200 g/m^2, which was about the half-unit mass of G4 and G5.

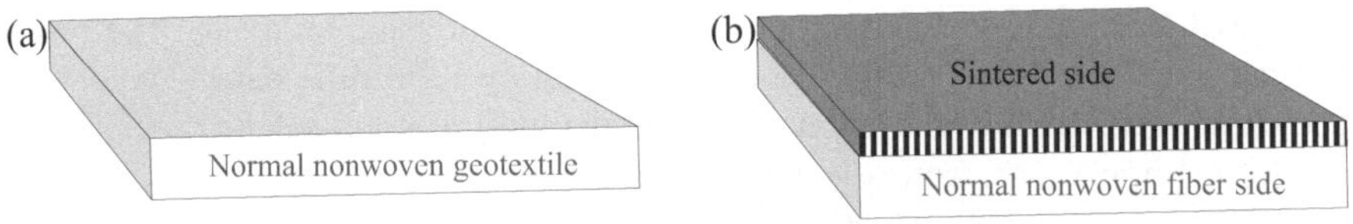

FIGURE 3.16 Schematic drawings of (a) normal nonwoven geotextile; (b) G2.

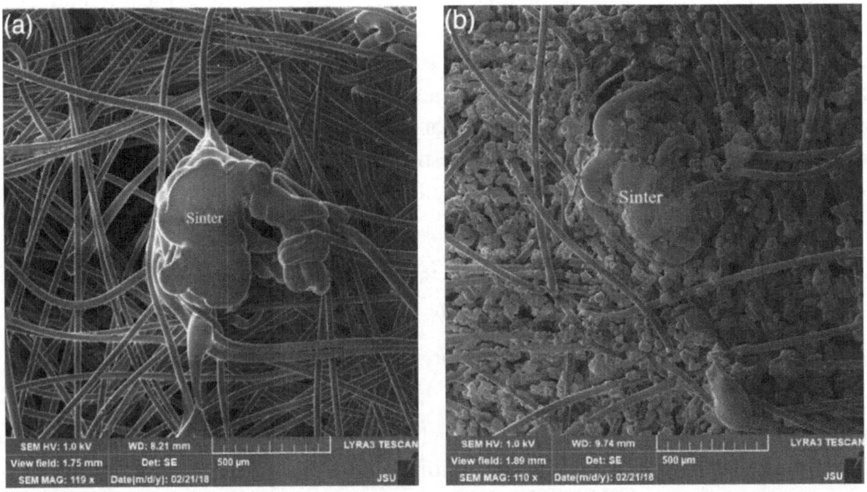

FIGURE 3.17 Scanning electron micrographs of needle-punched nonwoven geotextile: (a) G2 before MICP treatment; (b) G2 after MICP treatment.

3.2.3.1.2 Full-Contact Flexible Mold

Cylinder-shaped specimens were prepared by full-contact flexible mold (FCFM) to conduct an unconfined compression test and direct shear test. FCFM was made with the five different geotextiles listed in Table 3.3. The size of the UCS test mold was 38.1 mm in diameter and 76.2 mm in height. The size of the direct shear test mold was 63.5 mm in diameter and 25.5 mm in height. Each mold consisted of an outer part, a bottom, and a cover, as shown in Figure 3.16.

3.2.3.1.3 MICP Specimen Preparation

To test the mechanical properties of MICP-treated specimens, the UCS test, direct shear test, and four-point bending test were conducted in this study. For the UCS test specimen, 140 ± 5 g sand was uniformly mixed with a 45 mL bacteria solution and

TABLE 3.4
Properties of Geotextiles Used in This Study

	Mechanical Properties			Physical Properties			
Parameter	Tensile Strength (Grab) (N)	Trapezoidal Tear (N)	CBR Puncture (N)	Apparent Opening (mm)	Water Flow Rate (mm/s)	Thickness (mm)	Unit Mass (g/m²)
Test Method	ASTM D4632	ASTM D4533	ASTM D6241	ASTM D4751	ASTM D4491	ASTM D5261	ASTM D5199
G1	455.0	200.0	379.3	0.21	57.04	1.07	162.5
G2	1689.2	666.8	4,556.3	0.15	33.95	1.51	200.0
G3	912.0	356.0	2,335.8	0.18	36.67	1.91	262.5
G4	1,690.0	644.0	4,560.4	0.15	20.37	2.57	525.5
G5	1,691.0	623.0	4,560.4	0.15	20.37	2.91	562.5

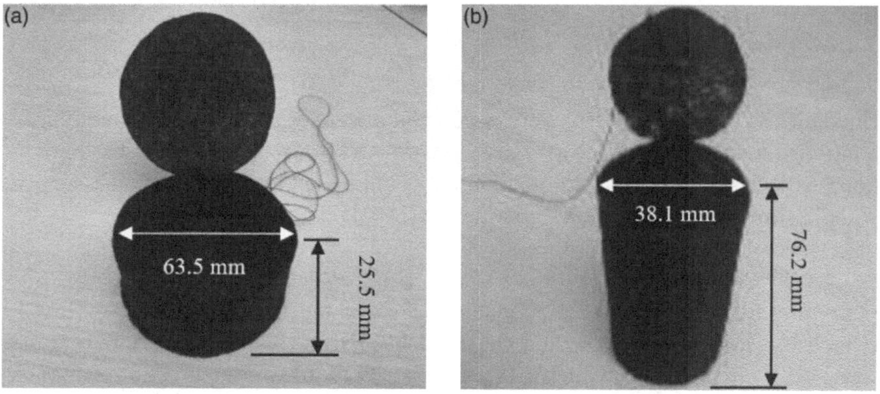

FIGURE 3.18 Photograph of (a) FCFM made of G2 for UCS tests; (b) FCFM made of G3 for direct shear tests; (c) rigid full-contact mold used for bending test specimen.

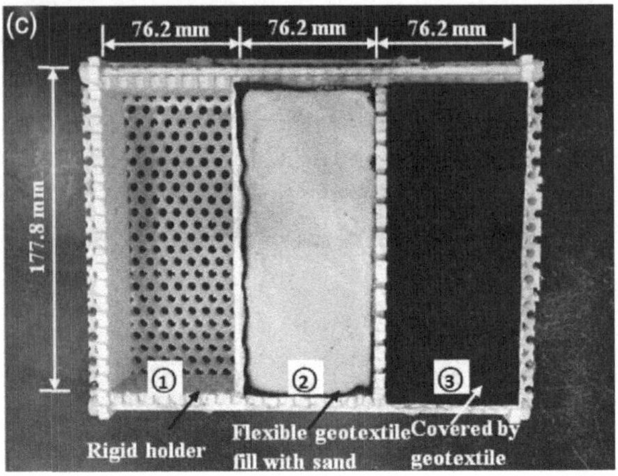

FIGURE 3.18 (Continued)

then compacted into the FCFM. For the bending test specimen, 900 ± 10 g was uniformly mixed with a 250 mL bacteria solution and then compacted into the RFCM. For the direct shear test specimen, 130 ± 5 g sand was uniformly mixed with a 40 mL bacteria solution and then compacted into the FCFM.

The immersing MICP method was used to soak soil specimens in the molds into the solution of 0.5 M Ca cementation media and react for 7 days under room temperature, as recommended by Martinez et al. (2013). The batch reactor is shown in Figure 3.4. The dimension of the batch reactor could be varied, depending on different requirements. No bacteria, growth medium, or cementation medium was added into or pumped out of the reactor during the 7 days of MICP reaction. The specimens were removed from the batch reactor at the end of 7 days. Full-contact flexible molds and rigid full-contact molds were disassembled, and the MICP-treated specimens were taken out and oven-dried for 48 h before testing.

3.2.3.2 Experimental Tests

3.2.3.2.1 Unconfined Compression Test

Unconfined compression tests were conducted under strain-controlled conditions at a uniform loading rate of 1.5%/min following ASTM D2166/D2166M-13 (International 2016). The specimens for unconfined compression tests were cylinder-shaped, with 38.1 mm diameter and 76.2 mm height (2H:1D ratio).

3.2.3.2.2 Four-Point Bending Test

Four-point bending tests were conducted to study the flexure behavior of MICP-treated bio-specimen. The specimens were located on two adjusted supports that were 152.4 mm apart from each other, and vertical load was applied on the top two supports in the middle of the specimen, with a 50.8 mm distance. Bending tests were in accordance with ASTM D6272–17 (International 1997). The four-point test system is shown in Figure 3.19.

FIGURE 3.19 Image of failed bio-specimen in four-point bending test system.

3.2.3.2.3 Direct Shear Test

Direct shear tests were conducted under the vertical loading–controlled condition of 50 kPa, 100 kPa, 150 kPa, and 200 kPa, at a uniform shearing rate of 1.0%/min, in accordance with ASTM D 3080 (International 2011). The specimens for direct shear tests were cylinder-shaped, with 25.5 mm height and 63.5 mm diameter.

3.2.3.2.4 Estimation of $CaCO_3$ Content

To determine precipitated $CaCO_3$ in the MICP-treated specimens and geotextiles, the acid washing method was used in this study. Crushed specimens and geotextiles peeled from MICP-treated specimens were oven-dried for 48 h to permit the dissipation of the excess moisture. The dry specimens and geotextiles were washed in HCl solution (0.1 M) to dissolve precipitated carbonates, then rinsed, drained, and oven-dried for 48 h. The difference between these two masses was considered to be the mass of the carbonates that were precipitated in the specimens or geotextiles (Rebata-Landa 2007). The $CaCO_3$ content in MICP-treated specimens was indicated as the percentage of the sample weight. The $CaCO_3$ content in geotextile was indicated as the weight of precipitated $CaCO_3$ in per unit area per unit thickness of the MICP-treated geotextiles sample.

3.2.3.2.5 Microstructure Analysis

To characterize the fiber structures of geotextiles before MICP reaction and the location of precipitated $CaCO_3$ in geotextiles after MICP reaction, images were captured by scanning electron microscopy (Lyra 3, TESCAN Inc). All untreated and MICP-treated geotextiles were flushed with tap water and oven-dried for 48 h before the test. Then the SEM samples of geotextiles were prepared by cutting into small pieces for scanning.

3.2.3.3 Effect of Properties of Geotextiles on the MICP Process

Grab tensile strength is defined as the maximum stress that a geotextile can experience while being pulled before failing. It also indicates the elongation property of geotextile. From Table 3.3, G1 had a tensile strength of 455 N, which was much lower than for the other four types of geotextiles, which means G1 had the lowest elongation. And G1 was the thinnest geotextile during these five types. When geotextile is used to make geotextile sand containers, the elongation properties of geotextile influence the deformability of geotextile sand containers, causing lower sand fill ratio (Dassanayake and Oumeraci 2012). Dassanayake and Oumeraci (2012) also reported that for thick nonwoven geotextiles, the effect of elongation was relatively lower. In this study, geotextiles were applied as flexible material to tailor FCFM for MICP-treated specimens, when the molds acted as containers for sand particles. For the UCS test, bending test, and direct shear test, the neat shape of specimens acts as an important role in accurate testing results. The tensile strength of geotextiles has a limited effect on bending test specimens because of the additional rigid holder of rigid contact flexible mold. But for UCS test and direct shear test specimens, the effect was nonnegligible. Especially for direct shear test specimens, the diameter–height aspect ratio (ratio of diameter to height) was 2.0; the soft fabric and inferior orthostatic of G1 caused the formation of irregular shape after MICP treatment

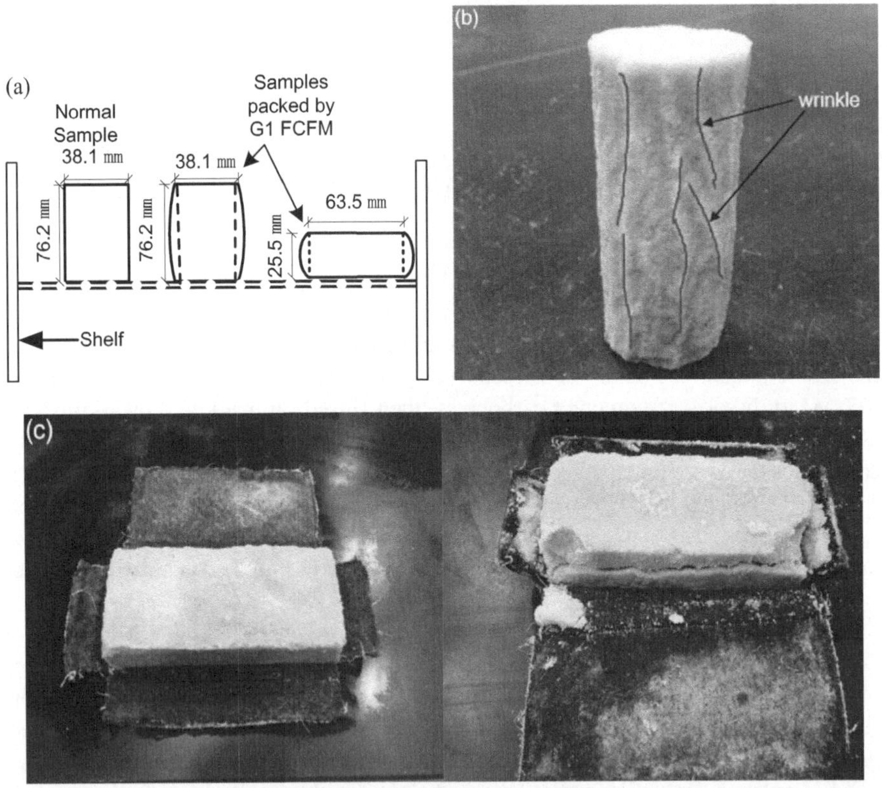

FIGURE 3.20 (a) Schematic drawings of normal sample and G1 FCFM packed specimen on shelf. (b) Sample with wrinkles on surface packed by FCFM made of G5 after MICP process. (c) Bending test specimen after MICP-treatment packed by FCFM made of G2 and G5 geotextile.

because of weak constraining forces. G1 was too soft to resist the normal stress from sand particles and cannot maintain specimens as regular shape. As shown in Figure 3.20a, the specimen prepared by G1 shows anomalous formation.

In this study, geotextiles were used as a container for sand particles and as a permeable medium for the cementation media to get into sand pores because of their porous structure in nature. The permeability of the water flow of geotextile mold is significantly influenced by the voids or pores available in the geotextile because the flow in the medium occurs only if these pores are interconnected (Fatt 1956). According to Komori and Makishima (1979), the mechanics of fluid flow depend on the width and depth of pores. The width of the pores determines the quantity of the fluid flow, and the depth of pores represents the free path of the fluid flow. Hence, the porous structure of geotextiles is important for filtration, drainage, and the separation function of geotextiles. Less thickness and higher water flow rate of geotextile can significantly increase the penetration of nutrients, urea, and calcium chloride inside the sand pores during the MICP process.

The thickness of geotextile represents the depth of pores, which determines the free path of fluid flow. The thickness of geotextiles kept increasing from G1 to G5, while the corresponding water flow rate has a general trend of decrease. Figure 3.21a shows the relationship of $CaCO_3$ content and UCS of MICP-treated samples as a function of the thickness of different geotextiles. The $CaCO_3$ content and UCS of MICP-treated samples decreased along with geotextiles' nominal thickness. Thicker geotextiles caused less permeation than less $CaCO_3$ formation and lower UCS. The water flow rate of geotextile represents water permeability, which is significant for

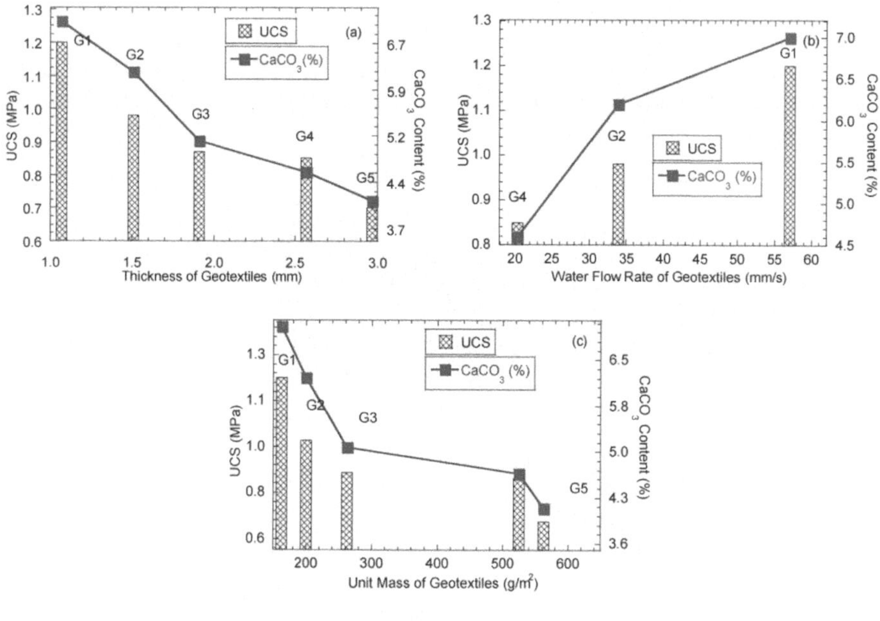

FIGURE 3.21 Median UCS and $CaCO_3$ content of MICP-treated specimen as a function of geotextile: (a) thickness, (b) unit mass, (c) water flow rate.

the application of geotextiles in this study. Figure 3.21b shows the relationship of $CaCO_3$ content and UCS of MICP-treated specimens as a function of the water flow rate. G1, G2, and G4 were selected to show this relationship, because the water flow rate of G3 was similar to that of G2, and the water flow rates of G4 and G5 were equal. The $CaCO_3$ content and UCS of MICP-treated specimens have a trend of increase along with a higher water flow rate. It can be explained that when geotextile has a low water flow rate, it will prevent cementation media from getting into MICP-treated specimens.

Unit mass represents the density of geotextile. Figure 3.21c shows the relationship of $CaCO_3$ content and the UCS of MICP-treated specimens as a function of the unit mass of different geotextiles. G1, G2, G3, G4, and G5 had an increasing unit mass in this study. $CaCO_3$ content and UCS decrease along with an increase of unit mass. A larger unit mass means the geotextile is denser, and the corresponding permeable pathway should be longer. For the concentration media, it would be the longer travel time through these geotextiles which would lead to the blocking of geotextiles' pores by the precipitated $CaCO_3$ inside the geotextiles during this travel time, limiting the MICP process. These would result in lower $CaCO_3$ content and lower strength for MICP-treated specimens, which could explain why bending test specimens prepared by G5 cannot be taken out unbrokenly after MICP reaction (see Figure 3.20c). Furthermore, thicker geotextiles with high unit mass were too stiff to result in wrinkles, as shown in Figure 3.20b on the surface of MICP-treated specimens. Especially for smaller-sized and cylinder-shaped specimens, this phenomenon would be more obvious.

3.2.3.4 Effect of Geotextiles on Mechanical Properties of MICP-Treated Specimens

The mechanical properties of MICP-treated specimens were significantly influenced by different types of geotextiles. Figure 3.22 shows the stress–strain curves obtained from unconfined compression tests on MICP-treated specimens prepared

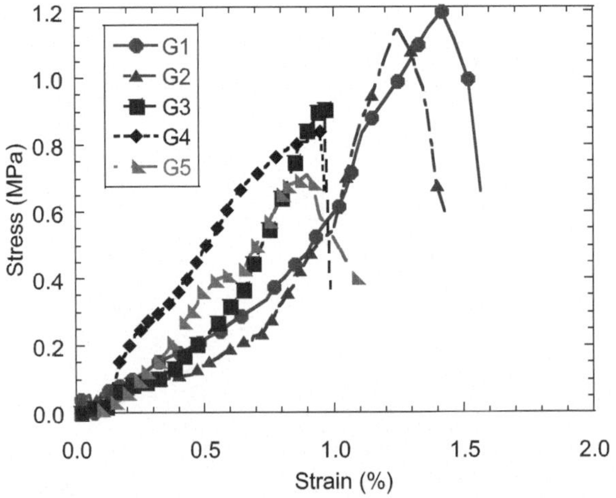

FIGURE 3.22 Unconfined compressive strength of MICP-treated specimen prepared using five different geotextile FCFM molds.

by five different geotextiles. The unconfined compression strength (UCS) of MICP-treated specimens varied when using different geotextiles, such as FCFM. When the FCFMs were made of G1, the peak strength of the MICP-treated specimen was about 1.2 MPa, whereas the peak strength of MICP-treated specimen decreased to 1.12 MPa, 0.90 MPa, 0.85 MPa, and 0.7 MPa when the geotextiles varied from G2 to G5. There was a nearly 42% decrease in peak strength (from 1.2 MPa in G1 to 0.7 MPa in G5) of MICP-treated specimens. This could be caused by the variety of properties of these five geotextiles. Muthukumaran and Ilamparuthi (2006) conducted laboratory studies on four different geotextile filters as used in geotextile tube dewatering. They found that the transition time varied much with the opening size of the geotextile. Their results indicated that at lower water contents, the formation of filter cake was only a function of the aperture opening size of geotextile and not on the gradation of sludge or its water content. The peak strain of G1 was the highest at 1.4%, and that of G5 was the lowest at 0.9%. The significant difference between G1 and G5 should cause the difference between the mechanical properties of specimens prepared by these two geotextiles. From these findings, the study proposed that G1 and G2 are more efficient in the MICP process.

Box plots of $CaCO_3$ content as a function of geotextiles (G1–G5) are shown in Figure 3.23. Each box shows the median value and ±25% of the $CaCO_3$ precipitation at the top and bottom of the box. The lines extending from the top and bottom of each box mark the 5% and 95% $CaCO_3$ content. The outlier is shown as an individual point. With the geotextile varying from G1 to G5, the $CaCO_3$ content of MICP-treated specimens decreased. The best precipitation effect exists in the specimens prepared through the FCFM of G1 and G2. The UCS of specimens can be more than 1.0 MPa, and more precipitated $CaCO_3$ can be produced, whereas for G3 to G5, the UCS is less than 1.0 MPa and precipitated $CaCO_3$ was produced poorly. The results indicated that less $CaCO_3$ content resulted in decreased UCS of MICP-treated specimens, which can be explained as the strength of MICP-treated material mainly depending on the part of $CaCO_3$ precipitation, which can bond the soil particles. When the amount of $CaCO_3$ induced in the sufficient MICP process was increased

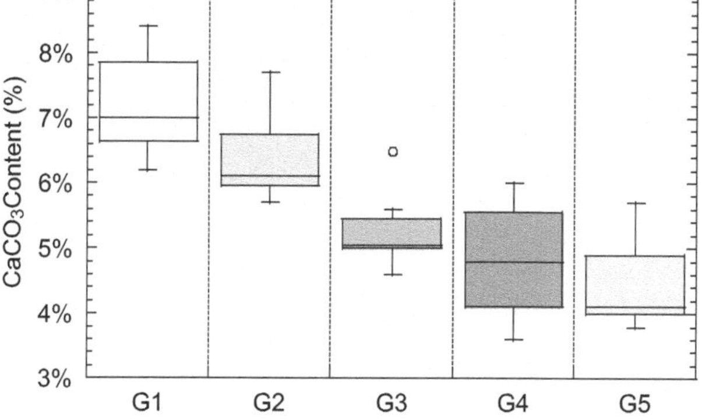

FIGURE 3.23 Box plot of $CaCO_3$ content of MICP-treated specimens as a function of different geotextiles.

and continually filled in the pore space, porosity decreased gradually, and UCS was improved with the increase in compactness.

Figure 3.24 shows the vertical load–displacement curves from four-point bending tests on the MICP-treated specimen using RFCM molds made of four different geotextiles, which are G1, G2, G3, and G4 geotextile. The MICP-treated specimens prepared by G5 were too soft to remove from RCFM molds (Figure 3.20). The peak bending load of MICP-treated specimens was 95 N when G1 was used in RCFM. The peak bending load decreased to 80 N, 75 N, and 57 N when the geotextile varied from G2 to G4, respectively. The results show that the peak bending load of MICP-treated

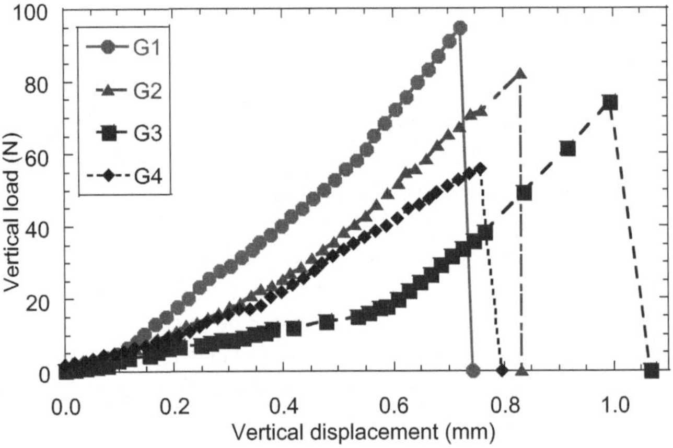

FIGURE 3.24 Load–displacement curves of MICP-treated specimen prepared by different geotextile RFCM molds from bending tests.

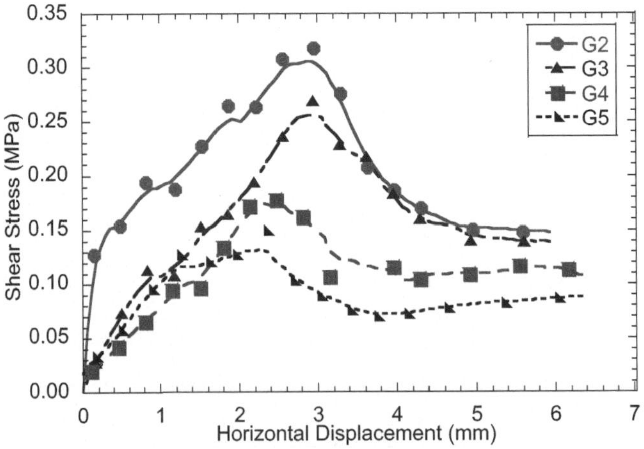

FIGURE 3.25 Shear stresses at normal stress of 50 kPa for MICP-treated specimens prepared by different geotextile molds in direct shear tests.

specimens decreased from 95 N to 57 N when the geotextile varied from G1 to G4, which is almost a 40% decrease. These reductions should be caused by the different properties of geotextiles.

Figure 3.25 shows shear stress as a function of horizontal displacement under the normal stress of 50 kPa for MICP-treated specimen prepared using the FCFM made of G2, G3, G4, and G5. The specimens prepared by G1 were too irregular-shaped to get direct shear test results (as shown in Figure 3.25a). The peak shear stress of the MICP-treated specimens prepared using FCFM made by G2 geotextile was about 0.3 MPa, while peak shear stress for MICP-treated specimens prepared using G3 geotextile is about 0.25 MPa. The difference between these two peak shear stresses was not obvious. But for specimens prepared by G4 and G5, the peak strength reduced to 0.18 MPa and 0.15 MPa, respectively. These findings indicate that the type of geotextile plays an important role in the immersing method for preparing MICP-treated specimens.

3.2.3.5 Singeing Geotextile

G2 was the only single-side singeing geotextile among the five geotextiles applied in this study. This singeing geotextile contained two different sides, one side of nonwoven fiber and another side of sintered nonwoven fiber.

The two-side geotextile achieves geotextile agglomeration on the surface, as well as maintaining the performance of the original nonwoven geotextiles, such as high strength and flexibility. For nonwoven geotextiles, including G1, G3, G4, and G5, $CaCO_3$ would hang wire bonding filaments easily during the MICP process. These bonding between filaments and $CaCO_3$ could result in sample damage when cutting molds off after MICP treatment. But for singeing geotextiles (e.g., G2), the filaments were singed into sinters after the singeing process, which would provide a relatively rough surface. The rough surface was applied as in-plane in the experiments. The sinters instead of filaments on the geotextile surface could prevent $CaCO_3$ crystal from adhering tightly to the geotextile, which made it easier to strip molds off after MICP treatment and keep specimens intact.

Among the five types of geotextiles in this study, G2 had a similar tensile strength, trapezoidal tear strength, and puncture resistance with G4 and G5, as shown in Table 3.3. The singeing nonwoven geotextile layer enhances the mechanical properties of geotextiles significantly. These enhancements made G2 strong enough to resist normal stress from sands and to contain regular-shaped specimens compared to G1. G2 had the secondary smallest thickness of 1.51 mm, which was much thinner than the last two geotextiles, G4 and G5. G2 was thin enough to avoid wrinkles on the surface of specimens compared to G4 and G5. Because of the smaller thickness, G2 had a higher water flow rate of 33.95 mm/s that can cause more penetration of cementation media and a much smaller unit mass of 200 g/m^2 that can prolong the effective MICP treatment duration before the blocking of pores in geotextile.

3.2.3.6 $CaCO_3$ Content

The precipitated $CaCO_3$ content in the bending test specimens prepared by different geotextiles through MICP was measured. As shown in Figure 3.26, the MICP-treated bending test specimen was divided into three parts, the lateral part of the length side,

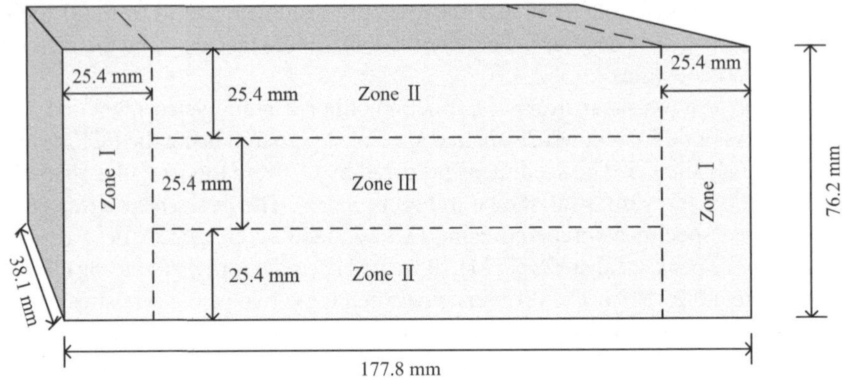

FIGURE 3.26 Three zones of MICP-treated specimen for $CaCO_3$ uniformity distribution.

TABLE 3.5
$CaCO_3$ Content (by Weight) of Three Different Zones of MICP-Treated Sample and $CaCO_3$ Content of Geotextiles after MICP Treatment

	$CaCO_3$ Content (%) of MICP-Treated Soil Specimens			$CaCO_3$ Content per Unit Area per Unit Thickness ($g/m^2/mm$) in Geotextiles
Geotextile	Zone I	Zone II	Zone III	
G1	7.5 ± 0.12	7.1 ± 0.09	6.8 ± 0.17	317 ± 16
G2	6.7 ± 0.13	6.4 ± 0.14	5.9 ± 0.08	324 ± 15
G3	5.6 ± 0.07	5.1 ± 0.11	4.9 ± 0.13	330 ± 9
G4	4.9 ± 0.09	4.7 ± 0.12	4.5 ± 0.10	337 ± 8
G5	4.5 ± 0.12	4.1 ± 0.06	3.9 ± 0.07	342 ± 8

the lateral part of the width side, and the central part. In each part, the $CaCO_3$ content was measured by the acid washing method. The $CaCO_3$ content was indicated as the percentage of the sample weight. The $CaCO_3$ content of three parts in bending test specimens prepared by five geotextiles is listed in Table 3.5. The $CaCO_3$ content of specimens using G1~G5 ranged from 6.8 ± 0.17 to 7.5 ± 0.12%, 5.9 ± 0.08 to 6.7 ± 0.13%, 4.9 ± 0.1 to 5.6 ± 0.07%, 4.5 ± 0.10 to 4.9 ± 0.09%, and 3.9 ± 0.07 to 4.5 ± 0.12% by weight, respectively. The $CaCO_3$ distribution results indicated that the precipitated $CaCO_3$ was distributed uniformly in MICP-treated specimens prepared by each kind of geotextile.

3.2.3.7 SEM Analysis

Nonwoven geotextiles are defined as products made of fibers, filaments, or other elements directionally or randomly oriented and mechanically, thermally, or chemically bonded (ISO 2015). A geotextile composed of filaments generally presents straight

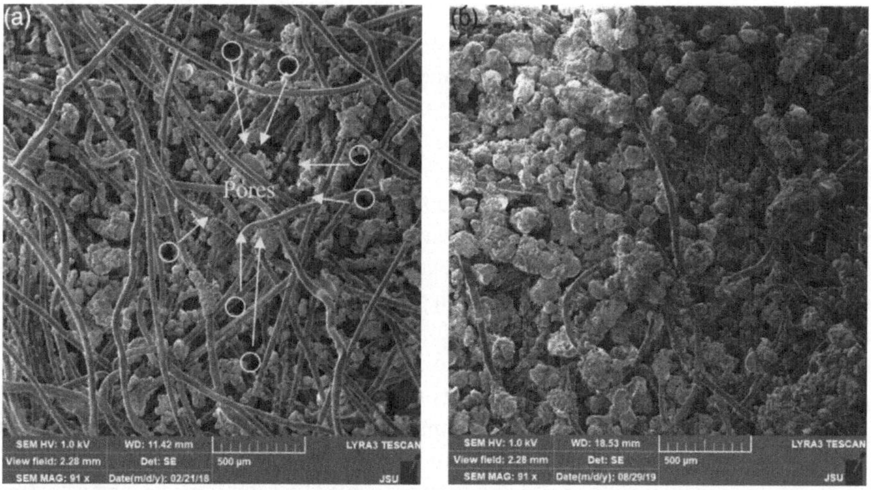

FIGURE 3.27 Scanning electron micrographs of needle-punched nonwoven geotextile: (a) G1, (b) G5 after 7 days' MICP reaction.

elements under an in-plane microscopic view of the geotextile, while a geotextile composed of staple fibers could present curvilinear elements (Silva et al. 2019). SEM images (Figure 3.27) were conducted on geotextiles G1 and G5 to characterize the fiber structures and $CaCO_3$ distribution in geotextiles after MICP treatment. As shown in Figure 3.27a, pores still could be observed on the geotextile G1 after 7 days' MICP reaction, while in Figure 3.27b, pores of geotextile G5 were almost totally occupied by $CaCO_3$ precipitation. These pores on geotextiles allowed nutrients, urea, and calcium chloride to penetrate the soil specimens. Once these pores were occupied by $CaCO_3$ precipitation, these blocks would prevent MICP reaction in the specimens. As listed in Table 3.4, $CaCO_3$ content per unit area per unit thickness in geotextile G5 is 342 ± 8 g/m^2/mm, which was higher than G1 of 317 ± 16 g/m^2/mm. These results also indicated that G5 got more blockage after MICP reaction, which may be the reason that MICP-treated specimens prepared by G5 had less $CaCO_3$ content and failed to be taken out from molds.

FCFM was developed to prepare MICP specimen because of its fibrous and porous structure. Many pores are present on the surface of the mold, which effectively increase the penetration of nutrients, urea, and calcium chloride into the sand pores and can result in increased contact between the cementation solution and the bacteria. Furthermore, geotextile provides a larger attachable area, which can attract a portion of the calcium carbonate precipitated on its surface and can reduce precipitation in the sand pore space and on the sample surface. For different types of geotextiles which were used to make FCFM and RFCM, the ability to increase penetration of the cementation media was quite different. The mechanical strength results of MICP-treated specimens showed the difference that resulted from different types of geotextile.

When comparing the properties obtained from different geotextile molds, the results showed that the mechanical properties of MICP-treated soil specimen prepared in FCFM made by G2 geotextile and G3 geotextile were similar. The unconfined compression strength of the MICP-treated specimen reached 1.12 MPa (G2) and 0.90 MPa (G3). The G2 geotextile had a tensile strength of 1,689.2 N, a thickness of 1.51 mm, and a unit mass of 200.0 g/m^2. The G3 geotextile had a tensile strength of 912.0 N, a thickness of 1.91 mm, and a unit mass of 262.5 g/m^2. Both geotextiles were strong enough to hold a soil sample during the MICP process to maintain the cylinder-shaped specimen. The two geotextiles had an apparent opening of 0.15 mm (G2) and 0.18 mm (G3), in which pores can promote the penetration of nutrients, urea, and calcium chloride into the sand pores to induce the MICP process and prolong the effective MICP treatment duration. These two geotextiles helped the MICP process get a higher efficiency among the five geotextiles in this study. As for the singeing geotextile, G2 had more advantages compared to others. The findings indicate that molds made by appropriate geotextile can provide a higher efficiency during MICP reaction. Besides, the soil engineering properties were greatly improved by selecting an appropriate geotextile to make FCFM. The type of geotextile used to make FCFM plays an important role in the MICP process.

3.2.4 Development of a Full-Contact Flexible Mold for MICP-Treated Sample

3.2.4.1 Sample Mold and Preparation

The geotextile used in this study was a polypropylene, staple-fiber, needle-punched nonwoven material. The geotextile had a grab tensile strength of 172.4 kg, an apparent opening size of 0.15 mm, a thickness of 1.5 mm, and a unit mass of 200 g/m^2. Direct shear tests were conducted on the FCFM specimens. As shown in Figure 3.28b, 125 g of washed Ottawa sand was uniformly mixed with 40 mL of bacteria and growth medium, which were then air-pluviated into an FCFM to reach an intermediate density condition (Dr in the range of 42–55%). Eighteen molds with sand samples on a shelf were immersed in 15 L of cementation medium in the completely stirred tank reactor. Small holes of 5 mm in diameter were punched in the shelf, and the spacing between these holes was 5 mm, which allowed the water to flow through the shelf. During the 7 days of reaction, only air was continuously pumped into the reactor. No bacteria, growth medium, or cementation medium was added to or pumped out of the reactor. After 7 days of reaction, the shelf was removed from the solution, each geotextile mold was cut, and the treated soil specimen was removed. The cylinder-shaped soil specimens had a diameter of 62 mm and a height of 26 mm. Direct shear tests were performed at vertical normal stresses of 20, 40, and 60 kPa, with a shearing rate of 1% per minute. The shear stress was recorded up to a shear strain of 7%.

Precipitated $CaCO_3$ in the soil specimens was determined using an acid-washing method, that is, by crushing the specimens, washing with an HCl solution (0.1 M) to dissolve precipitated carbonates, rinsing, draining, and oven-drying. The difference in weight between the original specimen and the treated specimen was the weight of the carbonates that were precipitated in the specimen.

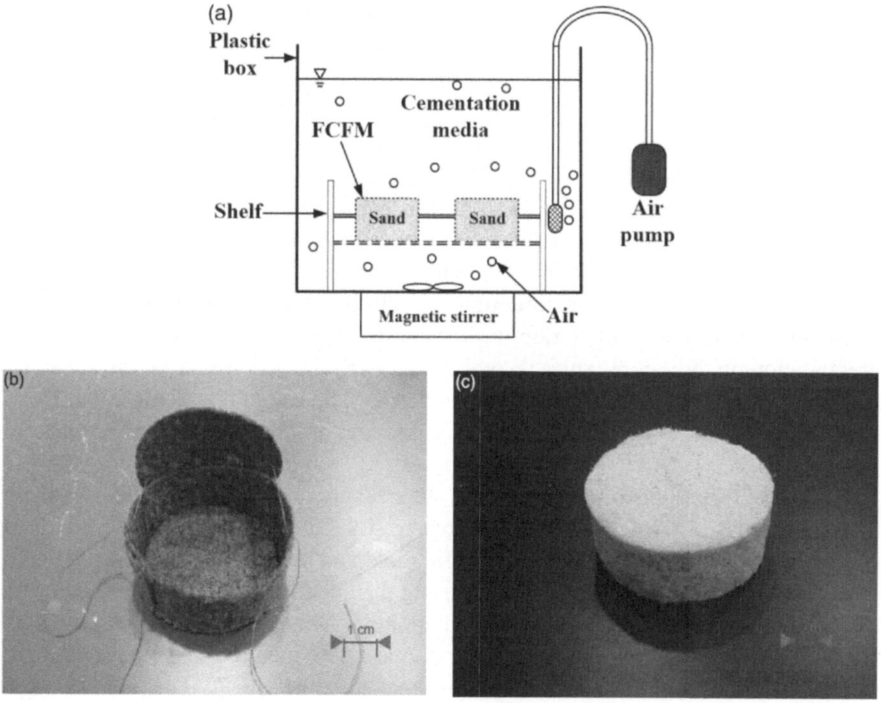

FIGURE 3.28 (a) Schematic drawings of the MICP batch reactor and (b) photograph of the full-contact flexible mold and specimens prepared for the direct shear test. The full-contact flexible mold was fabricated from a geotextile. A PVC shelf was used to support the full-contact flexible mold.

3.2.4.2 Results

The effect of FCFM on MICP-treated soil was compared to that of an opening rigid mold. The opening rigid mold was constructed of PVC. Porous holes were fabricated along the side wall to enhance the flow of water through the rigid mold. For the MICP-treated specimens prepared with a rigid mold, approximately 10 mm of hard crust formed in the upper and lower layers, and an annular shell formed around the soil specimens. However, the central parts of these samples remained loose. Because only the upper and lower layers of the sand were directly exposed to the cementation medium, calcium carbonate precipitation occurred and blocked the surface pores within a short amount of time. Precipitation on the upper and lower layers of the sand inhibited the cementation medium from flowing deeper into the core, thus limiting MICP within the soil specimens. Without pumping of the cementation medium, the contact surface area between the soil specimen and the cementation medium is the main factor affecting the MICP process and the engineering properties of the treated soil.

For the MICP-treated specimens prepared using FCFM, the soil specimen was fully supported and enclosed by the geotextile mold. This geotextile contains many pores, promoting the penetration of nutrients, urea, and calcium chloride to the bacteria in the sand pores, which is beneficial to the MICP process. A geotextile cover was added to reduce calcium carbonate precipitation that could potentially clog the pores in the upper layer of the soil specimen. The entire MICP-treated specimen was relatively uniform, with no obvious fragile layers. The MICP-treated soil specimens were like weak rocks, and a few of the samples were too hard to conduct direct shear tests. Furthermore, a large amount of calcium carbonate was generated on the geotextile surface. However, the fibrous structure of the FCFM ensured that calcium carbonate precipitation did not clog the mold, because it provided a large attachable area for calcium carbonate precipitation. The fibrous structure reduced the amount of calcium carbonate precipitation in the sand pore space as well as the amount of precipitation on the sample surface, which promoted the penetration of the cementation medium into the pore space of the sand core. The use of this flexible geotextile mold enhanced the MICP reaction in the sand cores and resulted in a more uniform precipitation.

Figure 3.29 provides the shear stress as a function of horizontal displacement under a normal stress of 20 kPa for untreated wet sand and for two MICP-treated specimens prepared using the opening rigid mold and the FCFM. The peak strength of the MICP-treated specimen prepared using the opening rigid mold was one order of magnitude higher than that of untreated sand. This finding is consistent with results reported by Chou et al. (2011), who found a 55% increase in the peak strength of the biocalcification of loose sand at a normal stress of 21 kPa. The peak strength of MICP-treated soil specimens prepared using the FCFM was nearly 100% higher than that of the opening rigid mold. By increasing contact between the cementation solution and the bacteria, the FCFM not only enhanced sample uniformity but also significantly improved the engineering properties of sand.

To evaluate the uniformity of the MICP process in the sand specimen, the $CaCO_3$ content in four areas of the unconfined compression test samples was measured.

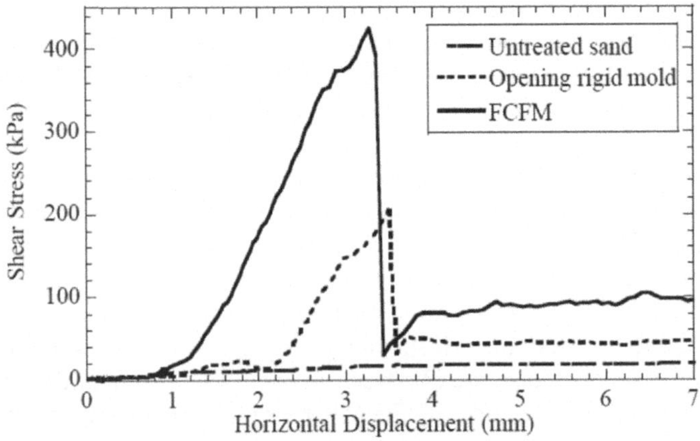

FIGURE 3.29 Comparison of stresses observed in direct shear tests at a normal stress of 20 kPa.

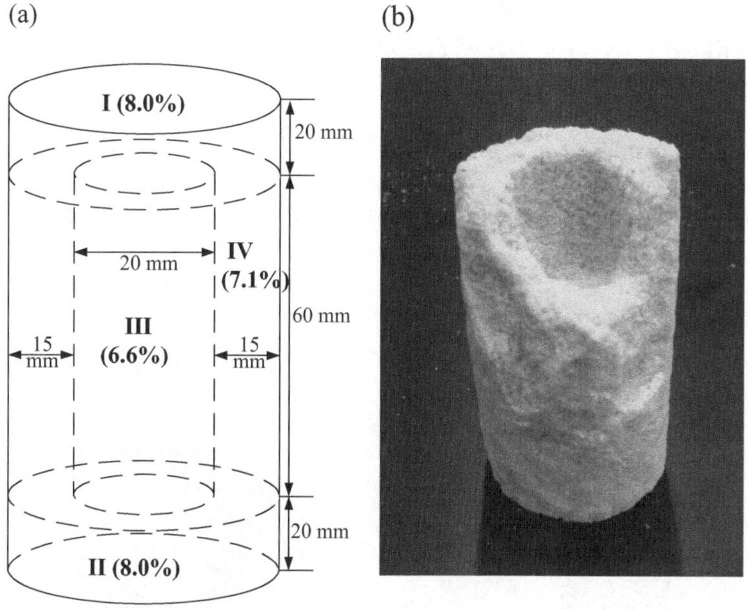

FIGURE 3.30 (a) Four different areas of one unconfined compression test sample for sample uniformity evaluation and (b) photograph of the outer $CaCO_3$ shell and the inner portion prepared using the full-contact flexible mold. The inner portions were manually removed to show the inner core.

Sample preparation was similar to that of direct shear samples, with the exception of the dimensions (51 mm in diameter and 102 mm in height). As indicated in Figure 3.30a, the four areas corresponded to the top (I) (20 mm thick), bottom (II) (20 mm thick), inner cylinder (III) (20 mm in diameter), and outer annular (IV) (15 mm thick). The $CaCO_3$ content in the four areas ranged from 6.6% to 8.0% (by weight). The $CaCO_3$ content in part III was lower than that of the other parts, while $CaCO_3$ contents in parts I and IV were higher.

A new full-contact flexible mold was developed to prepare MICP samples. Because of its fibrous structure, many pores are present on the surface of the full-contact flexible mold, which effectively increases the penetration of nutrients, urea, and calcium chloride into the sand pores and results in increased contact between the cementation solution and the bacteria. Furthermore, this geotextile mold provides a large attachable area, which can attract a portion of the calcium carbonate precipitated on its surface and can reduce precipitation in the sand pore space and on the sample surface. The strength results showed an increase in soil strength resulting from the use of the full-contact flexible mold when compared with the properties obtained for the rigid mold, indicating that soil engineering properties were greatly improved by using the full-contact flexible mold. An examination of the $CaCO_3$ content in different areas of the MICP-treated soil specimen indicated the extent of $CaCO_3$ distribution throughout the sample.

3.3 ENHANCING MICROBIALLY INDUCED CALCITE PRECIPITATION–TREATED SANDY SOIL THROUGH MULTIPLE TREATMENTS

3.3.1 Materials and Experimental Methods

The Ottawa sand used in this section is the same as that in Section 3.2.1. The bacteria and growth medium and cementation media followed are those in Section 3.1.1 and Section 3.1.2.

FIGURE 3.31 Image of (a) full-contact flexible mold. (b) Specimen prepared with full-contact flexible mold. (c) Specimens after MICP treatment.

3.3.1.1 Full-Contact Flexible Molds

All samples in this study were prepared by full-contact flexible mold, which was developed by Zhao et al. (2014b). This mold is made of geotextile that is a polypropylene, staple fiber, and needle-punched nonwoven material. The selected geotextile has grab tensile strength of 1,700 kN, grab elongation of 50%, trapezoidal tear of 670 kN, apparent opening size of 0.15 mm, water flow rate of 34 mm/s, thickness of 1.5 mm, and unit mass of 200 g/m^2 (Zhao et al. 2014b). This geotextile mold can stand alone and hold loose sandy soil, as shown in Figure 3.31. The mold can allow the cementation media to freely contact sand particles due to its large number of pores. The specimens made by full-contact flexible molds for unconfined compression tests were cylinder-shaped, with 38.1 mm diameter and 76.2 mm height. A specially designed frame was built for multiple full-contact flexible molds in MICP treatment (Figure 3.31c).

3.3.1.2 Immersing Method in Batch Reactor

All soil samples were treated in a completely stirred batch reactor. The reactor included a 25 L plastic box to contain soil samples and the cementation media, a magnetic mixer to keep the solution uniform, and an air pump to provide oxygen for the bacteria (Figure 3.32). In the immersing environment, each soil specimen can access the cementation media from the top, from the bottom, and from the surrounding of the molds to make uniform contact with the solution. A major feature of this method is to allow soil samples to fully immerse into the cementation media and to allow the cementation media to freely penetrate into the soil samples (Zhao et al. 2014b). As the cementation media permeates into the soil specimens, MICP can occur in sample pores, and the produced $CaCO_3$ can bond the sand particles together to improve

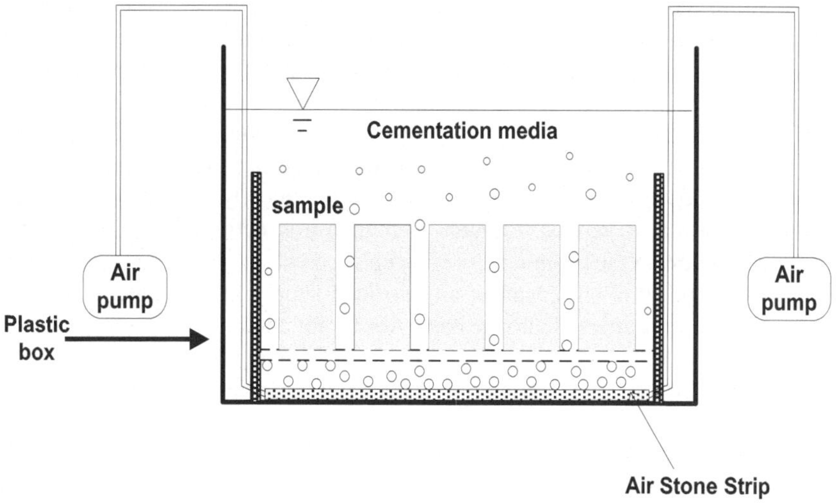

FIGURE 3.32 Batch rector component and setup.

engineering properties. Zhao et al. (2014b) found that the produced $CaCO_3$ was fairly distributed in the samples, and the samples could be easily taken out from the mold for different mechanical tests.

In the immersing environment, there is no hydraulic gradient to drive the flow through the soil specimen in MICP molds. When the chemical of the cementation media reacts under the catalysis of bacteria, the concentration of these chemicals is lower in the soil specimens, which causes the chemical substances to diffuse from a high-concentration area to a low-concentration area to continue MICP deeper in the soil specimens (Zhao et al. 2014a).

3.3.1.3 MICP-Treated Sample Preparation in Single Treatment

The 140 ± 5 g sand was uniformly mixed with 35 mL bacteria solution and then air-pluviated into the mold to reach a median dense condition (Dr in the range of 42~55%, and dry density of sand ranging from 1.58 to 1.64 g/cm^3). A total of 34 samples were placed on the shelf (Figure 3.31c). Then, the entire shelf was slowly immersed into the batch reactor filled with cementation media. Without adding any additional cementation media, bacteria, or growth media, MICP was maintained for 7~28 days. At the end of a single treatment, the shelf and the samples were removed from the reactor, and MICP-treated samples were removed from the mold by cutting the mold. Details of sample preparation can be found in Zhao et al. (2014b).

3.3.1.4 MICP-Treated Sample Preparation in Multiple Treatments

A batch reactor capable of holding 34 samples at a time treated all samples together during the initial MICP cycle, which lasted seven days, as described in Section 3.3.1 (6). After the first treatment cycle, additional MICP cycles were conducted on the same samples to achieve multiple treatments. Following each 7-day cycle, seven specimens were removed for unconfined compressive strength (UCS) testing. The remaining samples underwent further MICP treatment. The procedure for multiple treatments involved several steps performed sequentially. First, the treated samples were removed from the batch reactor, and the geotextile mold was carefully cut to extract the MICP-treated soil specimens. These specimens were then oven-dried at 105°C for 24 h to achieve a stable moisture content. After drying, the specimens were soaked in freshly prepared bacteria solution, ensuring the bacterial concentration remained consistent throughout all cycles. Subsequently, the soaked specimens were placed into new geotextile molds to maintain structural integrity. The used cementation media in the batch reactor was replaced with a freshly prepared solution of the same concentration, ensuring an adequate supply of reactants. Finally, the specimens were re-immersed into the batch reactor for another 7-day treatment cycle. This process was repeated for up to four cycles, enhancing the calcite precipitation and improving the soil stabilization properties of the samples over time.

3.3.1.5 Unconfined Compression Strength Test

The MICP-treated soil specimens for unconfined compression tests were cylinder-shaped, with a 2H:1D ratio. In single MICP treatment, the shelf was removed from the solution after 7 days of reaction. Each geotextile mold was cut, and the MICP-treated soil specimen was taken out. For multiple treatments, the reaction time for each treatment is 7 days. In this study, the longest treatment time is 28 days for quadruple MICP treatments. Unconfined compression strength tests were conducted under strain-controlled conditions at a uniform loading rate of 1.5%/min, in accordance with ASTM D2166.

3.3.1.6 Permeability Test

The falling head permeability testing method was used to test the hydraulic conductivity of MICP-treated sand following ASTM D5084–16a. For untreated Ottawa sand, constant head permeability testing method was used to conduct permeability test following ASTM D2434–68.

3.3.1.7 $CaCO_3$ Content

To determine the $CaCO_3$ content in MICP-treated sand, specimens were crushed using a mortar and oven-dried. The dry soil was washed in an HCl solution (0.1 M) to dissolve precipitated carbonates, then rinsed, drained, and oven-dried. The difference between the two weights is the weight of the carbonates that were precipitated in the specimen (Rebata-Landa 2007).

3.3.1.8 SEM Analysis

After MICP treatment, the soil specimen was examined with scanning electron microscopy (SEM) (Lyra 3) to examine the precipitation of calcium carbonate under microscale. The MICP-treated samples were oven-dried overnight at 105°C before test and mounted on the stubs with adhesive carbon conductive tabs. Then, prepared samples were imaged by secondary electron detection.

3.3.2 Effect of Reaction Time on MICP Single Treatment

Reaction time is one of key factors for the MICP process (Al Qabany et al. 2012). This chapter studied 7–28 days of reaction time in a single treatment. Figure 3.33 shows the stress–strain relationship of MICP-treated sample as a function of reaction time at a cementation media concentration of 0.25 M Ca and 0.5 M Ca. There is no significant improvement on the peak strength of MICP-treated sample when reaction time exceeds 7 days, as shown in Figure 3.33. In Figure 3.33b, the peak

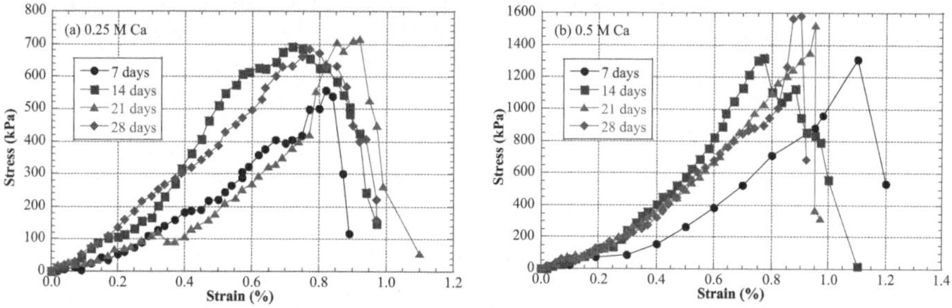

FIGURE 3.33 Stress–strain relationship as a function of reaction time with different cementation media concentration in single treatment: (a) 0.25 M Ca. (b) 0.50 M Ca.

strength of 7 days' MICP-treated sample is around 1,300 kPa, a finding similar to that in Zhao et al. (2014a), while 14 days' MICP-treated sample did not show any change on peak strength. The peak strength of MICP-treated sample only increased to around 1,500 kPa when the reaction time was extended to 21 and 28 days. This could be explained by the fact that the bacteria with negative surface charge had already been covered by $CaCO_3$ precipitation, and enzyme activity was reduced after 7 days (Zhao et al. 2014a).

3.3.3 Effect of Cementation Media Concentration with Multiple Treatments

The maximum MICP treatment time in this study was four times, which is a total of 28 days of treatment. After each MICP treatment, the unconfined compression strength, $CaCO_3$ content, hydraulic conductivity, and dry density of MICP-treated soil specimens were measured.

Figure 3.34 shows the stress–strain curve of multiple MICP-treated sands in different cementation media concentrations (0.25 M, 0.50 M, and 0.75 M), with up to four treatments. When the cementation media concentration was 0.25 M Ca, peak strength was around 600 kPa, with a failure strain at 0.9% for a single treatment. The peak strength increased to 1,300 kPa (an increase of 110% compared with 0.25 M Ca), with a failure strain of 1.1%, when the cementation media concentration doubled to 0.5 M Ca for a single treatment. When the concentration of the cementation media increased to 0.75 M Ca, the peak strength increased to 2,200 kPa (an increase of 69% compared with 0.5 M Ca), with a failure strain of 1.3% for a single treatment. These results indicate that the unconfined compression strength and failure strain of MICP-treated sand increase with cementation media concentration in a single treatment. However, the utilization of the cementation media decreases in high cementation media concentration, and a low cementation media concentration can improve cementation media utilization, which is similar to the findings in Nemati et al. (2005) and Soon et al. (2014).

A single MICP treatment has limited improvement on soil strength. Zhao et al. (2014a) found that the peak strength of MICP-treated sand with 1.5 M Ca cementation media concentration achieved only 2,100 kPa. Multiple MICP treatment can overcome the limitation of a single treatment. Figure 3.34a shows that the peak strength of a quadruple-treated sample was around 4,000 kPa, which is over 6.6 times that of a single-treated sample (600 kPa), 2.5 times that of double-treated sample (1,600 kPa), and 1.5 times that of triple-treated sample (2,700 kPa). Each treatment can increase around a 1,000 kPa peak strength of MICP-treated sample in 0.25 M Ca cementation media concentration. The failure strain of a quadruple-treated sample was around 1.6%, which is 1.3 times that of triple-treated sample (1.2%). This finding indicates that multiple MICP treatments with 0.25 M Ca cementation media concentration are still efficient after quadruple treatments.

Figure 3.34b shows the stress–strain curve of multiple MICP-treated sands up to four times in 0.5 M cementation media. It can be seen that the growth rate of UCS is reduced after second treatment. There was around 1,400 kPa increase in peak strength between single-treated sample (1,300 kPa) and double-treated sample

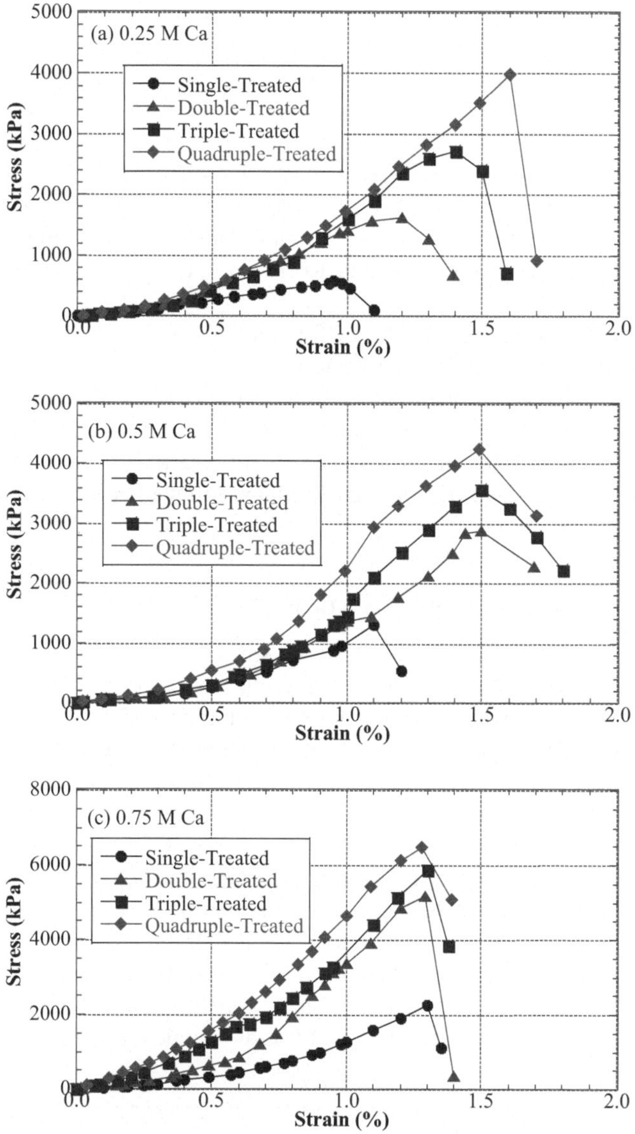

FIGURE 3.34 Stress–stain relationship of MICP-treated soil specimen in multiple treatments at different cementation media concentration: (a) 0.25 M Ca, (b) 0.50 M Ca, and (c) 0.75 M Ca.

(2,700 kPa), while, after second treatment, the increased strength reduced to 800 kPa in both triple-treated sample (3,500 kPa) and quadruple-treated sample (4,300 kPa). The less peak strength increasing rate is related to the reduction on void space of the soil specimen. The failure strain of quadruple-treated sample was around 1.5%, which is nearly the same as that of the triple-treated sample (1.5%) and double-treated sample (1.48%). The efficiency of the multiple treatment method decreased after second-time treatment in the 0.5 M cementation media.

When the cementation media increased to 0.75 M, the peak strength of the double-treated sample was around 5,100 kPa, which is over two times that of a single treated sample (2,200 kPa). However, the growth rate of the peak strength reduced dramatically after the second treatment. After the fourth treatment, peak strength (6,400 kPa) was only 1.25 times that of double-treated sample. Similar to the case of 0.5 M cementation media concentration, the efficiency of the multiple treatment method also decreased after second-time treatment in the 0.75 M cementation media concentration.

The failure stress and strain of MICP-treated sample both increased with the increase in cementation media concentration. MICP-treated samples exhibited as a brittle material according to their failure strain (<1.5%). Meanwhile, tangent Young's modulus can be calculated based on this stress–strain data (Table 3.7). Young's modulus increased with the increase in cementation media concentration and treatment time, which indicated that the MICP-treated sample became stiffer with more treatment.

3.3.4 Equivalent Reaction of Multiple MICP Treatments

Figure 3.35 shows a comparison of the stress–strain curves of multiple-MICP treatment samples under different cementation media concentrations. For clarification, only partial stress–strain curves were compared in Figure 3.35, where the curve ended at failure strain, and a trend line was added to each group of data. For each comparison, the quantities of $CaCl_2$ and urea (by weight) in the cementation media were the same, but the chemicals were applied with a different concentration of cementation media and treatment times. For example, double treatment of 0.25 M Ca had same quantity of $CaCl_2$ and urea with single treatment of 0.5 M Ca. Then, strength performance was compared between the two treatment procedures. Figure 3.35 (a and b) shows that a low cementation media concentration (0.25 M Ca) with multiple treatments can result in a better peak strength performance than for a high concentration (0.5 M Ca; 0.75 M Ca) in a single treatment. However, if the cementation media concentration is higher than 0.25 M Ca, the efficiency of multiple treatments

TABLE 3.6
Tangent Young's Modulus with Multiple Treatments

	Single-Treated	Double-Treated	Triple-Treated	Quadruple-Treated
Cementation Media Conc.	E_f* (MPa)	E_f (MPa)	E_f (MPa)	E_f (MPa)
0.25 M Ca	60 ± 6	120 ± 16	170 ± 28	230 ± 38
0.5 M Ca	120 ± 21	200 ± 35	260 ± 42	310 ± 42
0.75 M Ca	170 ± 14	380 ± 33	445 ± 34	476 ± 24

Note: *$E_f = \sigma_f/\varepsilon_f$; σ_f = peak strength; ε_f = failure strain

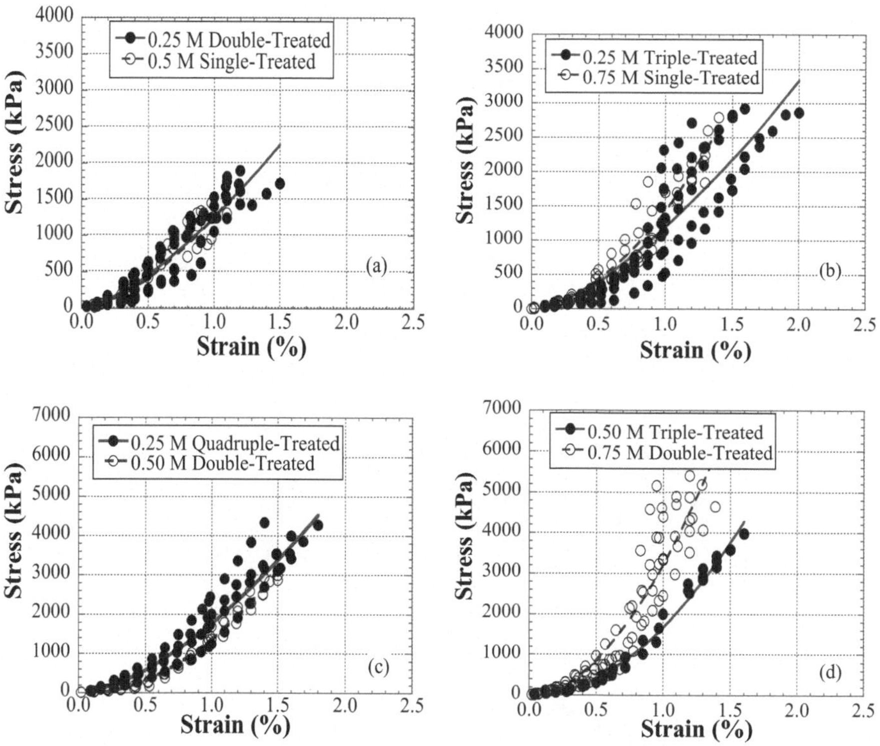

FIGURE 3.35 Comparison of stress–strain curves between (a) 0.25 M double-treated sample and 0.5 M single-treated sample, (b) 0.25 M triple-treated sample and 0.75 M single-treated sample, (c) 0.25 M quadruple-treated sample and 0.5 M double-treated sample, and (d) 0.75 M double-treated sample and 0.5 M triple-treated sample.

becomes limited after double treatment. Figure 3.35c shows that 0.25 M Ca cementation media concentration with four treatments can result in a similar peak strength performance at 0.5 M Ca cementation media concentration in double treatment. As shown in Figure 3.35d, the double-treated sample (0.75 M Ca) performed better than the triple-treated sample (0.5 M Ca). Therefore, the most efficient treatment time for a higher cementation media concentration (i.e., 0.5 M Ca and 0.75 M Ca in this study) is two times.

3.3.5 Effect of $CaCO_3$ Content

Figure 3.36 shows the $CaCO_3$ content of MICP-treated samples at different cementation media concentrations and treatment times. As shown in Figure 3.36, the $CaCO_3$ content in the MICP-treated samples increased with the number of treatment times. Meanwhile, UCS increased with more $CaCO_3$ content in the MICP-treated samples. Figure 3.36a shows the average $CaCO_3$ content to be 4.0%, and the corresponding UCS is 600 kPa after the first treatment in 0.25 M Ca cementation media

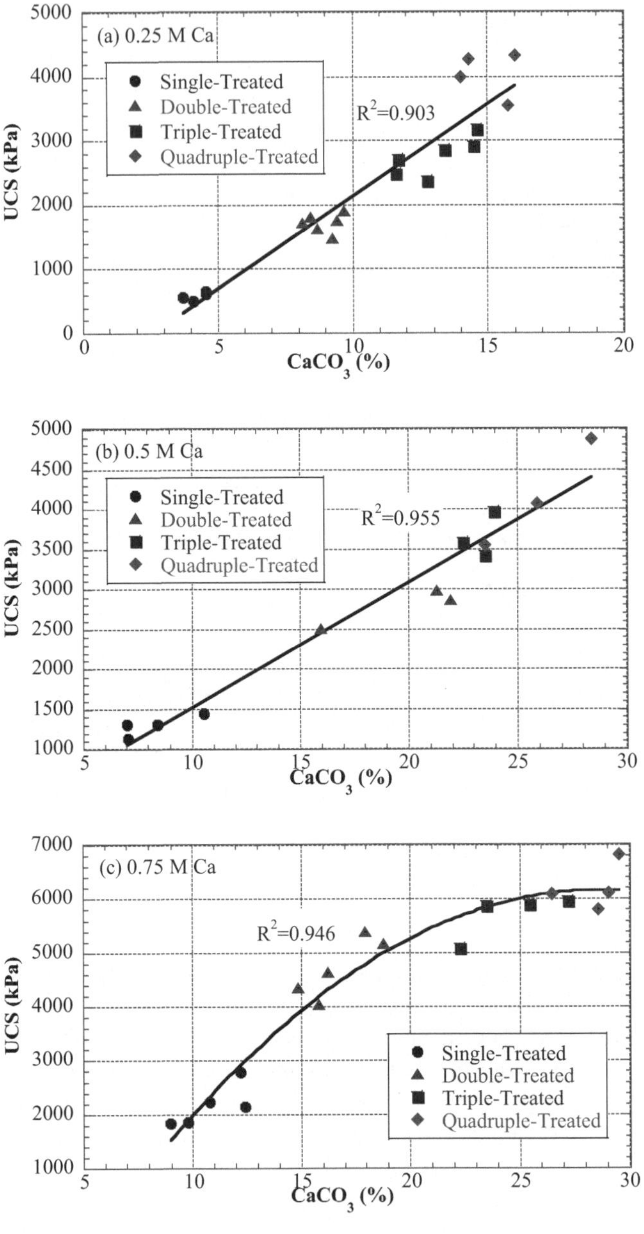

FIGURE 3.36 UCS of MICP-treated sand as a function of $CaCO_3$ content in different cementation media concentration: (a) 0.25 M Ca, (b) 0.50 M Ca, and (c) 0.75 M Ca.

concentration. $CaCO_3$ content increased to 8.2% after the second-time treatment. There is 5% more $CaCO_3$ (13.2%) found after the third treatment. However, only around 2% more $CaCO_3$ (15.1%) was produced after the fourth-time treatment. The strength of MICP-treated material mainly depends on the part of $CaCO_3$ precipitation which can bond the soil particles. This finding can be explained thus: (1) The void space between the sand particles is reduced after multiple treatments, and (2) when the void space is reduced, there is a better chance for effective $CaCO_3$ bond produced by multiple treatments.

In Figure 3.36b, the concentration of cementation media increased to 0.5 M. After the first treatment, the average $CaCO_3$ content was 8.1%, and the corresponding UCS was 1,300 kPa. $CaCO_3$ content increased to 19.6% after second-time treatment. Only 3% more $CaCO_3$ (22.7%) is found after the third-time treatment, and 2% more $CaCO_3$ (25.1%) after the fourth-time treatment. $CaCO_3$ precipitation increment and UCS growth rate reduced quickly under 0.5 M Ca cementation media concentration after double treatment. Figure 3.36c also demonstrates that the UCS of MICP-treated sample with 0.75 M Ca increased with $CaCO_3$ content, whereas $CaCO_3$ content has negligible effect on UCS when $CaCO_3$ content exceeds 25%. These results indicate that for the high cementation media concentration, the multiple treatments only have a prominent effect on the second-time treatment. When the soil sample is only MICP-treated for one time, some $CaCO_3$ precipitations are only attached to the surface of sand particles that may not help bond the sand particles (Cheng and Cord-Ruwisch 2012; Zhao et al. 2014a). Multiple treatments can increase the amount of $CaCO_3$ in the void space between sand particles. Meanwhile, it may produce a more effective $CaCO_3$ bond to improve the properties of MICP-treated samples.

3.3.6 Void Ratio and Dry Density

The calculated void ratio and measured dry density after each MICP treatment for different cementation media concentrations are shown in Figure 3.37. The void ratio was calculated assuming that the specific gravity of calcite is 2.7 (Cho et al. 1997). It is shown that the void ratio of MICP-treated sample decreases during MICP treatment times, and the dry density of MICP-treated sample increases with MICP treatment times. The void ratio of MICP-treated sample reduces with treatment times at the 0.25 M Ca cementation media concentration. But for the 0.5 M Ca and 0.75 M Ca cementation media concentrations, the reduction of void ratio along treatment times slows down after second-time treatments. This result shows a similar trend as that of the UCS result, which indicates that the efficiency of the multiple treatment method decreases after second-time treatment with 0.5 M Ca and 0.75 M Ca cementation

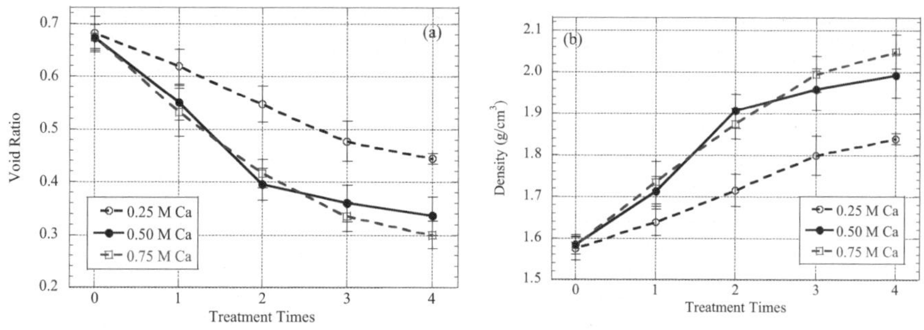

FIGURE 3.37 (a) Void ratio and (b) dry density of MICP-treated soil specimen versus treatment times at different cementation media concentration.

media concentrations. Meanwhile, Figure 3.37b indicates that the dry density of MICP-treated sample increases with treatment times. For 0.25 M Ca, 0.5 M Ca, and 0.75 M Ca cementation media concentration, the highest dry density can increase up to 1.83 g/cm^3, 1.99 g/cm^3, and 2.05 g/cm^3 after four MICP treatments, respectively.

3.3.7 Permeability of MICP-Treated Sand

Figure 3.38 shows the hydraulic conductivity of MICP-treated samples as a function of treatment times and void ratio. The hydraulic conductivity of untreated Ottawa sand was around 0.14 cm/s from the constant head permeability test. In Figure 3.38a, the hydraulic conductivity of MICP-treated sample decreases approximately two orders of magnitude after four MICP treatments. Meanwhile, the hydraulic conductivity of MICP-treated sample is also reduced more in a higher concentration of cementation media. As shown in Figure 3.38b, the hydraulic conductivity of MICP-treated sand decreases with a reduction in void ratio. This result is similar to the trend shown in Gyu et al. (2017).

3.3.8 SEM Analysis

The formation of MICP is examined by SEM, and there are noticeable differences between the size and the quantity of calcium carbonate precipitation for different cementation media concentration and treatment times. The dimensions of $CaCO_3$ precipitations and void space are shown in Figure 3.39. In Figure 3.39a, MICP-treated specimens with 0.25 M cementation can form small spherulitic calcium carbonate precipitations with an average radius of 17.76 μm, exhibited on the sand particle surfaces and inter-particle contact points. The average radius increases to 24.66 μm after quadruple treatment, and there is more precipitation at the interparticle contact point (Figure 3.39b). Meanwhile, void space is reduced from 140.83 μm to 55.52 μm after quadruple treatment, and $CaCO_3$ precipitation is distributed denser compared with single treatment, as shown in Figure 3.39b. From Figure 3.39c, $CaCO_3$ precipitation in MICP single treatment with 0.75 cementation media is still exhibited as a spherulitic crystal. In contrast, there is no remarkable spherulitic precipitation in the surface after quadruple treatment, and the sand is almost fully covered by calcium carbonate

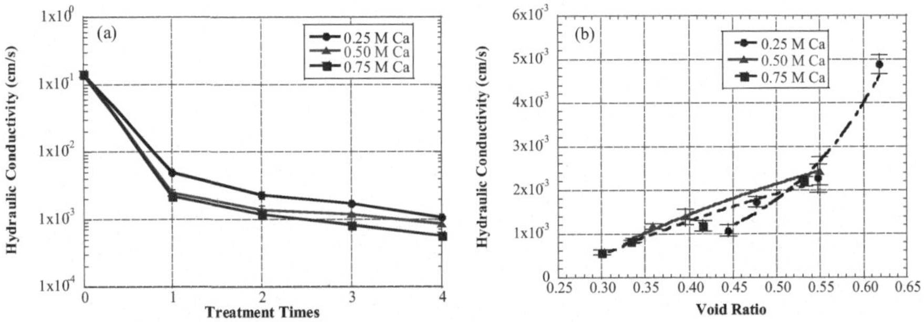

FIGURE 3.38 Hydraulic conductivity of MICP-treated sand as a function of (a) treatment times. (b) Void ratio in different cementation media concentration.

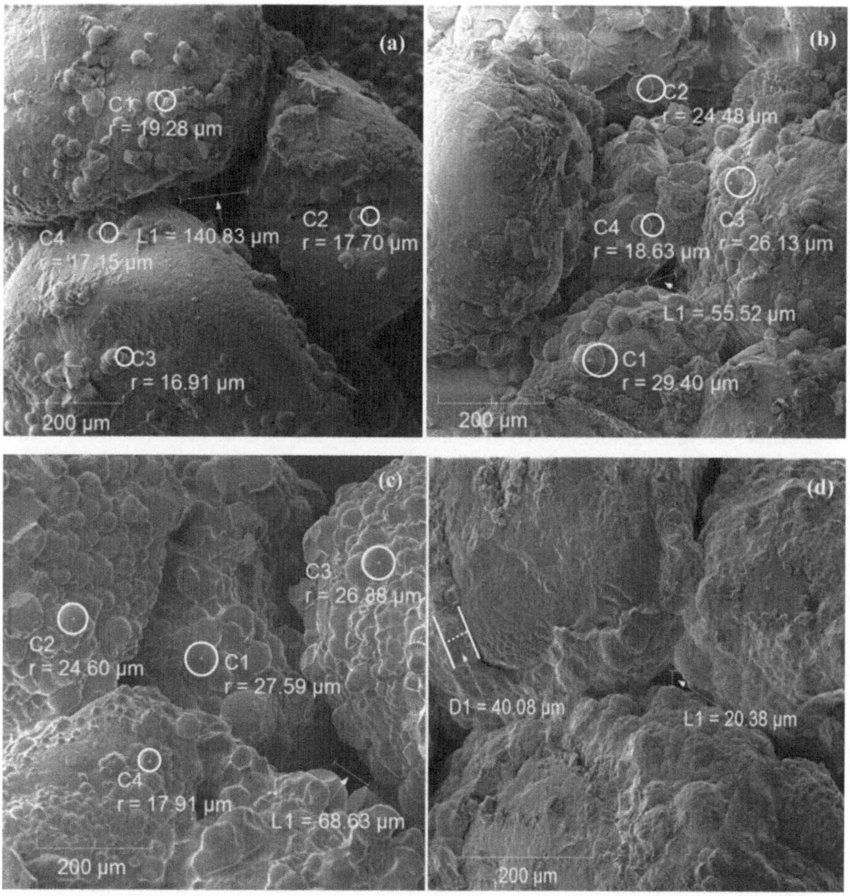

FIGURE 3.39 SEM image of MICP-treated specimen: (a) single treatment of 0.25 M cementation media concentration; (b) quadruple treatment of 0.25 M cementation media concentration; (c) single treatment of 0.75 M cementation media concentration; (d) quadruple treatment of 0.75 M cementation media concentration

precipitation with a 40.08 μm depth, as shown in Figure 3.39d. After a single treatment of 0.75 M cementation media, only a 68.63 μm void space remained for more precipitation, and void space was reduced to 20.38 μm after quadruple treatment.

3.4 FACTORS AFFECTING THE MICP PROCESS

3.4.1 Effect of Bacteria and Urease Concentration on the MICP Process

3.4.1.1 Materials and Methods

3.4.1.1.1 Bacteria and Cementation Media

The Bacteria and growth medium and cementation media followed are those in Section 3.1.1 and Section 3.1.2.

3.4.1.1.2 Precipitation Experiments

Four different bacteria concentrations (OD_{600} = 0.1, 0.3, 0.6, and 1.0) and initial urease concentrations (1.0, 2.5, 5.0, and 8.0 g/L) were set up in this experiment. The four urease concentrations were selected because the urease activity and electronic conductivity (EC) are similar to those of different bacteria concentrations. The experimental group list used in this study is summarized in Table 3.7. The precipitation experiments were carried out at 25°C in beakers, and no soil was involved in this experiment. The bacteria or urease solution (5 mL) was extracted by pipette and mixed with the cementation media (100 mL). Four glass slices were placed at the bottom of each beaker to collect the precipitations during the MICP/EICP process. A set of time-series experiments was conducted for each mixed solution. The mixed solutions were monitored at 12, 24, 72, and 168 h, and one piece of glass slice was then collected at each time for further SEM analysis.

Another group of experiments was conducted in a beaker only. The bacteria or urease solution (5 mL) was extracted by pipette and mixed with cementation media (100 mL). At sampling periods of 12, 24, 48, 72, 96, and 168 h, a 10 mL solution was sampled and filtered. The filtered solution was analyzed for Ca^{2+} concentration. The calcium carbonate precipitation deposited on the bottom of each beaker at different periods was collected for further XRD and FTIR analysis.

The calcium carbonate precipitations on sand particles were also investigated in this study. Ottawa silica sand (99.7% quartz) was used, and the soil sample was prepared with a full-contact flexible mold developed by Zhao et al. (2014a). The sand is uniformly graded, with a median particle size of 0.46 mm, and no fines were included. It was classified as poorly graded sand based on the Unified Soil Classification System. The 140 ± 5 g sand was uniformly mixed with a 45 mL bacteria or urease solution and then air-pluviated into the mold to reach a median dense condition (dry density of sand ranged from 1.58 to 1.64 g/cm^3). Then, the soil specimens were immersed into the cementation media and treated for 168 h. The batch reactor is shown in Figure 3.32. Soil samples are 1.5 in (38.1 mm) in diameter and 3.0 in (76.2 mm) in height. After the reaction, the samples were removed for SEM analysis. All samples were conducted at least in triplicate.

3.4.1.1.3 Analytical Procedures

Ca^{2+} concentration was measured by inductively coupled plasma-optical emission spectrometer (ICP-OES, SHIMADZU). The detection limitation of ICP-OES is one

TABLE 3.7
Summarized Group List in Precipitation Experiments

Group ID	Components
Group 1	OD_{600} = 0.1 and urease = 1.0 g/L
Group 2	OD_{600} = 0.3 and urease = 2.5 g/L
Group 3	OD_{600} = 0.6 and urease = 5.0 g/L
Group 4	OD_{600} = 1.0 and urease = 8.0 g/L

part per billion (ppb). The crystal morphologies of calcium carbonate were determined by Fourier transform infrared spectroscopy (FTIR, Spectrum Two, PerkinElmer), X-ray diffraction (XRD, MiniFlex 600, Rigaku), and scanning electron microscopy (SEM, Lyra 3, TESCAN Inc). The FTIR spectra were obtained for 50 scans at a resolution of 4 cm^{-1}. The XRD spectra were performed using a Cu target, and the scan was obtained at the exploration range $2\theta = 20–70°$ and a scanning velocity of 0.15 s/step. The precipitations were mounted on the stubs with adhesive carbon conductive tabs for SEM observation. The prepared SEM samples were imaged by secondary electron detection.

3.4.1.2 Effects of Bacteria and Urease Concentration on the Kinetics of $CaCO_3$ Precipitation

Calcium carbonate precipitation can be affected by many external environmental factors; among them, bacteria/urease concentration may be one of the most important factors (Al Qabany et al. 2012; Mortensen et al. 2011; Zhao et al. 2014a). The time-dependent behaviors of Ca^{2+} concentration in the mixed solution at different initial concentrations of bacteria and urease are shown in Figure 3.40. In the case of groups 1 and 2, Ca^{2+} concentration decreased faster in the urease-induced system in the first 72 h than that in the bacteria-induced system. Ca^{2+} concentration remained approximately 110 and 100 mmol/L at the end of 72 h in groups 1 and 2, respectively, indicating an incomplete reaction in the two groups. The faster decrease of Ca^{2+} concentration in the urease-induced system became more apparent in groups 3 and 4. Ca^{2+} concentration had been completely precipitated in the urease-induced system (8 g/L) of group 4, which was approximately 72 h earlier than that in the bacteria-induced system (OD 600 = 1.0). It was also clearly seen that the higher initial concentrations of bacteria and urease could yield faster decreases of Ca^{2+} concentration, which indicated that more $CaCO_3$ precipitations could be generated. However, the high concentration of bacteria and urease concentration doesn't represent the high urease activity. This conclusion is only correct for this study under the given experimental conditions. Meanwhile, the concentration of Ca^{2+} remained constant or decreased marginally in urease solution after 72 h, which suggested that urease used could be degraded by the surrounding chemicals. In contrast, Ca^{2+} concentration decreased gradually along with time in the bacteria-induced system. This conclusion is only correct for cases with lower urease/bacteria concentrations.

The modified exponential logistic model was used to fit the data of precipitated Ca^{2+} concentration according to Stocks Fischer et al. (1999), which is expressed as the following regression equation:

$$y_t = \frac{C}{1+e^{-k(t-t_0)}} \tag{3.1}$$

where *yt* is the amount of precipitated $CaCO_3$, C is the range of yt (precipitated $CaCO_3$) variation, k is the reaction rate constant, t is the reaction time, and t_0 is the time at the maximum (dc/dt). The consumption of Ca^{2+} in MICP/EICP is caused by $CaCO_3$ precipitation. The precipitated Ca^{2+} concentration was calculated with an initial Ca^{2+}

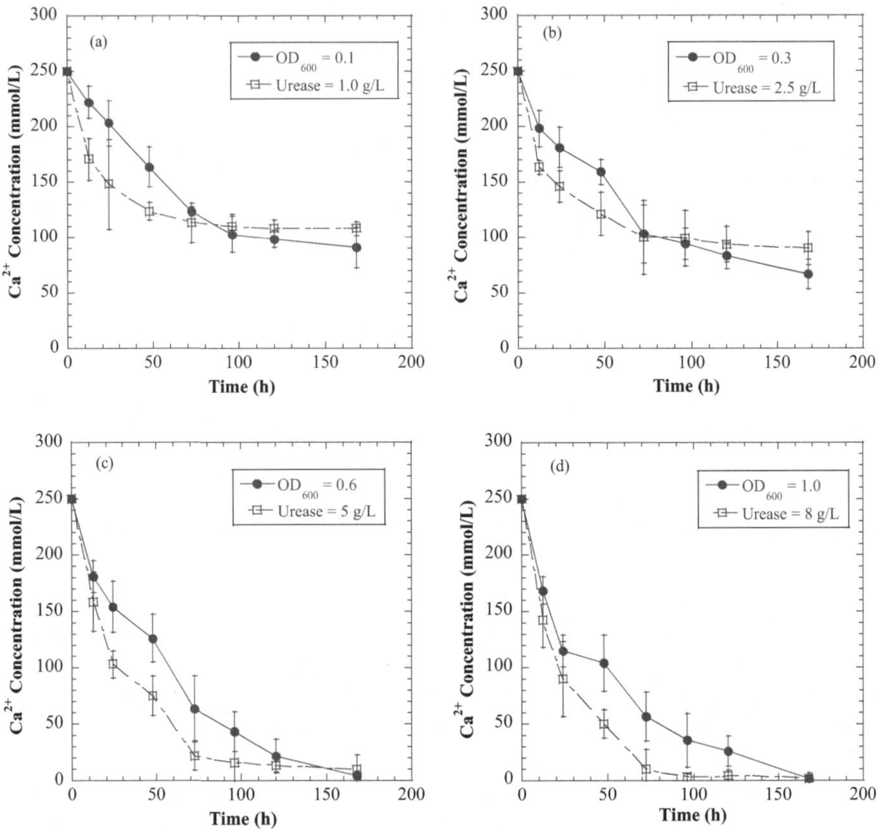

FIGURE 3.40 Temporal Ca^{2+} concentration in the mixed solution at different groups: (a) group 1, (b) group 2, (c) group 3, and (d) group 4.

TABLE 3.8

The Regression Parameters of Precipitated Ca^{2+} Concentration at Different Initial Bacteria and Urease Concentrations

Bacteria Concentration	C (mmol/L)	k (h^{-1})	t_0 (h)	R^2	Urease Concentration (g/L)	C (mmol/L)	k (h^{-1})	t_0 (h)	R^2
OD_{600} = 0.1	157.55	0.027	43.09	0.99	1.0	141.60	0.037	6.83	0.99
OD_{600} = 0.3	179.95	0.031	41.56	0.98	2.5	159.56	0.045	7.06	0.99
OD_{600} = 0.6	246.45	0.036	39.38	0.99	5.0	240.53	0.052	15.63	0.97
OD_{600} = 1.0	247.46	0.048	29.56	0.96	8.0	248.36	0.053	19.46	0.98

concentration of 250 mmol/L minus Ca^{2+} concentration in the solution. Figure 3.41 showed the fitting curves at different initial concentrations of bacteria and urease. The model was well-fitted with a confidence value of 95%. Each experiment data point on this graph was the average of at least triplicate samples. Table 3.8 summarizes the

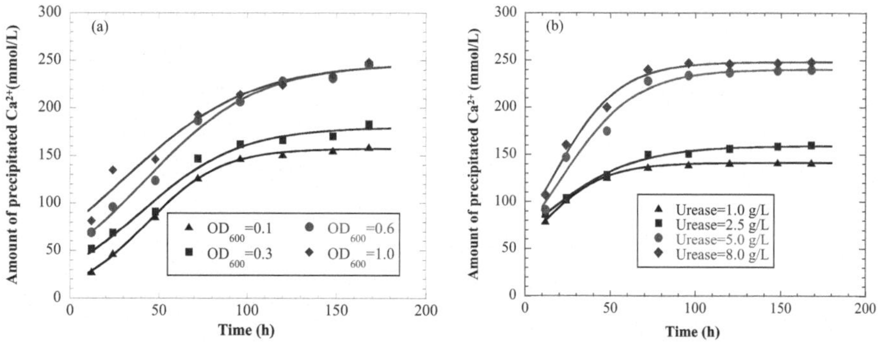

FIGURE 3.41 Exponential logistics model fitting curves of precipitated Ca^{2+} concentration at different initial (a) bacteria concentrations and (b) urease concentrations.

equation parameters, including C, k, t, t_0. In general, the higher initial concentrations of bacteria and urease had a greater rate constant, indicating a faster reaction rate. Moreover, the precipitated Ca^{2+} concentration had only slightly increased or remained constant in the urease solution after 72 h. This result is consistent with the finding of the remaining Ca^{2+} concentration in Figure 3.40. Stocks Fischer et al. (1999) found that the kinetics of bacteria-induced (*Sporosarcina pasteurii*) $CaCO_3$ precipitations at different initial Ca^{2+} concentration was well-fitted by this model.

3.4.1.3 Effects of Bacteria and Urease Concentration on the Morphology of $CaCO_3$ Crystals

Figure 3.42 shows the XRD patterns of $CaCO_3$ crystals at the end of each precipitation experiment (168 hours). The calcite phase was found in all XRD results, and the leading crystal face of calcite was C(104). An apparent vaterite phase was observed simultaneously in $CaCO_3$ crystals induced by different bacteria concentrations. The diffraction peaks happened at the diffraction angle of 24.9° and 27.1°, respectively, corresponding to vaterite face V(110) and V(112). It suggests that urease-induced $CaCO_3$ crystals favored the calcite phase, and bacteria-induced $CaCO_3$ crystals favored the formation of vaterite. The concentration of bacteria or urease did not have an apparent effect on the morphology of $CaCO_3$ crystals.

In order to better understand the formation processes of $CaCO_3$ crystals in the precipitation experiment, the $CaCO_3$ crystals obtained at different precipitation periods were used to conduct FTIR and SEM analyses. The FTIR spectra of $CaCO_3$ crystals at different times in bacteria and urease systems are shown in Figure 3.43 and Figure 3.44, respectively. There were no hydrated phases in all systems identified by FTIR absorption bands, because of the absence of absorption bands at 866 and 1074 cm^{-1} (Dyer and Viganotti 2016; Li et al. 2013; Xu et al. 2004). The presence of vaterite and calcite was confirmed by the observation of absorption bands at 747 and 1,085 cm^{-1} for vaterite and 711 cm^{-1} for calcite. These similar absorption bands of vaterite (717 and 1,085/1,084 cm^{-1}) and calcite (708/712 cm^{-1}) were also reported by Loste et al. (2003) and Li et al. (2013). There was only the vaterite phase existing

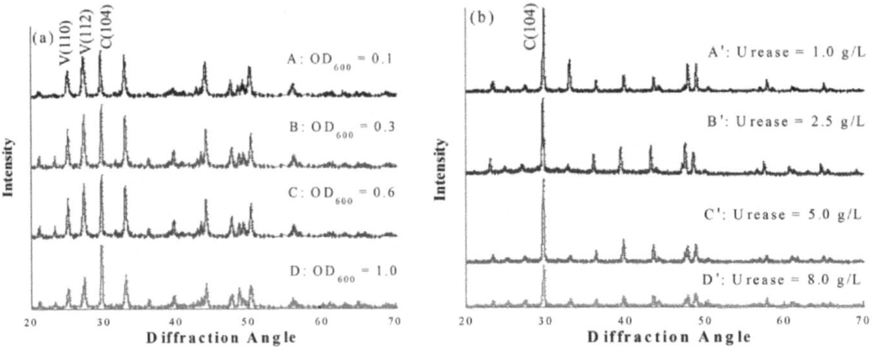

FIGURE 3.42 XRD patterns of $CaCO_3$ crystals at different initial (a) bacteria concentrations and (b) urease concentrations at 168 h.

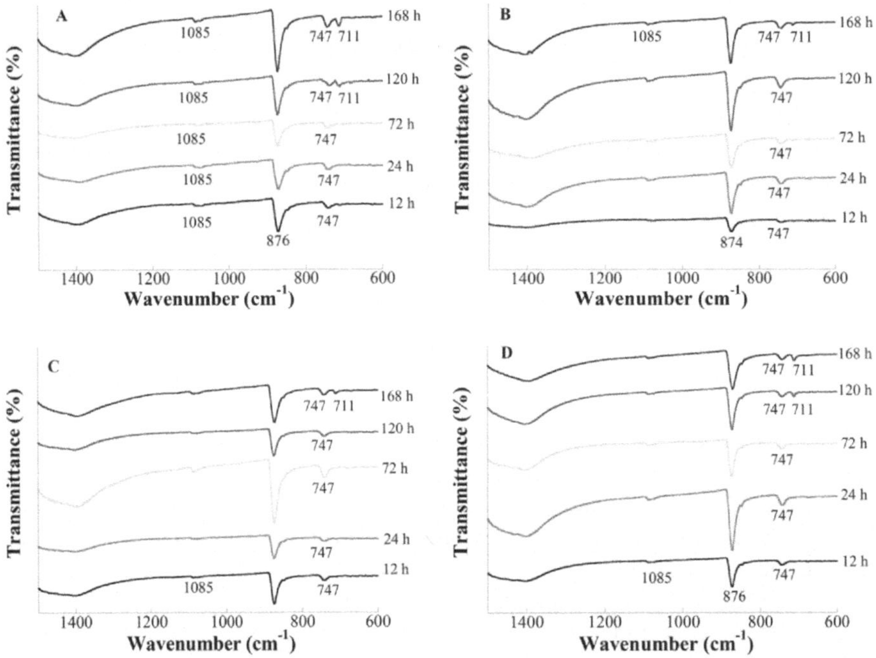

FIGURE 3.43 FTIR spectra of $CaCO_3$ crystals as function of time at different initial bacteria concentrations: (a) $OD_{600} = 0.1$, (b) $OD_{600} = 0.3$, (c) $OD_{600} = 0.6$, (d) $OD_{600} = 1.0$.

before 72 h in all bacteria concentration systems. The characteristic absorption band at 711 cm^{-1} was first observed at 120 h as OD 600 = 0.1 and 1.0, indicating that the calcite phase began to form in the bacteria solution. The calcite phase was formed in all bacteria concentrations at 168 h. This finding was consistent with the results of XRD pattern in the bacteria-induced system. In contrast, the calcite phase was

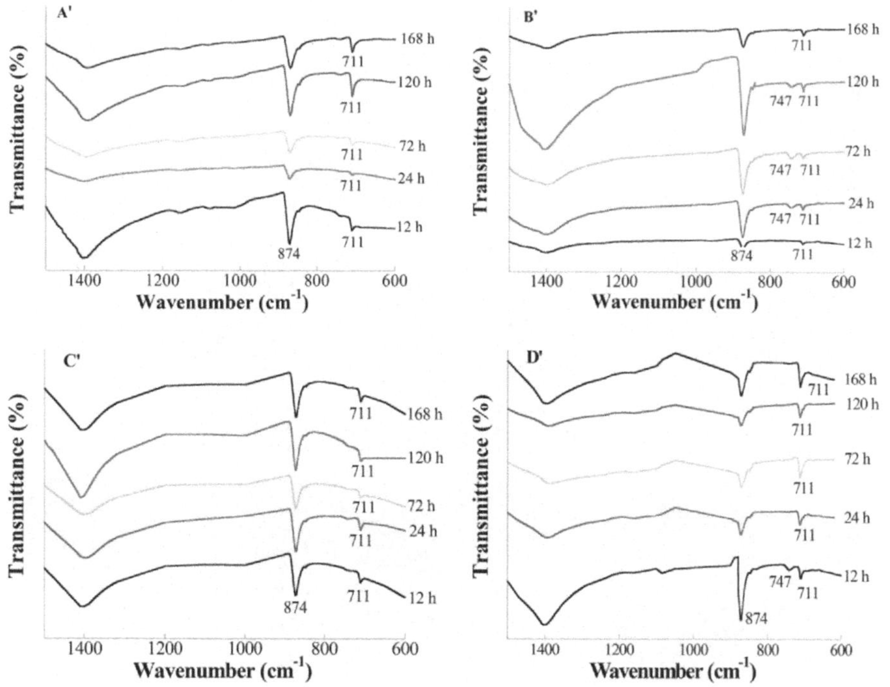

FIGURE 3.44 FTIR spectra of $CaCO_3$ crystals as function of time at different initial urease concentrations: (a) urease = 1.0g/L, (b) urease = 2.5g/L, (c) urease = 5.0g/L, (d) urease = 8.0g/L.

observed in the urease-induced system from the beginning for 12 h. There was no obvious absorption band at 747 and 1,085 cm^{-1} observed at the end of the experiments, indicating a limited vaterite phase produced in the urease-induced system with urease concentration between 1 and 5 g/L. This finding also corresponded to the previous XRD results. The study by Li et al. (2013) on calcium carbonate induced by carbonic anhydrase also revealed that the calcite phase was the primary formation after 24 h. Figure 3.43d presented that the vaterite phase was formed at 12 h in 8 g/L urease concentration. The fast precipitation rate caused by higher urease concentration might lead to an unstable reaction process, which could yield the formation of vaterite. The initial precipitation of the vaterite phase at very high urease activities was also reported by van Paassen (2009) and Al Qabany et al. (2012).

Figure 3.45 and Figure 3.46 show the SEM images of bacteria- and urease-induced system at different precipitation times, respectively, corresponding to those of FTIR spectra. In the bacteria-induced system, spherulitic crystals were observed at the beginning of the bacteria-induced system in all bacteria concentration, while the size of the spherulitic crystals tended to grow at 24 h. When OD_{600} = 0.1, rhombohedral crystal started to form at 72 h, and the precipitates were flower-like crystals at 168 h. Flower-like crystals were also observed at 72 h, as OD_{600} is higher than 0.3, and then large prismatic crystals with wrinkled surface were formed at 168 h, suggesting that the crystals consisted of vaterite and calcite phase.

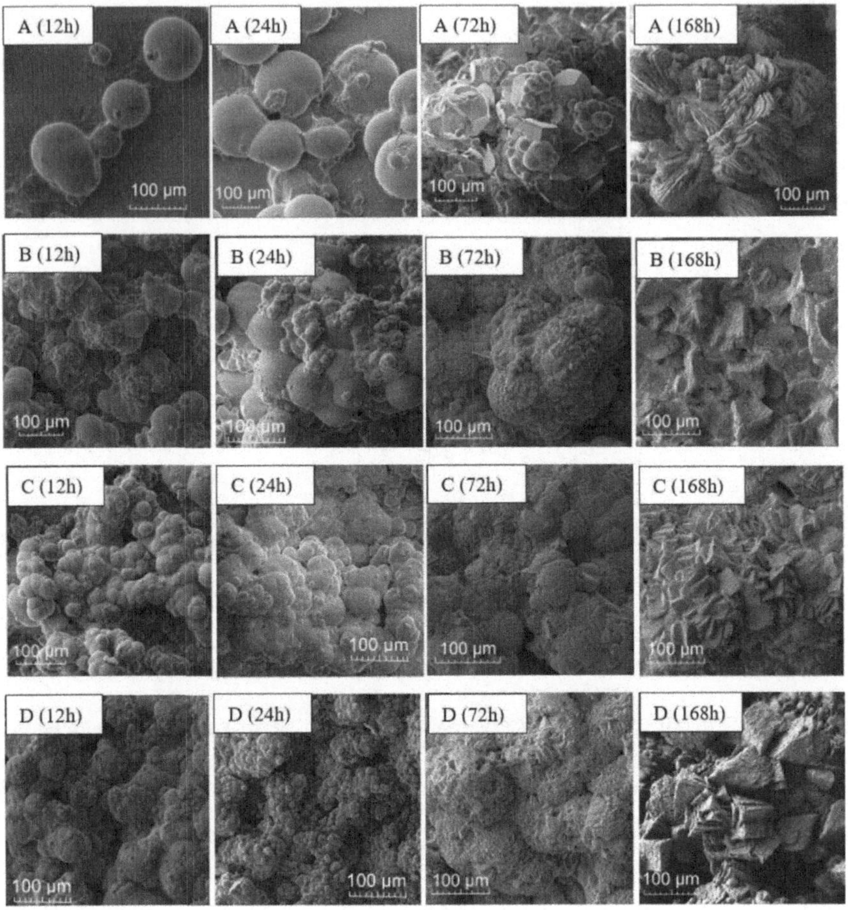

FIGURE 3.45 SEM image of $CaCO_3$ crystal at different bacteria concentrations: (a) $OD_{600} = 0.1$, (b) $OD_{600} = 0.3$, (c) $OD_{600} = 0.6$, (d) $OD_{600} =1.0$.

Overall, the comprehensive results of XRD, FTIR, and SEM indicated that the vaterite phase formed at the beginning and partly transformed to calcite phase in the bacteria-induced system. Sondi and Salopek-Sondi (2005) reported that *Sporosarcina pasteurii* could produce a large amount of acidic amino acids, for example, aspartic acid and glutamic acid, at the surface of the *Sporosarcina pasteurii* urease. The presence of aspartic acid in the solution tended to control the formation of vaterite, and higher concentrations of aspartic acid would favor the formation of vaterite (Tong et al. 2004). The amino acids of aspartic and glutamic acids were more like a nucleus during the formation of vaterite. That is why the vaterite phase was the preferential formation in the *Sporosarcina pasteurii*–induced system.

In the case of the urease-induced system, small prismatic crystals with wrinkled surface were observed at 12 h, as urease concentration is lower than 5 g/L. There were some spherulitic crystals found in the surface of crystals at 12 h in 8 g/L urease

FIGURE 3.46 SEM image of $CaCO_3$ crystal at different urease concentrations: (a) urease = 1.0g/L, (b) urease = 2.5g/L, (c) urease = 5.0g/L, (d) urease = 8.0g/L.

concentration, indicating a vaterite phase formed in the beginning. This finding corresponded to FTIR results. The crystal phase did not show remarkable differences along with time in 1.0 g/L urease concentration. Some large polyhedron shape was observed along with time as urease concentration was higher than 2.5 g/L, and spherulitic crystals disappeared with longer reaction time in 8 g/L urease concentration. Overall, the integrated results of XRD, FTIR, and SEM suggest that in the urease-induced system, the calcite phase was the major formation of the $CaCO_3$ crystals.

The formation of $CaCO_3$ crystals on sand particles was also studied in comparing with crystals on glass slices. The SEM images of $CaCO_3$ crystals on sand after 168 h at OD 600 = 0.6 and 0.25 M Ca cementation media are shown in Figure 3.46a. Both spherulitic and rhombohedral sharp crystals were found in the bacteria-induced system on sand surface after 168 h. Meanwhile, flower-like crystals were observed on sand particles, as the arrow indicates in Figure 3.46a, which is consistent with the $CaCO_3$ crystal precipitated on glass slices. For the urease-induced system (Figure 3.46b), $CaCO_3$ crystals precipitated on sand particles were majorly formed in a small prismatic shape, which is similar to that on glass slices.

Bacteria-induced $CaCO_3$ crystals were sparse in 12 h as OD_{600} of 0.1, but the size of individual particles at OD_{600} of 0.1 was larger than that at urease concentration of 1.0 g/L. More crystals precipitated and united together in both scale and density along with time. The shape of $CaCO_3$ crystals was spherulitic within the first 72 h, which may be caused by the acidic amino acids produced by bacteria. After that, the shape started to change, and rhombohedral crystals began to form. The results from FTIR also verified that the calcite phase appeared after 72 h. The results in Figure 3.32 reveal that the activities of bacteria reduced after 72 h. Also, bacteria with negative surface charge could attract Ca^{2+} ions and then wrapped by $CaCO_3$ crystals (DeJong et al. 2010). Therefore, less acidic amino acids may be produced in solution. Sondi and Salopek-Sondi (2005) reported that the vaterite phase favored a high concentration of acidic amino acids. The possible reducing concentration of acidic amino acids also explained that there were only small spherulitic crystals attached on the surface at 72 h; however, the acidic amino acid concentration test was not conducted in this paper. Meanwhile, there were no apparent spherulitic crystals on the surface at 168 h, and most crystals were found in cubic shape. The XRD results found both vaterite and calcite at 168 h, which may indicate that vaterite crystals undergo a dissolution and recrystallization process, thus transforming to calcite crystals. Li et al. (2013) used bacterial carbonic anhydrase to induce $CaCO_3$ in a NH_4HCO_3 and $CaCl_2$ solution. They also found that vaterite was gradually formed in the initial stage (~24 h), and then no vaterite appeared at 72 h. In contrast, crystals were smaller but more uniformly distributed in the urease-induced system than those in the bacteria-induced system. However, the shape and size of crystals did not have any obvious change until the urease concentration exceeded 5.0 g/L. The $CaCO_3$ crystals produced by the urease were almost in the calcite phase, which was confirmed by results reported by Yasuhara et al. (2012).

Much research had reported that both MICP and EICP processes can achieve a certain amount of $CaCO_3$ precipitation (Cheng et al. 2017; Chu et al. 2013; DeJong et al. 2006; Hamdan et al. 2013; Martinez et al. 2013; Zhao et al. 2014a). Among them, Zhao et al. (2014a) compared the mechanical properties of MICP-treated sand and EICP-treated sand. They found that MICP-treated samples (OD_{600} = 0.6) achieved an unconfined compression strength of 1.48 MPa, which was approximately four times than EICP-treated samples (5.0 g/L). From this study, the different morphologies of $CaCO_3$ crystals were found from MICP and EICP treatment, and it may explain the differences in mechanical behavior between MICP- and EICP-treated samples. Bacteria-induced $CaCO_3$ tended to achieve larger crystals and may provide better bonding for larger soil particles, like sand and gravel. However, the microbial growth–needed aerobic environment may limit deeper applications (Li et al. 2018a). For EICP, $CaCO_3$ crystals were smaller and more uniform, which may suit fine particles. However, the treatment of fine particles through EICP may not go deep, because the reaction time of EICP is short and the precipitated crystals will fill up the gap between particles and block the channel for further precipitation.

In this study, a series of experiments were conducted to investigate the effect of bacteria and urease concentration on the kinetics of Ca^{2+} concentration, and the morphology of $CaCO_3$ crystals obtained from MICP/EICP processes was analyzed by XRD, FTIR, and SEM. The results showed that higher bacteria and urease concentration could yield faster decreases of Ca^{2+} concentration, and the reaction ceased after 72 h in EICP processes. Meanwhile, OD_{600} of 0.6 or urease concentration of 5 g/L can achieve complete reaction in MICP or EICP processes in 0.25 M cementation media. The comprehensive results of XRD, FTIR, and SEM indicated that the vaterite phase was the major form of $CaCO_3$ crystals within 72 h, and the calcite phase was partly observed along with time in the bacteria-induced system. The vaterite phase favored the bacteria-induced solution. The integrated results of XRD, FTIR, and SEM suggested that in the urease-induced system, the calcite phase was the major form of $CaCO_3$ crystals. The crystals were smaller but more uniform in the urease-induced system than those in the bacteria-induced system.

3.4.2 Effect of Conductivity and pH on MICP Process

3.4.2.1 Materials and Methods

3.4.2.1.1 Bacteria and Cementation Media

The Ottawa sand used in this section is the same as that in Section 3.2.1. The bacteria and growth medium followed are those in Section 3.1.1. The concentration of the cementation media used in this section was 0.25 M Ca. The component of the cementation media is shown in Table 3.1.

3.4.2.1.2 Monitoring Method

In EC and pH tests, the bacterial suspension was mixed with a urea solution to study the EC and pH change due to urea hydrolysis. There are several factors that can affect EC and pH values, including reaction time, concentration of urea solution, and concentration of bacterial suspension. A series of different concentrations of urea solution and bacterial suspension was prepared (Table 3.9). The bacterial suspension (4 mL) was extracted by pipette from the cultivation tube and mixed with the urea solution (80 mL), as shown in Figure 3.47. Electric conductivity and pH meter were used to monitor the change of EC and pH values. Electric conductivity was monitored by a conductivity meter (Oakton, con 6 Acorn series) which has a detection limit of 0.1 μS/cm, and the pH of the solution was measured using an ACCUMET® AE 150 pH meter, which has a precision of 0.01. The mixed solutions were monitored at 5, 10, 20, 30, 40, 50, 60 min and 2, 3, 4, 5, 6, 12, 24, 48 h on the EC and pH values. The test condition was kept at the same for all the tests at a controlled temperature of 23°C. All solutions were opened to air, and there was no extra nutrition supplied to the mixed solution.

Urea concentration was also monitored during EC and pH tests using the colorimetric urea analysis method, which was developed by Knorst et al. (1997). A solution of 0.5 mL containing 4% (weight/volume) of p-dimethylaminobenzaldehyde and 4% (volume/volume) sulfuric acid in ethanol was prepared and added to 2 mL of urea solution. After 10 min, the absorbance of the solution was measured at 422 nm

TABLE 3.9
Concentration of Urea Solution and Bacterial Suspension in EC and pH Tests

Test Description	Concentration of Bacterial Suspension (OD_{600})	Concentration of Bacteria (Cells/mL)	Concentration of Urea Solution
Effect of bacteria concentration	0.3	1.67×10^7	0.25 M
	0.6	4.28×10^7	0.25 M
	0.9	7.44×10^7	0.25 M
	1.2	1.10×10^8	0.25 M
	1.5	1.49×10^8	0.25 M
	1.8	1.91×10^8	0.25 M
Effect of urea concentration	0.6	4.28×10^7	0.25 M
	0.6	4.28×10^7	0.50 M
	0.6	4.28×10^7	1.0 M

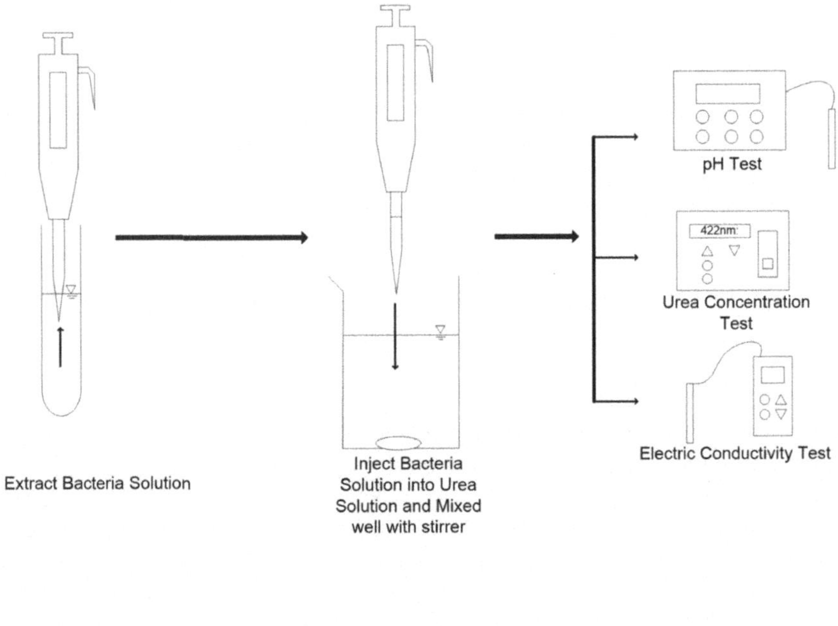

FIGURE 3.47 Schematic of pH test, urea concentration test, and electric conductivity test.

against a reagent blank using a spectrophotometer (Genesys 20, Thermo Scientific). A calibration curve was developed using standard urea solution before the test. All samples were conducted in triplicate.

3.4.2.1.3 Full-Contact Flexible Mold

The full-contact flexible mold (FCFM) was developed by Zhao et al. (2014b) and Li et al. (2015a). It was made with a geotextile which was a polypropylene, staple-fiber, needle-punched nonwoven material with grab tensile strength of 1,700 kN, thickness of 1.5 mm, opening size of 0.15 mm, water flow rate of 34 mm/s, thickness of 1.5 mm, and unit mass of 200 g/m^2.

3.4.2.1.4 Immersing Sample Preparation Method

MICP-treated specimens were prepared in a batch reactor (see Figure 3.32) to provide uniform calcite precipitation within the pore spaces. Sandy soil sample was supported by full-contact flexible mold with a cover, as shown in Figure 3.48a. Each of the sample had 140 g (±5 g) Ottawa sand to control the density of the specimens and mixed with 45 mL bacterial suspension (OD_{600} = 0.6). All specimens were immersed into the reactor tank with cementation media during the MICP process in this study. A major feature of this method is allowing soil samples to fully immerse into the cementation media and allowing the cementation media to freely penetrate into the soil (Li et al. 2015c; Zhao et al. 2014a).

3.4.2.1.5 Unconfined Compression Strength Tests for MICP-Treated Samples

The specimens made from FCFM for unconfined compression strength (UCS) tests were cylinder-shaped, with 2H:1D ratio (Figure 3.48b). After 7 days of reaction, the treated soil specimens were removed out of the batch reactor, and the geotextile mold was cut. According to Zhao et al. (2014a), a 7-day reaction time was needed for MICP-treated soil. UCS tests were conducted under strain-controlled conditions at a uniform loading rate of 1.5%/min, in accordance with ASTM D2166. All samples were conducted in triplicate.

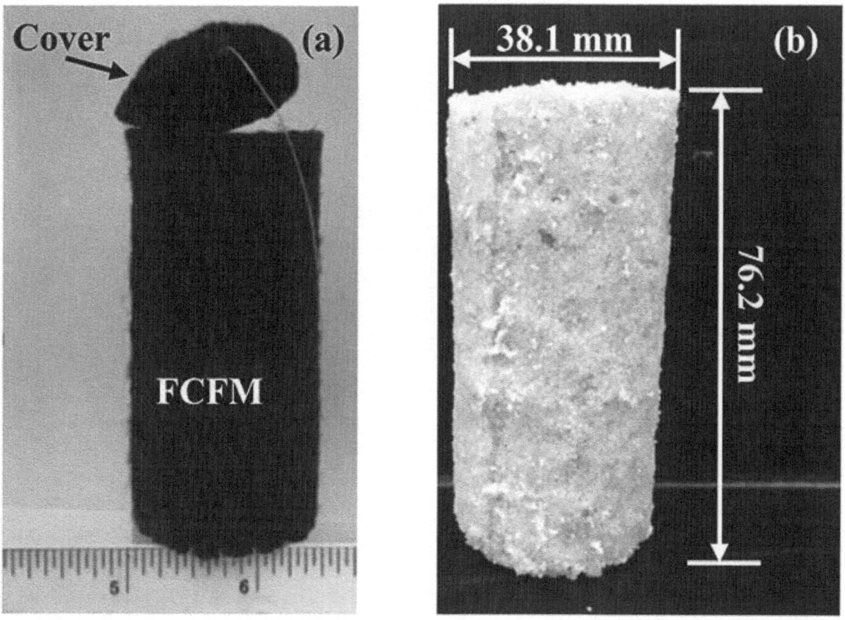

FIGURE 3.48 (a) Full-contact flexible mold (FCFM) made by geotextile. (b) MICP-treated specimen.

3.4.2.2 Study of Ureolytic Activities of Bacteria before MICP Treatment

When bacterial suspension was mixed with urea solution, urea was hydrolyzed. The concentration of urea solution would decrease as a sign of hydrolysis rate of urea. The time-dependent behaviors of hydrolyzed urea concentration in different initial urea concentrations are shown in Figure 3.49. The results show that the concentration of urea was hydrolyzed quickly within 24 h. For 0.25 M and 0.5 M urea solutions, urea had been fully hydrolyzed after 24 h. For 1.0 M urea solution, hydrolyzed urea concentration reached approximately 0.6 M after 24 h, and it continued to be hydrolyzed. Furthermore, it could be observed that higher initial urea concentration yielded faster hydrolysis rate of urea in the first 24 h. Jiang et al. (2016) also found a positive correlation between the ureolytic rate and urea concentration.

The temporal variations of the EC value of a mixed solution at different urea concentrations are shown in Figure 3.50a. Because the value of EC reflected the number of electrolytic ions in the solution, its evolution could indicate the rate of ureolytic reactions in equations (3.3)–(3.5). A higher urea concentration yielded higher EC values, which indicated that more urea was hydrolyzed. The detailed EC values within the initial 1 h are also shown in Figure 3.50a. The EC value increased over time in the first 1 h, and the EC value reached 1.62 mS/cm in 0.25 M urea solution and 1.73 mS/cm and 1.85 mS/cm in 0.5 M and 1.0 M urea solution, respectively. Whiffin (2004) reported a similar positive relationship between time and urease activity. In 0.25 M and 0.5 M urea solution, the value of EC remained constant or increased marginally after 24 h, which indicated that the urea solution had been hydrolyzed completely. However, the EC value of 1.0 M urea solution continued to increase after 24 h, indicating remaining urea in the solution. This was consistent with the results of the hydrolyzed urea concentration in Figure 3.49.

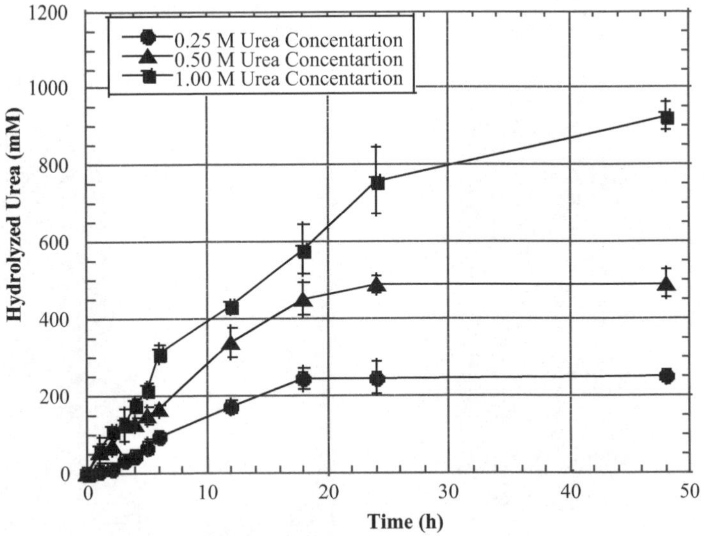

FIGURE 3.49 Hydrolyzed urea concentration as a function of time in different initial urea concentration (OD_{600} fixed at 0.6).

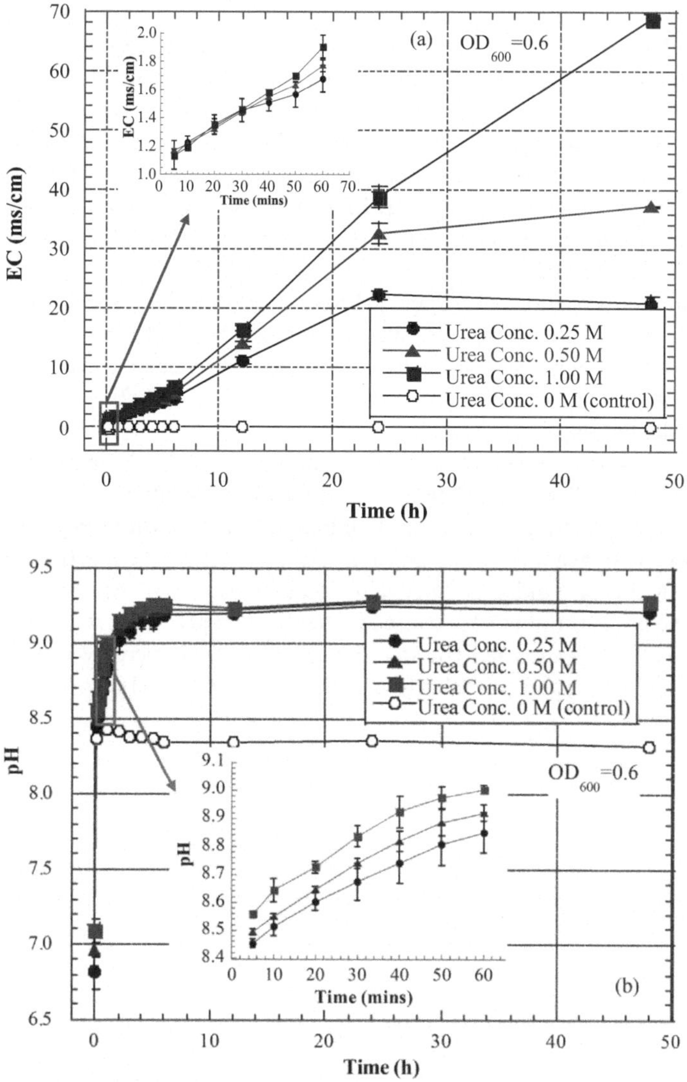

FIGURE 3.50 (a) Solution EC value and (b) solution pH value as a function of time in variable initial concentration of urea.

$$(NH_2)_2CO+H_2O \rightarrow 2NH_3+CO_2 \tag{3.3}$$

$$2NH_3 + 2H_2O \leftrightarrow 2NH_4^+ + 2OH^- \tag{3.4}$$

$$CO_2 + 2OH^- \leftrightarrow HCO_3^- + OH^- \leftrightarrow CO_3^{2-} + H_2O \tag{3.5}$$

The time-dependent behaviors of pH value for different concentrations of the urea solution are shown in Figure 3.50b. The pH value of the solution increased within the initial 1 h, which was attributed to urea hydrolysis (see equations (3.3)–(3.5)). The pH value of the mixed solution increased only slightly after 6 h, which indicated that the system had reached equilibrium. The pH value in 0.25 M, 0.5 M, and 1.0 M urea solution stabilized around 9.2–9.3 after 6 h, indicating that the effect of urea solution on pH value was negligible after 6 h. The trend of pH change was similar to that in the study of Jiang et al. (2016) with another bacteria, *B. megaterium*, mixed with 0.5 M urea solution. The relationship between EC values and hydrolyzed urea concentration is shown in Figure 3.51. In general, the EC value increased with the increase of hydrolyzed urea concentration. This result is consistent with EC values presented in Figure 3.50.

The temporal variations of EC and pH values for bacterial suspension in DI water (without urea solution) are also shown in Figures 3.50a and b, respectively. It was found that no distinguishable difference in EC and pH value occurred within 48 h. Therefore, it could be confirmed that the increase in EC and pH values was mainly caused by the hydrolysis of the urea solution.

Batch experiments at 0.25 M urea concentration were conducted with a range of cell concentration. The EC and pH values of the mixed solution as a function of different OD_{600} in a fixed urea concentration (0.25 M) are shown in Figures 3.52a and b, respectively. In general, the magnitude and rate of ureolytic reactions could contribute to the increase in EC and pH values (equations (3.3)~(3.6)). As shown in Figure 3.52, the higher initial OD_{600} caused higher EC and pH values within 60 minutes, which indicated that the ureolytic rate increased with the increase in bacteria concentration. Okwadha and Li (2010) found a linear relationship between ureolytic rate

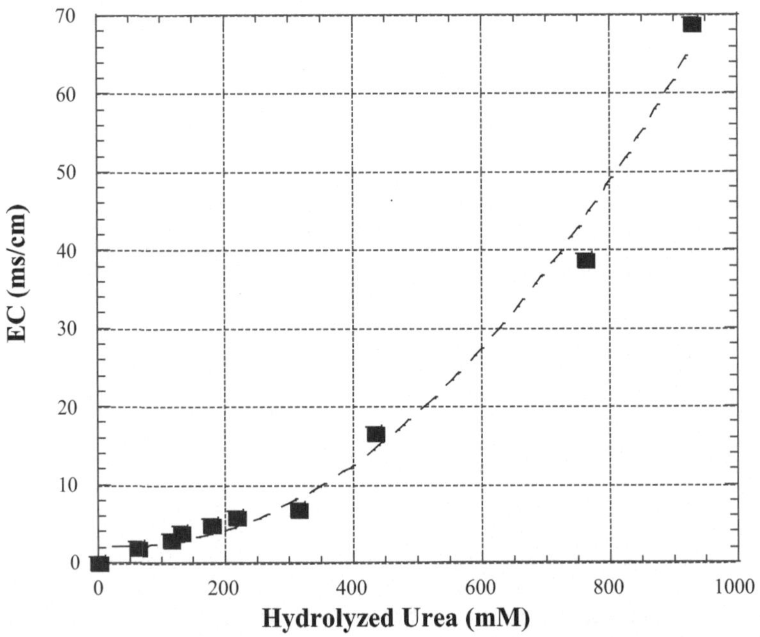

FIGURE 3.51 The relationship between EC values and hydrolyzed urea concentration.

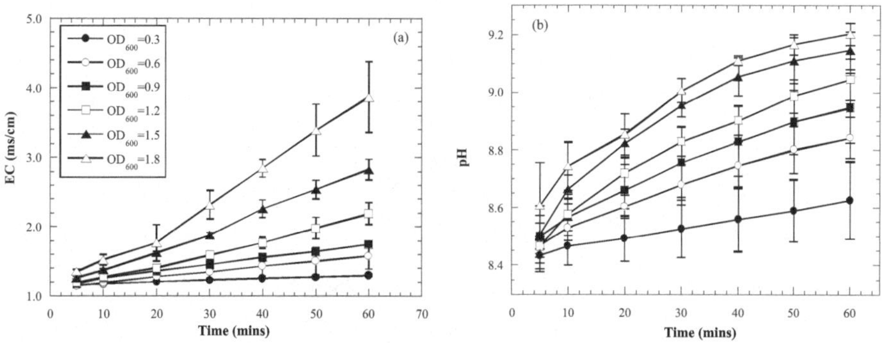

FIGURE 3.52 (a) Solution EC and (b) pH change value as a function of time with different concentrations of bacterial suspension. Legend is the same in (a) and (b), and urea concentration was fixed at 0.25 M.

and OD_{600} for the same bacteria. Zhao et al. (2014a) reported that a higher bacteria cell number in the aqueous system could maintain a higher ureolytic rate for this bacterium from 0.3 of OD_{600} to 1.5 of OD_{600} (1.66 × 107 cell/mL to 1.49 × 108 cell/mL, according to equation 3.6).

3.4.2.3 Study of the Relationship between Ureolytic Activities of Bacteria and the Unconfined Compression Strength of MICP-Treated Samples

The strength of sand sample was improved when $CaCO_3$ precipitated inside the sand and bonded the sand particles together, while the strength of MICP-treated sample is affected by many factors, and one of these factors is the rate of urea hydrolysis, which could be affected by the ureolytic activities of bacteria. EC and pH tests prior to MICP treatment could be used as an indicator to reflect the ureolytic activities of bacteria. In this study, after the evaluation of the ureolytic activities of bacteria, the left bacteria medium was mixed with sand sample and immersed into the commendation media to drive the MICP process. Then, a UCS test was conducted on these MICP-treated samples after 7 days' treatment. For EC and pH tests, the urea concentration was 0.25 M. The concentration of bacteria (OD_{600}) was fixed as 0.6, and the concentration of the cementation media was 0.25 M Ca in this section.

The relationship of EC and pH values in the first 60 min with UCS of MICP-treated samples is shown in Figure 3.53. This figure is divided into two phases (phase 1 and 2). In the case of phase 1, the unconfined compression strength increased from 120 to 600 kPa, and the EC and pH values ranged from 1.25 to 1.50 mS/cm and 8.65 to 8.82, respectively. Meanwhile, the unconfined compression strength tended to increase with the increase in EC and pH values. This result indicated that the ureolytic activities of bacteria had a remarkable influence on the UCS of the MICP-treated sample. It should be noted that although the OD_{600} value, which is the bacteria concentration, was the same for all the tests, the bacteria performed differently in EC and pH tests. Therefore, it was not enough to solely control bacteria concentration in

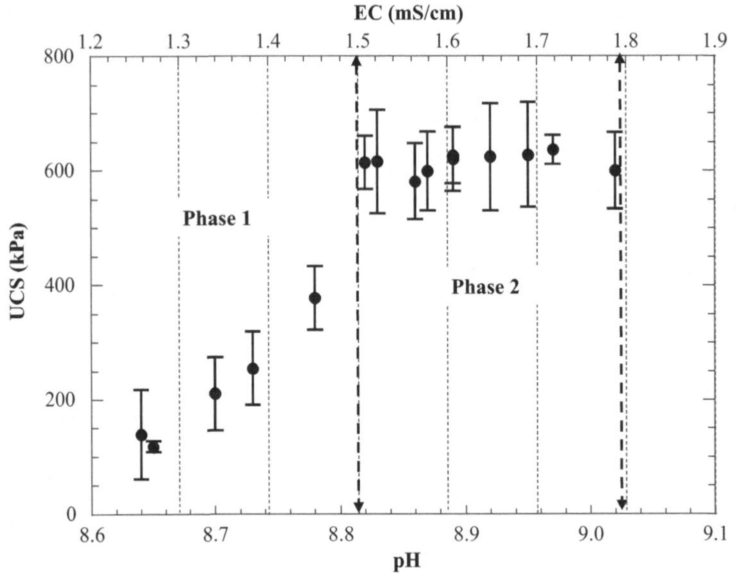

FIGURE 3.53 Effect of EC and pH values on UCS of MICP treatment sample. The EC and pH were tested in 0.25 M urea concentration, and the MICP-treated sample was treated in 0.25 M Ca. Bacteria concentration (OD_{600}) was fixed as 0.6 in all tests.

the MICP treatment, and the ureolytic activities of bacteria were more critical to the treatment result.

In contrast, in the case of phase 2, as EC and pH values ranged from 1.50 to 1.80 mS/cm and 8.82 to 9.02, respectively, the average UCS of the MICP-treated sample was maintained at around 600 kPa, which illustrated that the MICP treatment approached a consistent status. There was no significant variation on UCS with the increase in EC and pH values, which might be limited by the concentration of cementation media. Although a higher EC and pH values in urea hydrolysis test means better performance on strength improvement, the 0.25 M Ca cementation media with immersing MICP treatment could only achieve around 600 kPa UCS. Therefore, the optimum EC and pH values were found in phase 2, which could achieve a consistent result of UCS. This result indicated that EC and pH values could be a good indicator of the ureolytic activities of bacteria to maintain a consistent MICP treatment result. Moreover, these EC (1.5–1.8 ms/cm) and pH (8.8–9.2) values are tested within 60 mins. Timing is an essential factor during the test of ureolytic activity. The EC and pH values suggested in this study only work for 60 min testing result. Therefore, other OD conditions (such as $D_{600} = 0.3$) may reach the same zone with enough reaction time, but the treated sample cannot achieve the same or higher UCS. Zhao et al. (2014b) reported that UCS increased with an increase of initial bacteria concentrations. The result in Figure 3.52 is just an example of using EC and pH tests as an indicator of ureolytic activities in a certain treatment condition. For other treatment methods and treatment conditions, it is highly recommended to establish an EC and pH indicator before applying MICP treatment in large scale.

3.5 MICROSCALE ANALYSIS OF MICROBIALLY INDUCED CALCITE PRECIPITATION IN SANDY SOIL

MICP has gained much attention in soil improvement studies, where it can enhance the physical properties of sandy soil. Small-scale sand cylinder tests were conducted to investigate the formation and failure of calcium carbonate precipitation bonding between individual sand particles. Bonding formation by precipitation was examined by scanning electron microscopy. Energy-dispersive X-ray spectroscopy analysis verified the existence of calcium, carbon, and oxygen, which could form $CaCO_3$ after MICP treatment. Focused ion beam milling was applied to study the interior structure of calcium carbonate precipitation during the MICP process.

3.5.1 Microanalysis Sample Preparation

MICP-treated samples were removed from the reactor tank for UCS testing to failure. Failed specimens were prepared for microscopy by oven-drying overnight at 105°C before mounting on SEM stubs with adhesive carbon conductive tabs. The fracture surfaces were sputter-coated with gold for SEM analysis and FIB/SEM analysis.

Specimens were imaged and ion beam–milled in a TESCAN LYRA3. Secondary electron image was acquired at 5 kV. Certain fracture surfaces were milled with a Ga+ ion beam current of 10 nA at 30 keV to expose cross sections of sand particle bonding. The cross section surface was then polished with a low ion beam current of 100 pA at 25 keV for 1 h prior to SEM imaging. Elemental analysis was obtained with a Thermo Scientific NORAN (System 7) energy-dispersive spectrometer (EDS).

3.5.2 Unconfined Compression Strength

Unconfined compression strength (UCS) tests were carried out to investigate the mechanical performance of MICP-treated sample. Table 3.10 shows the UCS and dry density results for MICP-treated samples over different numbers of treatment cycles. The UCS and dry density of MICP-treated samples increased with treatment cycles, indicating that more calcium carbonate precipitation was generated through repeated treatment cycles. The UCS of MICP double-treated sample was more than twice that of MICP single-treated sample, which indicated that the increase in treatment

TABLE 3.10
Unconfined Compression Strength (UCS) of MICP-Treated Samples over Different Numbers of Treatment Cycles for a Cementation Media Concentration of 0.5 M Ca

	Single Treatment (kPa)	Double Treatment (kPa)	Triple Treatment (kPa)	Quadruple Treatment (kPa)
UCS (kPa)	1,200 ± 130	2,700 ± 250	3,500 ± 280	4,300 ± 670
Dry density (g/cm^3)	1.72 ± 0.12	1.88 ± 0.08	1.96 ± 0.10	1.98 ± 0.05

cycles not only induced more calcium carbonate precipitation but also increased the effective bonding. However, the relative increase in UCS was less for triple- and quadruple-treated samples. This was due to the reduction on the void space of sample with each treatment, which would reduce the surface area available for bonding.

3.5.3 Carbonate Bonding of Sand Particles

Figure 3.54a shows a SEM image of untreated Ottawa sand with individual sand particles exhibiting smooth surfaces. After the unconfined compression strength test, the fracture surface of a crushed MICP-treated sample was observed using SEM and EDS to analyze the microstructure of $CaCO_3$. Figure 3.54b shows sand particles bonded by calcium carbonate precipitation after a single MICP treatment. There were many dome-like calcium carbonate precipitates attached to the surface of sand particles that did not participate in bonding. However, Table 3.10 shows that the strength and density of multiply MICP-treated sand increased with each treatment. Nontreated samples (loosely sand) that have a little strength are not shown in Table 3.10.

Multiple treatments increase calcium carbonate precipitation content and bond area between sand particles. Figure 3.55a shows that, after four processing treatments, calcium carbonate precipitation wrapped around sand particles so that the sand particles could be better bonded together. MICP-treated specimens could bear higher loads with more cementation bonding. The failure mode of MICP-treated sand was bond particle failure. Figure 3.55a shows that some sand grains were detached from the calcium carbonate bond, leaving a smooth concave region, impressions of missing sand grains.

3.5.4 X-Ray Emission Spectrometry

An EDS X-ray spectrum, acquired while scanning the entire image frame shown in Figure 3.55a, is shown in Figure 3.55b. EDS analysis confirms the presence of

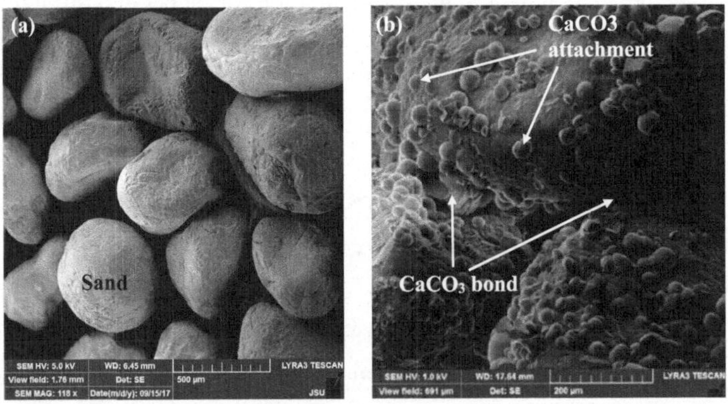

FIGURE 3.54 SEM image of (a) untreated Ottawa sand and (b) fracture surface of MICP single-treated Ottawa sand.

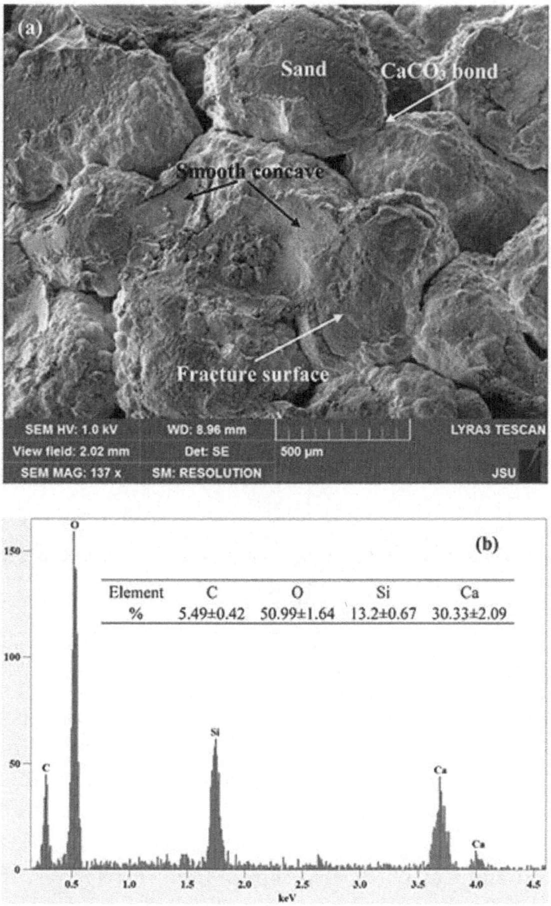

FIGURE 3.55 (a) Fracture surface of MICP quadruple-treated Ottawa sand. (b) EDS analysis of quadrable-treated Ottawa sand.

carbon, oxygen, and calcium, the elements that make up $CaCO_3$, in the MICP-treated sand. The silicon peak and part of the oxygen peak in the spectrum were from the Ottawa sand itself (known to be 98% SiO_2).

3.5.5 MICP-Treated Four Times

Fracture surfaces were observed after samples were MICP-treated four times. Figure 3.56 shows the details of the fracture surface of $CaCO_3$ in a quadruple-treated sample. This image captures the exposed surface of a bond particle failure for quadruple-treated sand. There were several interface lines exposed on the fracture surface, which illustrated the evolution of $CaCO_3$ during multiple treatments (Figure 3.56b). The $CaCO_3$ precipitated on the surface of sand particles first, and newly formed $CaCO_3$ attached to prior carbonate during each treatment cycle. After

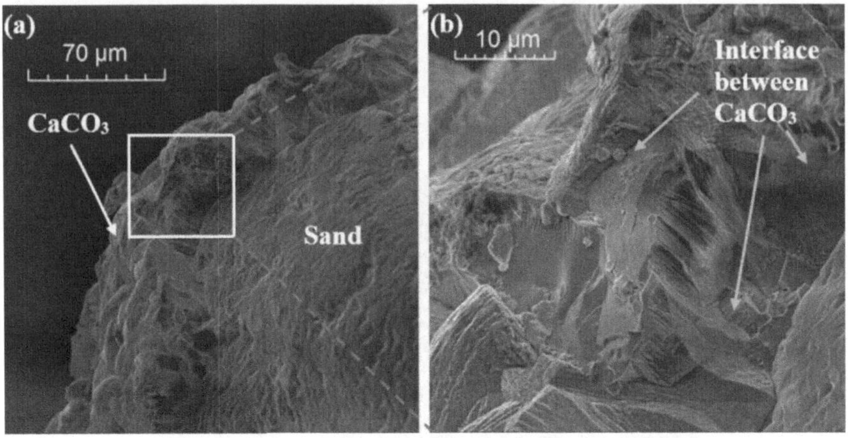

FIGURE 3.56 (a) Fracture surface showing precipitation and (b) at a higher magnification.

quadruple treatment, the $CaCO_3$ layer expanded and converged to a shell-like precipitation which wrapped around the sand particles.

3.5.6 Cross Sections of Bonding

Figure 3.57 shows a SEM image of the ion beam sectioned inner core of a region of the quadruple-treated sand specimen. The left side of Figure 3.57a shows part of a bare sand particle, whereas the right side shows a portion of the $CaCO_3$ layer on top of the sand particle. The cross section through the $CaCO_3$ layer was milled to about a 30 μm depth (yellow box). Figure 3.57b of the milled region shows a clear interface line between sand and $CaCO_3$. It also shows three interfaces in $CaCO_3$ (arrows), indicating that $CaCO_3$ precipitation was accumulated layer by layer with multiple treatments, with the final layer forming the precipitation shell that wraps around the sand particles. Thus, the $CaCO_3$ contact surface area expanded, and the bonding of the sand particles became stronger.

3.5.7 Remnant Bacteria

In each layer, there were some inactive bacteria surrounded by $CaCO_3$ (see circled region in Figure 3.58b). Incidentally, some curtain-like FIB milling artifacts are shown in the black box of Figure 3.58b. SEM imaging provided insight into the remnant bacteria between the sand particles. Figure 3.58a shows the inactive bacteria attached to the surface of the precipitation. After FIB milling to 30 μm, the SEM image of the exposed section provided a much clearer image of inactive bacteria inside the $CaCO_3$ in Figure 3.58b. The bacteria with negative surface charge were attracted to Ca^{2+} and were trapped by the $CaCO_3$ precipitation.

After a single MICP treatment, many $CaCO_3$ hemispheres were found on the surface of the same particles, reducing the possible bonding area of the calcite on the sand particles (Figure 3.54b). Multiple treatments increased $CaCO_3$ content and

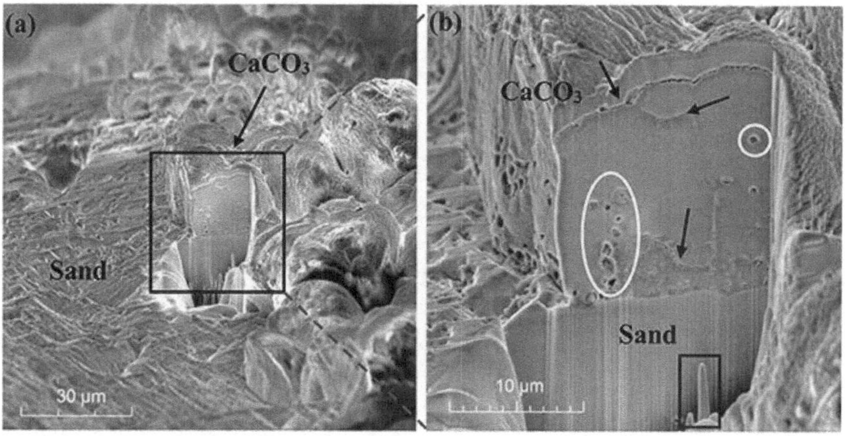

FIGURE 3.57 (a) FIB/SEM image of inner precipitation exposed by ion milling and (b) at higher magnification.

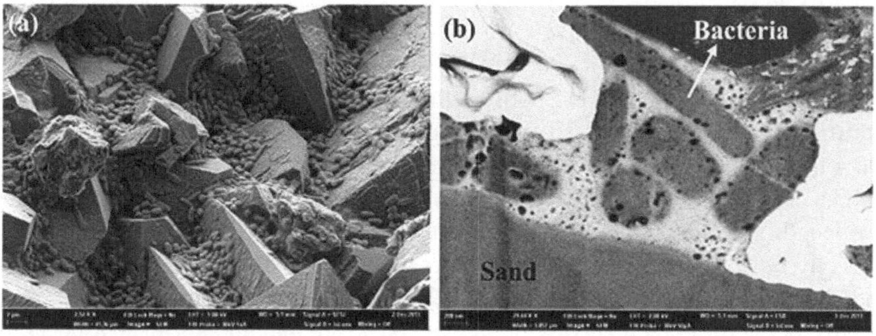

FIGURE 3.58 Inactive bacteria (a) on surface of the precipitation product and (b) wrapped by the precipitation.

enhanced $CaCO_3$ bonding (Figure 3.56). Conventional SEM can only analyze the surface of the sample. After ion beam milling of a small cross section at the surface, a SEM image could be obtained for the deeper layers (Figure 3.57). Repeating this procedure produces a cross section of the internal structure. However, during this manual ion beam trimming process, the interfacial bonding between cells and the mineral was damaged. The automated trimming process, along with simultaneous SEM observation, was useful in studying the interior structure of the MICP-treated soil specimen. More $CaCO_3$ precipitated after each additional treatment cycle, and the milling function of the focused ion beam exposed a cross section inside the MICP layer-by-layer deposition of $CaCO_3$ (Figure 3.57b). The effect of additional treatment cycles could be observed along the cross-sectional face.

4 Application of MICP in Construction Materials

4.1 APPLICATION OF MICP-REINFORCED SAND

4.1.1 Materials and Methods

4.1.1.1 Materials

4.1.1.1.1 Sand

The Ottawa sand used in this section is the same as that in Section 3.2.1.

4.1.1.1.2 Bacteria and Cementation Media

The bacteria and growth medium and cementation media followed are those in Section 3.1.1 and Section 3.1.2.

4.1.1.1.3 Fiber

As one of the most used fibers both by the concrete and geotechnical engineering industries, Fibermesh 150 was used throughout the laboratory investigations. It is a 100% uniform homopolymer polypropylene multifilament fiber with a specific gravity (Gs) of 0.91 (Fibermesh 150). It is chemically inert, with high acid salt resistance. The length and thickness of the fibers used in this study are 12 mm and 0.1 mm, respectively, with an aspect ratio of 120 between the length and the thickness of the fiber (shown in Figure 4.1b). Consoli et al. (2009b) used similar fibers of different lengths for the reinforcement of sand. It was concluded that fibers with an aspect ratio above 300 experienced strain-hardening behavior, which caused significant mechanical problems, such as fracture failures during shearing of the soil and a significant decrease in the ductility behavior of the fiber. The aspect ratio of the fiber used in this study was lower than the upper limit determined in the previous studies, indicating that it should provide efficient reinforcing performance (Consoli et al. 2009b).

4.1.1.2 Sample Preparation

4.1.1.2.1 Full-Contact Flexible Mold

The samples for unconfined compression tests were prepared with full-contact flexible molds (FCFM), which are made of geotextile. The geotextile is a polypropylene, staple-fiber, and needle-punched nonwoven material. The fibers are needle-punched to form a stable framework that retains dimensional stability relative to each other. The selected geotextiles have grab tensile strength of 172.4 kg, grab elongation of 50%, trapezoidal tear of 68 kg, apparent opening size of 0.15 mm, water flow rate of 34 mm/s, thickness of 1.5 mm, and unit mass of 200 g/m^2. This geotextile has many pores, promoting the penetration of chemicals from the cementation media to sand pores, which is beneficial to the MICP process. The size of the unconfined

DOI: 10.1201/9781003519638-4

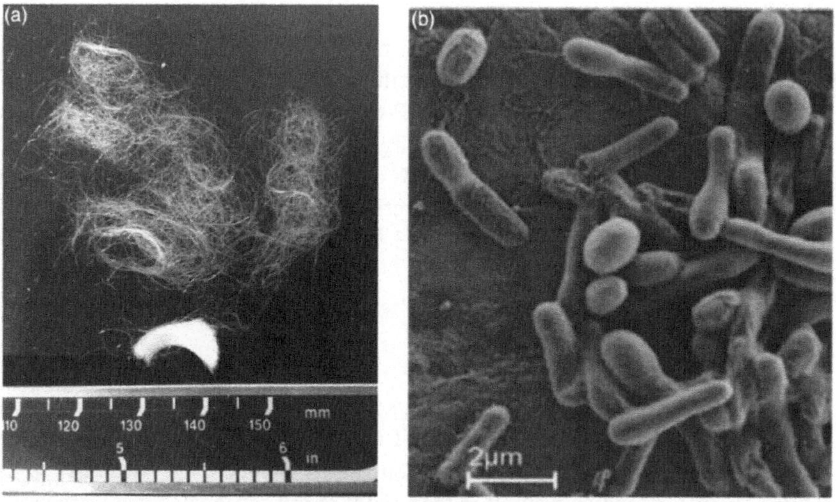

FIGURE 4.1 Image of (a) synthetic fibers and the (b) bacteria *Sporosarcina pasteurii* used in this study.

Source: Image of bacteria adopted from ssbang.sdsmt.edu.

compression test mold is 51 mm diameter and 102 mm height. The mold consists of an annular part, a bottom, and a cover.

4.1.1.2.2 Batch Reactor

All samples were prepared in a completely stirred tank reactor. The reactor included a plastic box to contain soil samples and cementation media, a magnetic mixer to keep the solution uniform, and an air pump to provide oxygen for the bacteria (Figure 3.32). A major feature of this method is to allow soil samples to fully immerse into the cementation media and to allow the cementation media to freely penetrate the soil samples instead of using a pump to inject the cementation media.

As the cementation media permeates into the soil specimens, MICP can occur in sample pores, and the produced $CaCO_3$ can bond the sand particles together to improve its engineering properties. Many studies have used a pump to inject cementation media into sample pores to promote the MICP process in samples. In this study, the samples were prepared with the full-contact flexible mold and immersed into the cementation media in a batch reactor. The chemical substances freely diffuse into the sample space under the action of the magnetic stirrer. In the batch reactor, there is no hydraulic gradient to drive the flow through the soil specimen. The hydraulic conductivity of the MICP-treated sand is in the range of 0.001 cm/s, which is permeable to the cementation media. When the chemical of the cementation media reacts under the catalysis of bacteria, the concentration of these chemicals is lower in the soil specimens, which causes the chemical substances to diffuse from a high-concentration area to a low-concentration area to continue the MICP deeper in the soil specimens.

4.1.1.2.3 *Curing Conditions*

To investigate the effect of environmental factors (humidity and temperature) on MICP-treated samples, four different curing conditions were considered. After removing from the batch reactor, the MICP-treated samples were cured under four different conditions, including (1) air-drying at room temperature (20°C), (2) sealing into ziplock bags then storing in a 30°C constant-temperature oven, (3) sealing into ziplock bags then storing in a −18°C refrigerator, and (4) directly soaking into water. All these samples were cured for 7 days, before unconfined compression tests began.

4.1.1.3 Experimental Method

4.1.1.3.1 *$CaCO_3$ Content*

To determine the precipitated $CaCO_3$ in the soil specimens, specimens were crushed using a mortar and oven-dried. The dry soil was washed in an HCl solution (0.1 M) to dissolve the precipitated carbonates, then rinsed, drained, and oven-dried. The difference between the two weights is the weight of the carbonates that were precipitated in the specimen.

4.1.1.3.2 *Unconfined Compression Tests*

The specimens of FCFM for unconfined compression tests were cylinder-shaped, with a 51 mm diameter and a 102 mm height. A geotextile cover was used for the top and the bottom (Figure 4.2). The 330 g washed Ottawa sand was uniformly mixed with 85 mL of bacteria and growth media and then added into the mold. The mixture was air-pluviated into the mold to a median dense condition (resulting in a D_r ranging from 42% to 55%, porosity ranging from 38% to 40%, and dry density of sand ranging from 1.58 to 1.64 g/cm^3). A total of 12 molds with sand samples on the shelf

FIGURE 4.2 Photograph of specimens after the reactions in the full-contact flexible mold. The full-contact flexible mold was fabricated by geotextile. PVC shelves were used to support the full-contact flexible molds.

were immersed into 20 L cementation media in a completely stirred tank reactor. After 7 days of reaction, the shelf was removed from the solution, and each geotextile mold was cut and the treated soil specimen was removed. Unconfined compression tests were conducted under strain-controlled conditions at a uniform loading rate of 1.5%/min, in accordance with ASTM D2166.

4.1.1.3.3 Scanning Electron Microscopy and Energy-Dispersive X-Ray Spectroscopy

The formation of MICP was examined by scanning electron microscopy (SEM). With the bacteria concentration (OD_{600}) of 0.6 and cementation concentration of 0.5 M, the samples were prepared in the reactor for 7 days and were air-dried at room temperature for 7 days. Subsequently, these samples were putter-coated for SEM analysis. The chemical composition of the samples was also analyzed using an energy-dispersive X-ray analyzer (EDX).

4.1.2 Factors Affecting the Improvement of MICP-Treated Soil Catalyzed by Bacteria and Urease

4.1.2.1 MICP Sample Preparation

The 330 g sand was uniformly mixed with an 85 mL bacteria solution (or urease solution) and then air-pluviated into the mold to reach a median dense condition (D_r in the range of 42%~55%, and dry density of sand ranging from 1.58 to 1.64 g/cm^3). These samples were placed on the shelf (Figure 3.32). Then, the entire shelf was slowly immersed into the batch reactor filled with cementation media. Without adding any additional cementation media, bacteria, or growth media, the MICP would be maintained for a desired period (reaction time). At the end of the reaction time, the shelf and the samples would be removed from the reactor (Figure 4.2). Then, the MICP-treated samples were removed from the mold by cutting the mold, and stored in a curing room. Details of the sample preparation can be found in Zhao et al. (2014b).

4.1.2.2 Effect of MICP on Stress–Strain Behavior

The stress–strain curve obtained from unconfined compression tests on the MICP-treated samples, catalyzed by *S. pasteurii* under different bacteria concentration (OD_{600}), is shown in Figure 4.3. With cementation concentration of 0.5 M and five different OD_{600}, the standard sands were prepared in the MICP reactor for 7 days and air-dried at room temperature for 7 days before the UCS tests. As shown in Figure 4.3, the UCS of MICP-treated samples increased with bacteria concentration (OD_{600}). When OD_{600} increased from 0.3 to 0.6, there was almost a 200% increase in UCS from 0.44 MPa to 1.36 MPa. With three, four, or five times more bacteria, UCS increased by 237%, 324%, and 378%, respectively. This finding indicates that bacteria plays a key role on MICP, that is, (1) producing enzyme to hydrolyze urea and (2) acting as nucleation sites for the formation of calcium carbonate crystals (mainly calcite) (Fujita et al. 2000). More bacteria in the solution can promote more enzymes and provide more nucleation sites for the MICP. As shown in Figure 4.3,

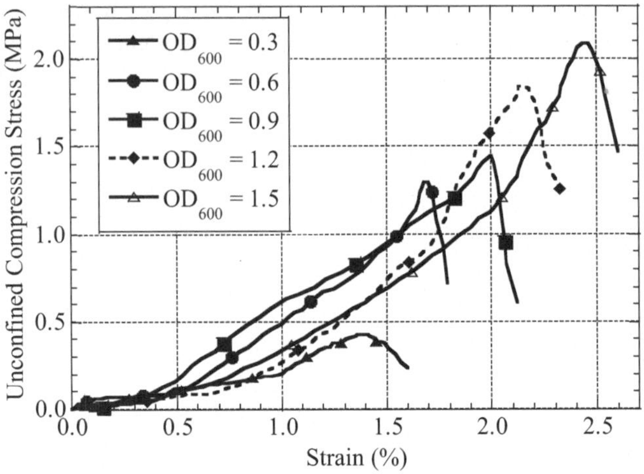

FIGURE 4.3 Unconfined compression stress of MICP-treated samples catalyzed by *S. pasteurii* under different bacteria concentration.

UCS does not increase significantly until it reaches 0.5% strain. This is because there were some sand particles and $CaCO_3$ crystals on the surface of the samples, leading to an uneven surface. Before it reached 0.5% strain, the main effect of UCS was flattening and compacting. Although the MICP-treated samples can reach a fairly uniform state in the full-contact flexible mold, there was still slightly different $CaCO_3$ content distributed within each sample (Zhao et al. 2014b), which may cause UCS to fluctuate during the test. At least four samples were prepared for each case to produce a representative condition.

The SEM image in Figures 4.4a and b shows that the $CaCO_3$ crystals were mainly irregular bulk, similar to observations made by Al Qabany et al. (2012). However, the crystal size (10~100 μm) was obviously larger than the size indicated in their results, probably due to the different sample preparation method. This study used a full-contact flexible mold to prepare the soil specimen by immersing the sample into the cementation media. These MICP-treated sand specimens could be taken out from the mold for different mechanical tests. With the full-contact flexible mold preparation, $CaCO_3$ in the MICP sand specimen was more evenly distributed in samples (Zhao et al. 2014b). Al Qabany et al. (2012) used plastic syringes and injected the cementation solution. The disturbance of the liquid may impact the formation of crystals. In this study, entire samples were completely immersed into the cementation media, which may be helpful in the formation of large crystals. Most crystal particles were dispersed and uniformly distributed in the gap between sand particles, and sometimes there was an overlapping structure on the crystal surface. As shown in Figure 4.4c–f, the results of EDX analysis demonstrated that the dominant minerals were SiO_2 and $CaCO_3$, and Ca, O, C were the main elements in the mineral precipitation. Studies by van Paassen (2009) and Al Qabany et al. (2012) confirm that the crystals observed were actually $CaCO_3$ crystals precipitated in silica sand.

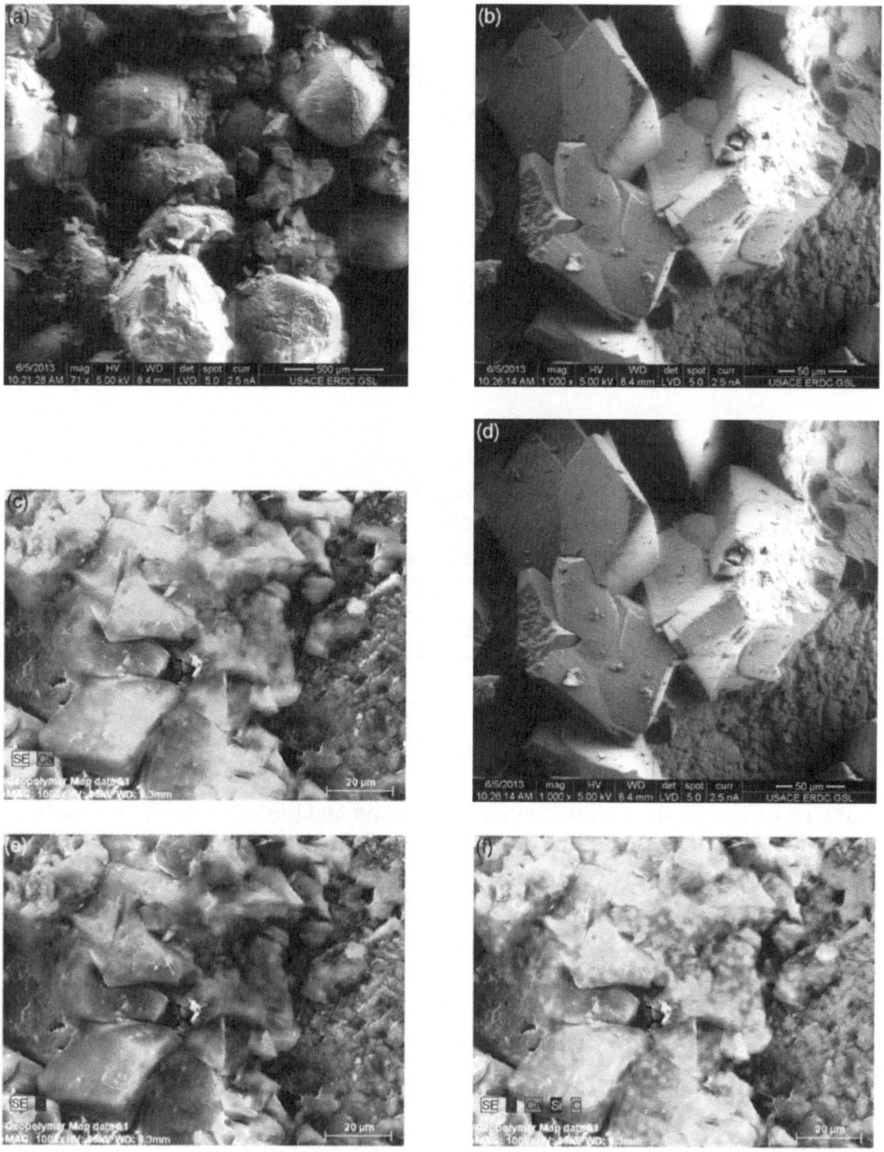

FIGURE 4.4 SEMs of MICP-treated sand catalyzed by *S. pasteurii* and energy-dispersive X-ray image of precipitation on sand surface: (a) sand and $CaCO_3$ crystals, (b) $CaCO_3$ crystals, (c) element Ca, (d) element C, (e) element O, and (f) elements of O, Ca, Si, C.

4.1.2.3 MICP Catalyzed by *S. pasteurii*

4.1.2.3.1 Effect of Bacteria Concentration

Five different bacteria concentrations on OD_{600} were used for the MICP-treated soil catalyzed by *S. pasteurii*. Table 4.1 shows the urease activity, specific urease activity, $CaCO_3$ content, and UCS under different OD_{600}. OD_{600} represents the

TABLE 4.1
Urease Activity, Specific Urease Activity, $CaCO_3$ Content, and UCS under Different OD_{600} of MICP-Treated Samples Catalyzed by *S. pasteurii*

OD_{600}	0.3	0.6	0.9	1.2	1.5
Urease activity	1.14	3.31	5.27	6.90	8.85
Specific urease activity	3.81	5.52	5.85	5.75	5.90
$CaCO_3$ content (%)	4.88~5.19	7.21~7.88	8.57~9.45	10.14~11.36	13.09~14.44
UCS (MPa)	0.34~0.53	1.28~1.43	1.36~1.60	1.76~2.04	2.01~2.22

Note: Urease activity (mM hydrolyzed urea/min) and specific urease activity (mM hydrolyzed urea/min/OD).

concentration of bacteria, urease activity reflects the catalytic ability of bacteria to hydrolyze urea, and the specific urease activity illustrates the catalytic ability of unit bacteria to hydrolyze urea. As OD_{600} value increases, urease activity gradually rises. But the specific urease activity only increased rapidly when the OD_{600} increased from 0.3 to 0.6. After OD_{600} reached 0.6, the increase in specific urease activity became slow. Although *S. pasteurii* can constitutively produce high levels of urease theoretically (Mörsdorf and Kaltwasser 1989), Whiffin (2004) found that no correlation existed between urease activity and biomass. He indicated that urease was not produced constitutively and was affected by many factors, resulting in the variation of specific urease activity. The box plot of UCS as a function of bacteria concentration (OD_{600}) is shown in Figure 4.5a. Each box shows the median value and ±25% of the UCS population as the top and bottom of the box. The lines extending from the top and the bottom of each box mark the minimum and maximum UCS. The outlier is shown as an individual point. With more bacteria, the $CaCO_3$ content and the median UCS of MICP-treated samples increase, obviously. The increase in $CaCO_3$ content results in an increased strength in the MICP-treated samples. The range of UCS under different OD_{600} is similar, except when OD_{600} is 1.5. The increase in UCS is consistent with the increase in urease activity and $CaCO_3$ content (Table 4.1).

4.1.2.3.2 Effect of Cementation Media Concentration

To investigate the effect of cementation media concentration on MICP-treated sand, standard sand samples were prepared under four different cementation media concentrations with the OD_{600} of 0.6. According to Table 4.1, both the MICP-treated efficiency and specific urease activity obviously increased when OD_{600} reached 0.6. The chemical compositions of the four different cementation media concentrations are summarized in Table 4.2.

The results in Table 4.2 show that the UCS and $CaCO_3$ contents increase with increased urea and calcium ions in the cementation media during the MICP process. When cementation media concentration increases from 0.25 Ca M to 0.5 M Ca, the median UCS increases from 0.13 MPa to 1.36 MPa, which is almost ten times. However, when cementation media concentration increases from 0.5 M Ca to 1.5 M Ca, the median UCS becomes double from 1.36 MPa to 2.13 MPa. This observation

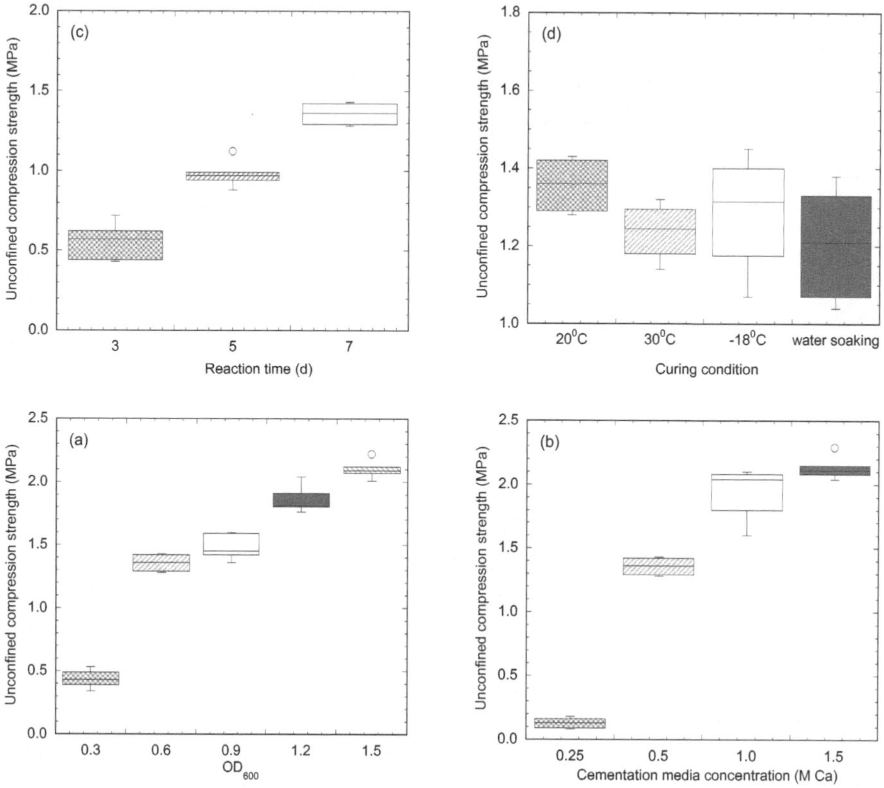

FIGURE 4.5 Box plot of unconfined compression strength of MICP-treated samples catalyzed by *S. pasteurii* as a function of (a) OD_{600}, (b) cementation media concentration, (c) reaction time, and (d) curing conditions.

shows that urea and calcium ions in the cementation solution have not been fully utilized when its concentration is higher than 1 M Ca. This is probably because the MICP process was limited by enzyme quantities and reached the maximum ability of urea hydrolysis, or the high concentration of calcium chlorine in cementation solution reduced the urease activity of the bacterial enzyme (van Paassen 2009; Whiffin 2004). Although high concentrations of cementation media can improve MICP treatment effect and mechanical properties to some extent, only a portion of calcium ions was precipitated by MICP. The results indicate that the sole increase in concentration of cementation media is a limited method to enhance MICP.

4.1.2.3.3 Effect of Reaction Time

MICP through bacteria is a complex biochemistry and transient process. The reaction time is one of the key factors for MICP. Standard MICP-treated sand samples after 3 days, 5 days, and 7 days were prepared respectively to study the influence of reaction time on MICP and soil properties. The MICP samples were not strong enough to be removed from the molds after 1-day reaction. The bacterial concentration (OD_{600}) was 0.6, the specific urease activity was 5.5 mM hydrolyzed urea/min/OD,

TABLE 4.2
$CaCO_3$ Content and UCS under Different Cementation Media Concentrations of MICP-Treated Samples Catalyzed by *S. pasteurii* and Urease

Cementation Media Concentrations	*S. pasteurii*[1]		Urease[2]	
	$CaCO_3$ Content (%)	UCS (MPa)	$CaCO_3$ Content (%)	UCS (MPa)
0.25 M Ca	1.90~2.02	0.08~0.18	–	–
0.5 M Ca	7.21~7.88	1.28~1.43	5.12~5.55	0.58~0.62
1.0 M Ca	9.56~11.12	1.60~2.10	6.03~7.02	0.57~0.76
1.5 M Ca	12.14~13.39	2.04~2.13	7.36~8.90	0.79~0.81

Note:
[1] $OD_{600} = 0.6$;
[2] urease concentration = 10 g/L.

and the cementation concentration was 0.5 M Ca. Both the MICP-treated efficiency and specific urease activity obviously increased when the OD_{600} reached 0.6. Urea and calcium ions have been fully utilized when cementation media concentration is 0.5 M Ca. After reaction for 3 days, 5 days, or 7 days in the MICP batch reactor, the samples were air-dried at room temperature.

Figure 4.5c shows an increase in median UCS from 0.56 MPa to 0.98 MPa, and then to 1.36 MPa, when the MICP reaction time was extended from 3 days to 7 days. The increase in the rate of UCS from 3 days to 5 days is greater than from 5 days to 7 days. This may be explained by the fact that the bacteria had already been wrapped by calcite crystal and enzyme activity was reduced after 5 days. Moreover, the precipitated $CaCO_3$ reduced pore volumes of samples and even blocked the pores, which reduced the effective contact between the cementation media and urease, thereby bringing down the MICP bonding efficiency.

4.1.2.3.4 Effect of Curing Conditions

To study the effect of curing conditions on MICP-treated samples, four different curing conditions were studied. After removing from the batch reactor, the MICP-treated samples were cured either by air-drying at room temperature (20°C), sealing into ziplock bags then storing in 30°C constant-temperature oven, sealing into ziplock bags then storing in −18°C refrigerator, or directly soaking into water. All these samples were cured for 7 days, before unconfined compression tests began. OD_{600} of 0.6 and cementation concentration of 0.5 M Ca were used for the MICP-treated standard sand samples. The box plot of UCS as a curing condition is presented in Figure 4.5d.

The result shows that the highest UCS is 1.36 MPa in air-dried condition, then 1.24 MPa in 30°C and 1.29 MPa in −18°C, and the lowest one is 1.21 MPa in water soaking. The curing temperatures of −18°C and 30°C are used to study the effect of temperature on MICP-treated samples. The purpose of curing in air and water is to study the effect of moisture content on MICP-treated samples. Figure 4.5d shows that curing temperatures have a less than 10% impact on the UCS of MICP-treated sand, while the moisture content has only an 11% impact on the UCS of MICP-treated sand. It is shown that once the MICP-treated samples are removed from the batch reactor, the MICP process in sand is substantially terminated, and the precipitated $CaCO_3$ is relatively stable. The engineering properties of MICP-treated sand are relatively stable and are not significantly affected by environmental conditions.

4.1.2.3.5 Effect of Sand Type

Samples of standard sand and Mississippi local sand were prepared under the same MICP conditions as OD_{600} of 0.6 and cementation concentration of 0.5 M Ca. After 7 days of reaction, the samples were air-dried at room temperature.

The result shows that the $CaCO_3$ content and UCS of standard sand are higher than that of Mississippi local sand, which indicates that soil permeability has a great influence on MICP (Table 4.3). The D_{10} of standard sand is 0.31 mm, and that of Mississippi local sand is 0.20 mm. The permeability of standard sand is higher than for the Mississippi local sand, which is conducive for the cementation media to penetrate specimens and promote the MICP process. Rebata-Landa (2007) conducted research on the effect of soil type on the MICP and indicated that soil type had two kinds of limitations on MICP. In very fine soil particles, the MICP is hindered by permeability, while in coarse soils, a thin layer of mineral precipitation cannot cement particles together. In this chapter, the D_{50} values of standard sand and Mississippi local sand were 0.46 mm and 0.33 mm, respectively, which were close, and thus, the grain size should have little effect on MICP. In addition to soil permeability, the differences in mineralogy may also affect the chemical reaction network.

4.1.2.4 MICP Catalyzed by Urease

S. pasteurii plays a key role in the MICP process, mainly because it can produce urease, which can catalyze urea hydrolysis as a specific enzyme (Jabri et al. 1995). In fact, similar enzymes can be manually extracted from many natural plants (El-Shora 2001; Kerr et al. 1983). Bacterial growth and division are affected by many factors (e.g., nutrients, pH, etc.) (van Paassen 2009; Whiffin 2004), and it is difficult to control its activity during the MICP process (Yasuhara et al. 2011). In this study, urease was directly used in MICP-treated sand instead of using the bacteria *S. pasteurii*. Similar impact factors were studied to determine the possibility of using urease instead of bacterial cells in practical applications of MICP.

TABLE 4.3
$CaCO_3$ Content and UCS under Different Sand Type of MICP-Treated Samples Catalyzed by *S. pasteurii*

Sand Type	Effective Size D_{10} (mm)	$CaCO_3$ Content (%)	UCS (MPa)
Standard sand	0.31	7.21~7.88	1.28~1.36
MS local sand	0.20	4.98~5.11	0.49~0.53

4.1.2.4.1 *Effect of Urease Concentration*

Like the experiment of bacteria concentration effect on MICP-treated soil, four different urease concentrations (2.5 g/L, 5 g/L, 10 g/L, and 15 g/L) were selected to study the effect of urease concentration on MICP. The four urease concentrations were selected because urease activity and electronic conductivity are like the OD_{600} of 0.3, 0.6, 1.2, and 1.5. Standard sand was used, and the concentration of the cementation solution was 0.5 M Ca. After 7 days of reaction, the samples were air-dried at room temperature.

After 7 days of reaction, the sand treated by 2.5 g/L urease was still loose and could not form a firm treated sample. Similar to the results of bacteria-treated soil, as urease concentration increased from 5 g/L to 15 g/L, the UCS of MICP-treated samples increased (Figure 4.6a). But the UCS of bacteria-treated samples is higher than that of urease-treated samples under similar urease activity. When the urease activity of bacteria is 6.90 mM hydrolyzed urea/min of UCS of bacteria-treated sand (1.79~1.94 MPa), it is almost five times the UCS (0.33~0.43 MPa) of urease-treated sample at the urease activity of 6.36 mM hydrolyzed urea/min. This may be explained

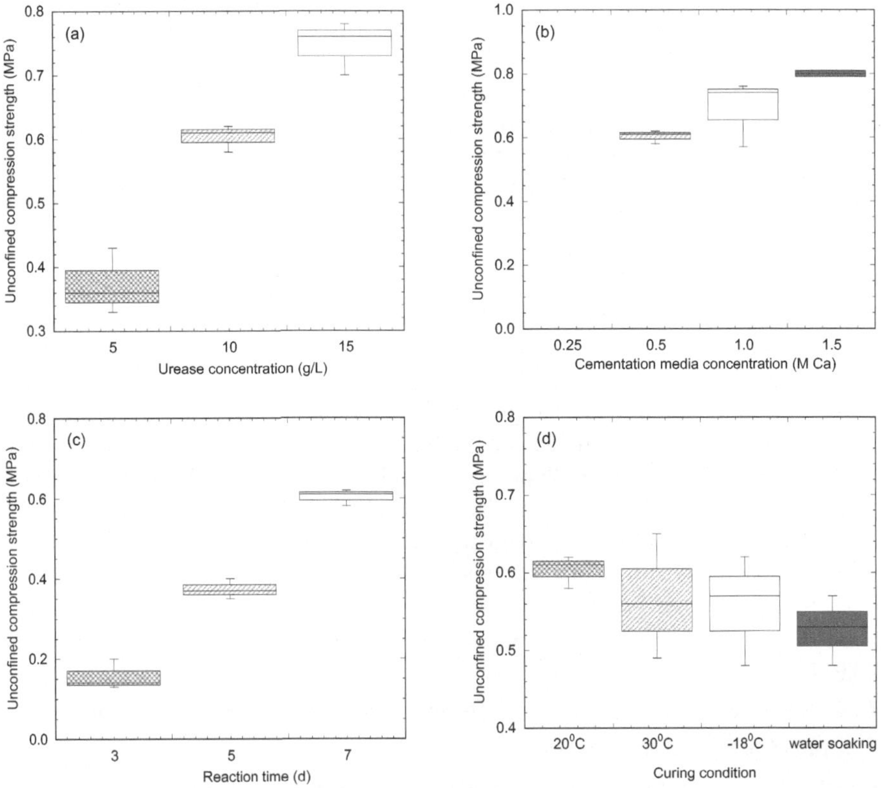

FIGURE 4.6 Box plot of the unconfined compression strength of treated samples catalyzed by urease as a function of (a) OD_{600}, (b) cementation media concentration, (c) reaction time, and (d) curing conditions.

by the fact that the bacteria kept growing and reproducing during the MICP process and continuously produced new urease, which improved the urease activity. However, the urease activity of MICP by urease decreases with its degradation and/or is inhibited in the long term. Moreover, the manual extraction of urease from plants is a complex process, and thus, production is relatively low and costly. Compared to the MICP catalyzed by bacteria, urease-treated soil may not be suitable for large-scale application due to the low efficiency and high cost of urease. Table 4.4 also indicates the correlation between $CaCO_3$ content and UCS for four urease concentrations. With an increase in urease concentration, both $CaCO_3$ content and UCS gradually increased. At similar $CaCO_3$ content, the UCS of the bacteria-treated sample is higher than that of the urease-treated soil (Table 4.2). When $CaCO_3$ content ranged from 7.21 to 7.88%, the UCSs of the bacteria-treated samples were 1.28~1.43 MPa, which is much higher than the 0.58~0.62 MPa of urease-treated samples, with 5.12~5.55% $CaCO_3$ precipitated. With the increase in $CaCO_3$ content, this trend was obvious. The UCSs of the bacteria-treated samples were 2.04~2.13 MPa, with 12.14~13.39% $CaCO_3$ content, while the UCSs of urease-treated samples were 0.79~0.81 MPa, with 7.36~8.9% $CaCO_3$ content. During the formation of $CaCO_3$ crystals, bacteria itself acted as the nucleating material, or another organic matter had an effect on the mineralization that resulted in different $CaCO_3$ crystals formed, and thus, the engineering properties of these samples varied (Stocks Fischer et al. 1999).

4.1.2.4.2 Effect of Cementation Media Concentration

To investigate the effect of cementation media concentration on MICP-treated sand catalyzed by urease, standard sand samples were prepared under four different cementation media concentrations. Urease concentration was 10 g/L, and urease activity was 11.5 mM hydrolyzed urea/min. The urease concentration was selected because the efficiency of MICP treatment is the highest (Table 4.4). Treated standard sand samples were air-dried at room temperature after 7 days of reaction.

Because the treated sample was still loose with a cementation media concentration of 0.25 M Ca, there is no UCS result for this cementation media concentration. Like the bacteria-treated soil, the urease-treated soil increases $CaCO_3$ content and UCS with more urea and calcium ions in the cementation solution (Figure 4.6b). Under the same cementation media concentration, the median UCS of bacteria-treated samples (1.28~1.36 MPa) is twice that of urease-treated samples (0.58~0.62 MPa), although the urease activity of bacteria is 3.31 mM hydrolyzed urea/min, far lower than the

TABLE 4.4
Urease Activity, $CaCO_3$ Content, and UCS under Different Urease Concentration of MICP-Treated Samples Catalyzed by Urease

Urease Concentration (g/L)	2.5	5	10	15
Urease activity (mM hydrolyzed urea/min)	3.15	6.36	11.51	14.86
$CaCO_3$ content (%)	—	4.43~5.02	5.12~5.55	7.09~7.32
UCS (MPa)	—	0.33~0.43	0.58~0.62	0.70~0.78

urease of 11.5 mM hydrolyzed urea/min. $CaCO_3$ content is also higher in bacteria-treated samples than in urease-treated samples. The efficiency of MICP catalyzed by bacteria is higher than that of urease.

4.1.2.4.3 *Effect of Reaction Time*

Standard sand samples were prepared to study the effect of reaction time on the engineering properties of urease-treated samples. Urease concentration was 10 g/L, and cementation concentration was 0.5 M Ca. After 3 days, 5 days, or 7 days of reaction, the samples were air-dried at room temperature.

As shown in Figure 4.6c, the results show that UCS increases with more reaction time for the urease-treated soil. Compared to the bacteria-treated soil, the median UCS of urease-treated soil increased less. This is probably because the effect of MICP catalyzed by urease is not as good as that of MICP catalyzed by bacteria. There may still be a large-volume pore space between the sand particles, and thus, MICP was still unhindered in the urease-treated samples. However, a comparison of samples treated by bacteria (OD_{600} = 0.6) and urease (10 g/L), indicates the median UCS of urease-treated sample is 0.58 MPa after 7 days of reaction, which is close to 0.59 MPa after 3 days of reaction for bacteria-treated samples. The efficiency of MICP by urease is much lower than the one by bacteria.

4.1.2.4.4 *Effect of Curing Conditions*

Using the same four curing conditions as the bacteria-treated samples, the urease-treated samples were cured for 7 days before unconfined compression tests began. The urease-treated samples were prepared with urease concentration of 10 g/L and ccementation concentration of 0.5 M Ca.

As shown in Figure 4.6d, median UCS of urease-treated samples curing in room temperature is slightly higher than that of 30°C and −18°C. The median UCS in the four curing conditions varied from 0.48 to 0.62 MPa. The impact of curing conditions on the urease-treated samples is limited. The engineering properties of MICP-treated sand are relatively stable and are not significantly affected by environmental conditions.

4.1.2.5 Relation between $CaCO_3$ Content and UCS

The relation between $CaCO_3$ content and UCS of treated samples catalyzed by *S. pasteurii* and urease is shown in Figure 4.7. The presence of higher $CaCO_3$ content with more mineral precipitation in the treated samples causes further strength improvement. The UCS of samples treated by bacteria is always higher than that of urease-treated samples at the same $CaCO_3$ content, because the efficiency of urease is quite lower than the one by bacteria. Moreover, the existing enzyme in the urease-treated samples may eventually degrade and/or be inhibited, while the bacteria can produce newer urease to replace the old ones. In terms of improvement efficiency, the treated sample catalyzed by bacteria is better than that catalyzed by urease.

MICP utilizing urea hydrolysis is a complex biochemical process, especially when it takes place between sand particles, for the improvement of soil engineering properties. There are many factors that may affect this process. Some of the factors studied in this paper include bacteria concentration (urease concentration), cementation

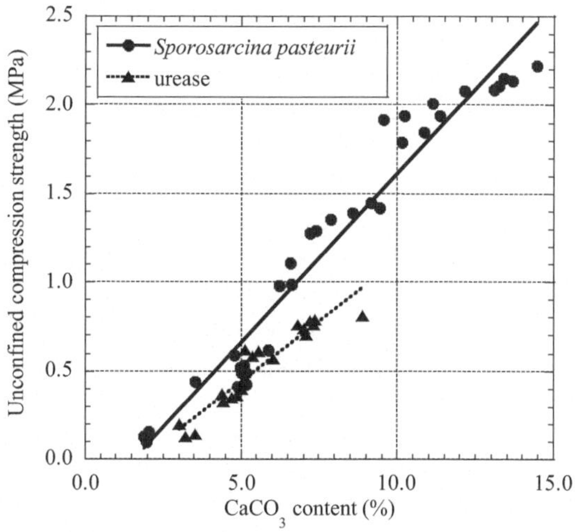

FIGURE 4.7 Relation between $CaCO_3$ content and unconfined compression strength of treated samples catalyzed by *S. pasteurii* and urease.

media concentration, reaction time, sand type, and curing conditions. This study highlights the effects of various factors on the MICP process under gravity-driven infiltration condition without additional pressure (such as pumping or injecting). All the samples were prepared with full-contact flexible molds to ensure full contact between the entire sample and the cementation media. In addition, based on the influence of these factors, a comparison between urease and whole bacteria cell in the MICP soil remediation was made.

The result of the unconfined compression test shows that all the studied factors, except curing conditions, have an obvious effect on the MICP process and UCS of treated sand, and the variation trend is similar between using bacteria and urease. With the OD_{600} ranging from 0.3 to 1.5 or urease concentration from 5 to 15 g/L, the UCS of MICP-treated samples increased. When cementation media concentration rose from 0.25 Ca M to 1.5 M Ca, the UCS of samples treated by bacteria or urease increased. Probably because there was no extra injection of cementation media, the MICP rate was slow, and the UCS still increased in 7 days. MICP performed on standard sand was better than on MS local sand. The UCS of MICP-treated samples catalyzed by *S. pasteurii* and urease linearly increased with the $CaCO_3$ content.

MICP utilizing urease powder is simple and can easily be controlled. On the other hand, bacteria need to experience the culture step before being used in the MICP, and its activity is difficult to control, owing to its division and growth during the MICP process. But the new urease can also be produced during MICP by bacteria, leading to a higher efficiency than MICP by urease powder. That is why the $CaCO_3$ content and the UCS of bacteria-treated samples are always higher than in urease-treated samples. The findings of this study indicate that the MICP by bacteria is better for engineering application due to its high improvement efficiency and low cost.

4.1.3 Influence of Fiber Addition on Mechanical Properties of MICP-Treated Sand

4.1.3.1 MICP Sample Preparation and Testing Methods

The 130 g of sand and fibers (0%, 0.1%, 0.2%, 0.3%, or 0.4% by weight of dry sand) were uniformly mixed with a 45 mL bacterial solution and then air-pluviated into the FCFM mold to reach a median dense condition (D_r in the range of 44%~55%). The percentage of fiber is based on the previous work of Shao et al. (2014). Preliminary test was conducted, with a gradual increment of fiber content mixing with the 130 g of sand. The soil strength increased with the fiber content, until 0.4% fiber content was reached. After fiber content, soil strength started to decrease. In the preparation of all samples, the required bacteria solution was first added into dry sand to prevent the fibers from floating in the soil matrix, and then the proposed content of fibers was mixed in small increments by hand to obtain a uniform mixture. It is important to ensure that all fibers are mixed thoroughly (Shao et al. 2014). A total of 12 molds with sand samples on the shelf were immersed into a 10 L cementation media in a completely stirred tank reactor (Figure 4.2). Small holes of 5 mm in diameter were punched in the shelf, and the spacing between these holes was 5 mm, which allowed the water to flow through the shelf. During the 7 days of reaction, only air was continuously pumped into the reactor. No bacteria, growth medium, or cementation medium was added to or pumped out of the reactor. After 7 days of reaction, the shelf was removed from the solution, each geotextile mold was cut, and the treated soil specimen was removed. The MICP sample preparation method is followed (Shao et al. 2014). After 7 days of reaction, the strength improvement of MICP-treated soil reached steady state (Shao et al. 2014).

Unconfined compression tests were conducted under strain-controlled conditions at a uniform loading rate of 1.5%/min, in accordance with ASTM D2166. Consolidated drained triaxial compression tests were conducted under 100, 200, 300 or 400 kPa cell pressure at a constant axial strain rate of 0.02% strain/min, where the pore water pressure was controlled near 0 kPa. The tests were halted after axial strain had reached 15%.

4.1.3.2 Effect of Fiber on Unconfined Compression Behavior of MICP-Treated Sand

The stress–strain curves obtained from UCS tests on MICP-treated sand specimens with different fiber contents (0.1%, 0.2%, 0.3%, and 0.4% by weight of sand) is shown in Figure 4.3a. Although the MICP-treated samples can reach a uniform state in the full-contact flexible mold, there was still slightly different $CaCO_3$ contents distributed within each sample, which may cause strength to fluctuate during the tests (as shown in Table 4.5). At least four samples were prepared for each case to produce a representative condition. With the addition of fiber, the unconfined compression strength increased to 0.3% fiber content. When fiber content reached 0.4% by weight, unconfined compression stress–strain curves deteriorated.

Figure 4.8 shows the effect of fiber on the UCS of MICP-treated sand. All the strengths (peak strength q_u, strength at half failure strain $q(\varepsilon_f/2)$, and residual strength q_{ru} at 15% strain) increase with up to a fiber content of 0.3%. When 0.1%, 0.2%,

TABLE 4.5
Effect of Fiber Content on Unconfined Compression Strength, Failure Strain, and Elastic Modulus of MICP-Treated Samples

Fiber Content (%)	q_u (kPa)	ε_f (%)	$q_u(\varepsilon_f/2)$ (kPa)	E_{50} (MPa)	E (MPa)	q_{ur} (kPa)
0.0	68 ± 8	1.6 ± 0.4	32 ± 8	4.3 ± 1.7	3.7 ± 0.7	0.0
0.1	90 ± 5	2.0 ± 0.3	53 ± 5	4.9 ± 0.3	4.6 ± 0.7	50 ± 6
0.2	167 ± 2	3.5 ± 0.1	85 ± 7	4.9 ± 0.7	4.8 ± 0.1	104 ± 10
0.3	164 ± 6	3.7 ± 0.5	87 ± 11	4.8 ± 1.3	4.4 ± 0.6	99 ± 7
0.4	102 ± 6	3.1 ± 1.1	61 ± 9	4.0 ± 0.8	3.4 ± 0.5	70 ± 15

Note: q_u is peak strength, ε_f is failure strain, $q_u(\varepsilon_f/2)$ is strength at half peak strain, E_{50} is elastic modulus at 50% of q_u (= $q_u(\varepsilon_f/2)/(\varepsilon_f/2)$), E is elastic modulus (= q_u/ε_f), and q_{ur} is residual strength.

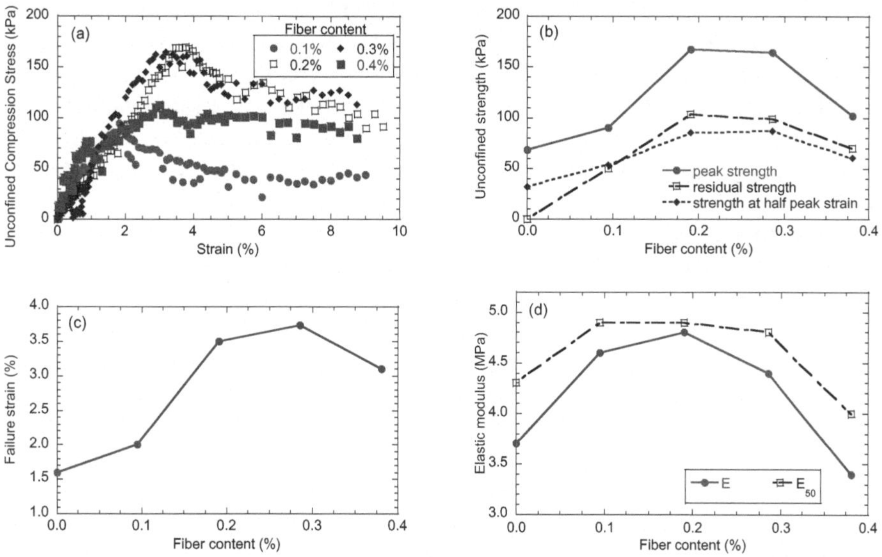

FIGURE 4.8 Effect of fiber content on (a) unconfined compression stress, (b) unconfined compression strength, (c) failure strain, and (d) elastic modulus of MICP-treated samples.

0.3%, or 0.4% fiber was added, UCS increased by 132%, 245%, 240% and 149%, respectively. The peak strengths of MICP-treated sand at a fiber content of 0.2–0.3% are more than two times higher than that without the addition of fiber. The strength at half failure strain of MICP-treated sand at fiber contents of 0.2–0.3% are also approximately two times higher than that without the addition of fiber. The residual strengths of MICP-treated sand at a fiber content of 0.2–0.3% are more than half of their peak strengths, while residual strength is 0 kPa for that of MICP-treated sand without fiber. When local cracks appear on MICP-treated sand with fiber, fibers across the cracks take the tension within the soil due to the fiber–soil friction, which effectively impedes further development of cracks and, thus, improves the resistance of soil to the force applied. These results indicate that fiber plays an important role on the unconfined compression strength of MICP-treated sand, and 0.2–0.3% fiber (by weight of sand) is the most suitable fiber content.

The effect of fiber on the failure strain of MICP-treated sand under unconfined condition is shown in Figure 4.8c. It shows that failure strain increases with the addition of fibers. As 0.1%, 0.2%, 0.3%, or 0.4% fiber was added, the failure strain of MICP-treated sand increased 125%, 219%, 233%, and 194% of that without fiber, respectively. The results indicate that the ductility of MICP-treated sand can be significantly improved by adding a suitable number of fibers (0.2–0.3%). This finding is consistent with the results of past studies.

The effect of fiber on the elastic modulus of MICP-treated sand under unconfined conditions is shown in Figure 4.8d. Young's modulus at the unconfined compression strength (E) and Young's modulus at 50% of the unconfined compression strength (E_{50}) increase with fiber content up to 0.2%. Young's modulus E reaches peak value at 0.2% fiber content, and E_{50} reaches peak value at 0.1% fiber content, which are 130% and 112% of that of MICP-treated sand without fiber, respectively. However, when the fiber contents exceed 0.2%, E and E_{50} slightly decrease. At 0.4% fiber content, E and E_{50} are lower than the elastic modulus of MICP-treated sand without fiber. Therefore, the strengthening effect of fiber is mainly influenced by extending the elastic strain range, rather than by increasing the elastic modulus, which is on behalf of rigidity.

When fiber content was increased up to 0.4%, there was a drop in shear strength and elastic modulus in the UCS test. The reason is that fiber impedes the uniformity of bacteria distribution in the soil specimen. The non-uniformly distributed bacteria impact the distribution of calcite precipitation in the soil mixture, and thus the strength of soil specimen. In the sample preparation, sand and fiber were mixed with the bacteria solution and then air-pluviated into the full-contact flexible mold. The individual bacterium is a long narrow shape with approximately 2 μm in length (as shown in Figure 4.1b). As the bacteria are alive and moving around in the pore among the soil mixture, fiber can impede bacteria movement. The more the fiber content, the more bacteria are obstructed by the fiber. The formation of calcite precipitation is around fiber, instead of around soil particles. The less bonding of soil particles by calcite precipitation causes weak strength at the higher fiber content (0.4% of fiber content).

4.1.3.3 Effect of Fiber on Triaxial Compression Behavior of MICP-Treated Sand

The stress–strain relationship is important to determine the mechanical behavior of modified soils (Consoli et al. 2009a; Heineck et al. 2005). Therefore, through studying the shear stress–strain curves, a better understanding can be developed as to how fiber reinforcement affects the behavior of the MICP-treated sand (Figure 4.9). It is well-known that the effectiveness of the inclusion of fibers into soil is greatly affected by the fiber content (Consoli et al. 2011; Shao et al. 2014). Moreover, the amount of confining pressure applied on MICP-treated sand has a direct impact on the shear

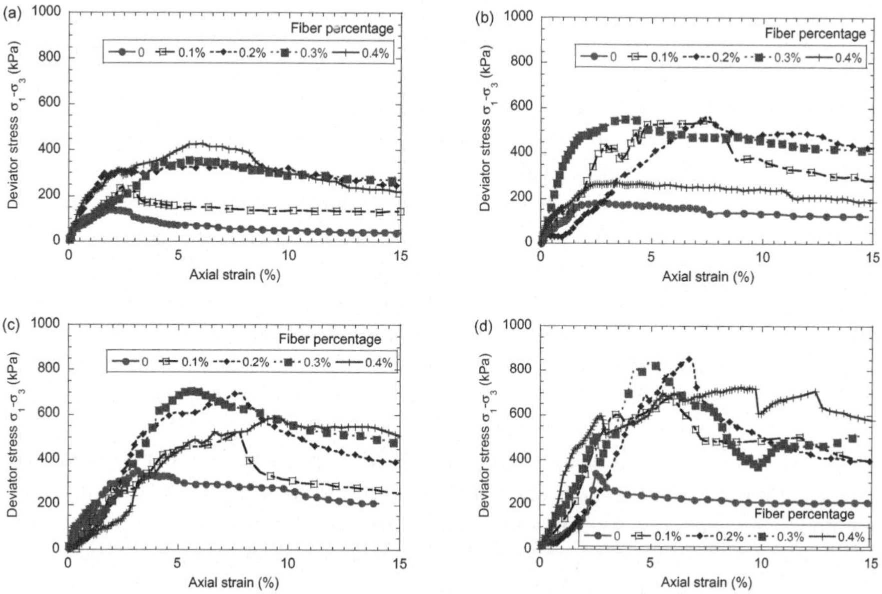

FIGURE 4.9 Shear stress–strain curves of MICP-treated sand with different fiber contents and confining pressure: (a) 100 kPa, (b) 200 kPa, (c) 300 kPa, and (d) 400 kPa.

strength of fiber-reinforced soils. Therefore, this study was particularly focused on determining the effects of fiber content and different confining pressures on the mechanical behavior of reinforced soils. Figure 4.9 shows the stress–strain curves obtained from triaxial compression strength tests on MICP-treated sand specimens mixed with different fiber contents (0%, 0.1%, 0.2%, 0.3%, and 0.4%) under confining pressures of 100 kPa, 200 kPa, 300 kPa, and 400 kPa.

As shown in Figure 4.9, the presence of fiber affected the trend of stress–strain curves. The fiber content tended to increase the ultimate shear strength of specimens substantially until fiber content of 0.3%, compared with MICP-treated sand-only. The inclusion of fiber up to a fiber content of 0.3% in MICP-treated sand makes less shear strength reduction. Yetimoglu et al. (2003) reported that the inclusion of fibers in sand provided great potential to reduce the brittleness of soils and improve its ductility and to limit the loss of post-peak strength. One of the main advantages of fiber reinforcement in sand is improving the ductility of sand (Babu et al. 2011; Yetimoglu et al. 2003). As shown in Figure 4.9, MICP-treated sand mixed with fibers exhibits a more ductile behavior than MICP-treated sand alone and limited reductions in post-peak shearing resistance until fiber content of 0.3%. MICP-treated sand mixed with fibers can maintain the shear stress with continuing deformation even at large strains, suggesting that these materials are very ductile.

The increase of fiber content to 0.4% inhibits the movement of bacteria in the soil matrix. The non-uniformly distributed bacteria impact the distribution of calcite precipitation in the soil mixture, and thus the strength of soil specimen. Therefore, there is no consistent relationship between the confining pressure and shear strength of the MICP-treated samples prepared with 0.1% and 0.4% by weight fiber content.

Shear strength of MICP-treated sand under 100 kPa, 200 kPa, 300 kPa, and 400 kPa confining pressures is shown in Figure 4.9a. When 0.1%, 0.2%, and 0.3% fiber (by weight) were added, the shear strengths of MICP-treated sand were improved under the four confining pressures. The lower the confining pressure (100 kPa), the greater the fiber contents of MICP-treated sand, and the greater the shear strength of MICP-treated sand at the same confining pressure. When fiber content is 0.2% or 0.3%, the shear strength of MICP-treated sand shows a good relationship with confining pressure. Compared to the shear strength of MICP-treated sand without fiber, the improvement of shear strength was observed to be as high as 310% for the MICP-treated sand prepared with a fiber content of 0.2% at a confining pressure of 200 kPa. Therefore, fiber reinforcement is more effective for samples at a fiber content of 0.2–0.3% for MICP-treated sand.

Figure 4.10b shows the Mohr circle and failure envelope of MICP-treated sand mixed with 0.2% fiber content under four confining pressures. From the envelopes obtained, cohesion and friction angle can be found as the intercepts and the slopes of the envelopes, respectively. Figure 4.10c shows the effect of fiber content on the shear strength parameters of MICP-treated sand. When fiber content is 0.1%, 0.2%, 0.3%, or 0.4%, cohesion increases by 236%, 273%, 311%, and 238% of that of MICP-treated sand without fiber, respectively. It indicates that cohesion can be considerably improved by the addition of fibers. With an increase in fiber content of samples

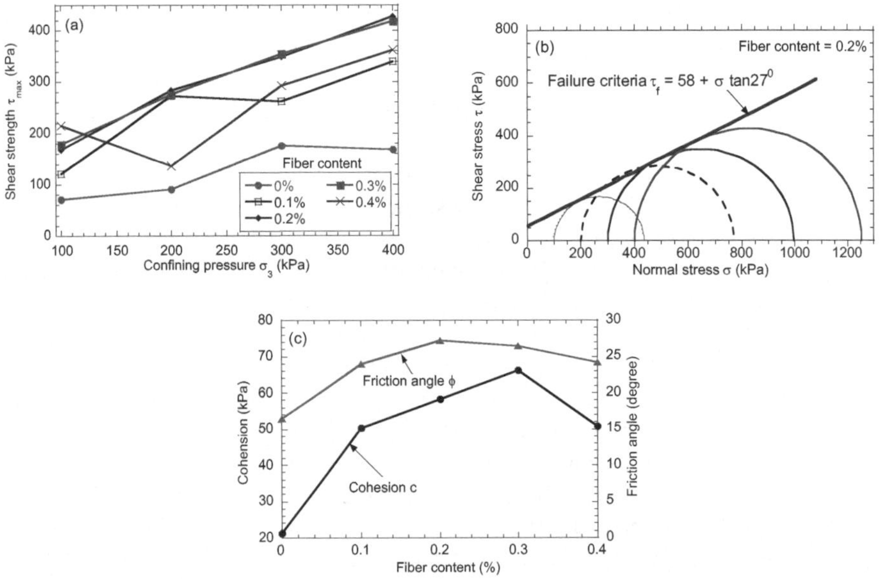

FIGURE 4.10 Shear strength for MICP-treated sand: (a) as a function of cell pressure at different fiber contents under triaxial compressive conditions, (b) Mohr circle and failure envelope, and (c) effect of fiber content.

until 0.3% of fiber content, the interface between fiber and soil particles increases, and hence, the friction between fiber and soil particles increases, which makes it difficult for soil particles surrounding fibers to change in position from one point to another and thereby improves the bonding force between soil particles, that is, cohesion of soil. Cementation by the MICP process also increases the cohesion of soil (Burbank et al. 2013; Chou et al. 2011; DeJong et al. 2010).

When fiber content is 0.1%, 0.2%, 0.3%, or 0.4%, the angle of internal friction was improved to 24°, 27°, 26°, and 24°, respectively, compared with that of MICP-treated sand without fiber with a friction angle of 16°. It indicates that the angle of internal friction can also be improved by the addition of fibers. The peak friction angle occurred at fiber content of 0.2%, with the value of 27°. This finding is consistent with the results of previous studies (Consoli et al. 1998; Jiang et al. 2010; Zaimoglu and Yetimoglu 2012). Consoli et al. (1998) found an improvement in both cohesion and angle of internal friction induced by 0.3% fiber addition for cemented sandy soil. Jiang et al. (2010) noted an improvement of both cohesion and angle of internal friction induced by the addition of fibers and summarized the optimal fiber content as 0.3% for clayey soil. When fiber content was 0.4%, the decreasing trend of shear strength parameters is like in the UCS findings, indicating that too much fiber is not appropriate for MICP-treated sand. This is because obtaining a homogeneous distribution of bacteria is difficult at high fiber contents.

The failure strain and elastic modulus of MICP-treated sand with fibers under four confining pressures are shown in Table 4.6. This table shows that failure strain increased with the addition of fibers. The average failure strain of MICP-treated sand without fiber is 2.6%, with a maximum value of 3.1%, while the average failure strain of MICP-treated sand with fiber is 6.2%, with a maximum value of 9.4%. The failure strain and elastic modulus increase along with confining pressure, which is consistent with previous studies. Consoli et al. (1998) noted improvements in failure strains induced by 0.3% fiber addition under 20 kPa, 60 kPa, and 100 kPa confining pressures for cemented sandy soil.

TABLE 4.6
Effect of Fiber Content on Failure Strain and Elastic Modulus of MICP-Treated Sand under Triaxial Compressive Conditions

Cell Pressure (kPa)	Fiber Content 0.0%		0.1%		0.2%		0.3%		0.4%	
	ε_f (%)	E (MPa)	ε_f (%)	E (MPa)	ε_f (%)	E (MPa)	ε_f (%)	E (MPa)	ε_f (%)	E (MPa)
100	2.0	7	2.4	10	6.1	6	5.7	6	6.1	11
200	2.8	7	7.3	7	7.5	8	4.1	14	3.4	13
300	3.1	11	7.8	7	7.9	9	5.4	13	9.4	9
400	2.5	13	5.9	12	6.7	13	4.9	17	9.0	13

Note: ε_f is failure strain, and *E* is elastic modulus.

4.1.3.4 Distribution of $CaCO_3$ Precipitation

The SEM image in Figure 4.12a shows that $CaCO_3$ crystals were mainly an irregular bulk, similar to the observations made by Al Qabany et al. (2012) and Zhao et al. (2014a). The crystals were verified as calcium carbonate using energy-dispersive X-ray analysis and were composed of elements of O, Ca, and C (Figure 4.12b). The $CaCO_3$ crystals tend to participate in the contact segments between sand particles, which results in a better bonding effect among sand particles than uniform distribution on sand particles. As a result of cementation, the strength of MICP-treated sand is improved (Burbank et al. 2013; DeJong et al. 2010). In addition, the reduction in void space due to $CaCO_3$ precipitation is a secondary strengthening factor (DeJong et al. 2010).

To evaluate the uniformity of the MICP process in the specimen, the $CaCO_3$ content in four areas of the unconfined compression test samples was measured. As indicated in Figure 3.30a, the four areas corresponded to the top (I) (18 mm thick), bottom (II) (18 mm thick), inner cylinder (III) (17 mm in diameter), and outer annular (IV) (15 mm thick). The $CaCO_3$ content in the four areas ranged from 6.6% to 8.0% (by weight). The $CaCO_3$ content in part III was lower than that of the other parts, while the $CaCO_3$ contents in parts I and IV were higher. Figure 4.11b shows the sand particles surrounding fibers, which was taken after soil specimen failure in unconfined compression strength test.

4.1.3.5 Discussion on the Optimum Content of Fiber Content

When fibers are mixed uniformly with sand, that is, 0.1% or 0.2% by weight of sand, the cohesion and friction between fibers and sand particles increase with fiber content. This is due to the increase in the interface between them, which improves cohesion, angle of internal friction, UCS, and failure strain of MICP-treated soil. However, if the amount of fiber mixed with soil is too much, that is, 0.4% by weight

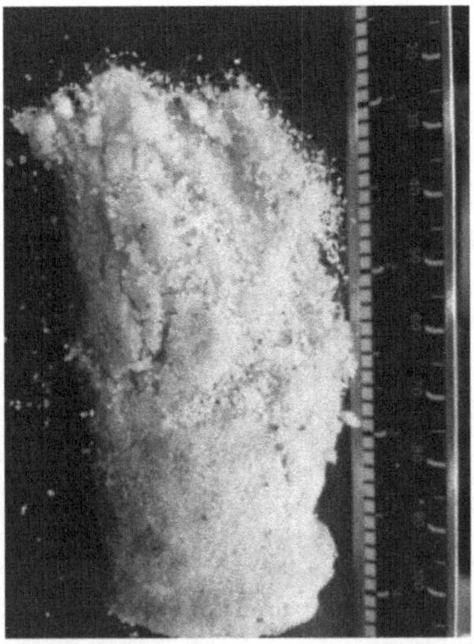

FIGURE 4.11 Photograph of the sand particles connected by fiber and $CaCO_3$ precipitation.

FIGURE 4.12 SEM image of MICP-bonded sand particles and energy-dispersive X-ray image of precipitation on sand surface: (a) sand, $CaCO_3$ crystals, and *S. pasteurii*; (b) element Ca; (c) element C; and (d) element O.

Note: Fiber is 100 µm in diameter, which is too big to show in the SEM image with a resolution of 1 µm.

of sand, the fibers will inhibit the movement of bacteria in the soil matrix, which leads to an uneven distribution of bacteria in soil and thus results in non-uniform calcite precipitation. This consequently reduces the effectiveness of improvement in compressive strength and shear strength parameters. This is similar to the previous finding by Jiang et al. (2010) based on their research on fiber-reinforced clay.

Hejazi et al. (2012) summarized the different optimum fiber contents for fiber-reinforced soil. Their results indicated a large optimum content scope ranging from 0.1 to 0.4% for synthetic fibers and ranging from 0.5% to 1% for natural fibers. The phenomenon of lump forming is the most important criteria for identifying the optimum fiber content. Based on the mechanical properties investigated by both UCS tests and triaxial tests in this study, including strength, cohesion, angle of internal friction, and failure strain, the optimal fiber content scope was found to be 0.2–0.3% by weight of dry sand. The optimum fiber content depends on the characteristics of the fiber, including type, chemical composition, length, diameter, strength, failure strain, and modulus of elasticity.

The addition of fiber in the MICP process has the potential to increase the ductility of the MICP-treated sand and to prevent the loss of post-peak strength. In this chapter, a series of unconfined compression tests and triaxial shear strength tests were carried out on MICP-treated sand mixed with fiber contents of 0.1, 0.2, 0.3, and 0.4%, in order to study the effect of randomly distributed discrete fiber on the mechanical properties of MICP-treated soil and to investigate the corresponding mechanisms. MICPs were catalyzed by *Sporosarcina pasteurii* at a concentration of 4.3×10^7 cells/mL. The concentration of the cementation media was 0.5 M. All the samples were prepared with full-contact flexible molds to ensure full contact between the entire sample and the cementation media.

Under unconfined compression and triaxial conditions, shear strength and failure strain were significantly improved, while elastic modulus showed a slight improvement. The results indicate various improvements achieved by adding fibers, compared with the corresponding value of MICP-treated sand without fiber. Under an unconfined compressive condition, 233% increase in failure strain, 240% increase in UCS, and 130% increase in elastic modulus were achieved. Under triaxial compressive conditions, 270% increase in failure strain, 253% increase in failure shear strength, 131% increase of elastic modulus, 311% increase in cohesion, and 51% increase in friction angle were noted. The UCS and shear strength increased gradually with an increase in fiber content up to 0.3%. In addition, MICP-treated sand reinforced with fiber exhibited more ductile behavior and limited reductions in residual strength. The optimum fiber content in the MICP-treated sand was found to be 0.2–0.3%. The UCS of MICP-treated sand at fiber content of 0.2–0.3% is more than two times higher than that without the addition of fiber. The residual strengths of MICP-treated sand at the fiber content of 0.2–0.3% are near 100 kPa, while the residual strength is 0 kPa for that of MICP-treated sand without fiber.

Fiber content had positive effects on improving the shear strength parameters (angle of internal friction and cohesion) of the mixtures. The cohesion and angle of internal friction of fiber-reinforced sand prepared at different ratios of fiber increased by 29–45 kPa and 7.6–11°, respectively. The inclusion of fibers improves the ductility of the soil by preventing the loss of post-peak strength. The strengthening effect of fiber is mainly achieved by the cohesions and frictions between fibers and sand particles, while the improvement in ductility is mainly achieved by the tension and large failure strain of fiber in the cracks. The failure strain of MICP-treated sand at a fiber content of 0.2–0.3% is nearly three times higher than that without the addition of fiber. The strengthening effect of fiber is mainly characterized as an extension of the elastic strain scope, rather than advancing elasticity modulus, which is different from the strengthening effect of calcium carbonate precipitation content.

4.2 APPLICATION OF BIO-BRICKS TREATED WITH MICP

4.2.1 Experimental Methods

4.2.1.1 Four-Point Bending Test

Four-point bending tests, as shown in Figure 4.13a, were conducted on bio-bricks to study their flexure behavior. The samples were located on two adjusted supports that were 152.4 mm apart from each other, and vertical load was applied on the top

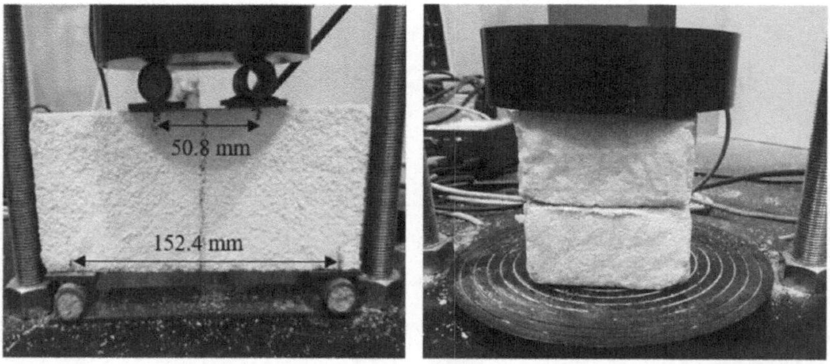

FIGURE 4.13 (a) Four-point bending testing system. (b) Brick compressive strength testing system.

two supports in the middle of the specimen, with a 50.8 mm distance. Following the testing method of ASTM D6272, the vertical load was conducted under strain control conditions at a uniform loading rate of 1.5%/min until the beam failed.

4.2.1.2 Brick Compressive Strength Testing

Brick compressive strength tests (see Figure 4.13b) were conducted on bio-bricks. Each test sample was cut into halves along the width direction before testing. These two halves were stacked along the surface perpendicular to the cutting surface. The testing procedure applied on the stacked halves followed the testing method of ASTM C67, and the vertical load was subjected to strain control conditions at a uniform loading rate of 1.5%/min.

4.2.1.3 Accelerated Erosion Testing

The accelerated erosion testing carried out on the bio-bricks was like the testing method outlined by Walker (2002). The erosion setup of this study is shown in Figure 4.14a. A jet of water was sprayed onto the surface of a brick at a constant pressure of 200 kPa and a constant distance of 470 mm. The erosion depth was recorded in Figure 4.14b at frequent intervals in a 60 min period. The maximum erosion rate of the sample was determined by the following:

$$\text{Maximum erosion rate} = \frac{\textit{maximum erosion depth}}{\textit{height of brick}} \times 100\% \tag{4.1}$$

4.2.1.4 Resistance to Water Absorption

The procedure of ASTM C67 was followed in this study to determine the water absorption of bio-bricks. Three representative intact samples were used for each test. The dry mass (m_d) of the bio-bricks was measured at the beginning of this procedure. Then, the bio-bricks were submerged in water at ambient temperature for 24 h. The

FIGURE 4.14 (a) Accelerated erosion setup. (b) Measuring the eroded depth of bio-bricks.

samples were taken out and weighed immediately as saturated mass (m_s). The water absorption (W_a) of the bricks was calculated using equation (4.2):

$$W_a = \frac{m_s - m_d}{m_d} \times 100\% \tag{4.2}$$

where m_d is the dry mass of bio-bricks, m_s is the saturated mass of bio-bricks, and W_a is water absorption of the bio-bricks. Resistance to water absorption was determined as the average water absorption of three bio-brick samples.

4.2.1.5 $CaCO_3$ Content

The precipitated $CaCO_3$ content of bio-bricks caused by MICP treatment was determined by acid washing method. The samples were weighed and washed with an HCl solution (0.1 M) to dissolve precipitated carbonates. The weight of specimens was measured after rinsing, draining, and oven-drying. The difference in weight before and after acid washing was considered as the weight of the carbonates that were precipitated during MICP treatment.

4.2.1.6 SEM Analysis

The microstructure of the bio-bricks was directly observed by scanning electron microscopy (SEM) (Lyra 3, TSECAN Inc.). The tested samples were oven-dried overnight at 105°C and mounted on the stubs with adhesive carbon conductive tabs before testing.

4.2.2 Enhancement of Erosion Resistance of Bio-Bricks through Fiber and Multiple MICP Treatments

4.2.2.1 Materials and Sample Preparation

4.2.2.1.1 Sands

The Ottawa sand used in this section is the same as that in Section 3.2.1.

4.2.2.1.2 Reinforcing Fiber

Two kinds of fiber, as shown in Figure 4.15, were used as reinforcement additions in the soil. Fibermesh 150 synthetic fiber (see Figure 4.15a) was used in this study, and the same fiber was used by Shao et al. (2014) and Li et al. (2015c). This fiber is a 100% homopolymer polypropylene multifilament fiber, and it is a very light material, with a specific gravity (Gs) of 0.91. It is chemically inert and has uniform characteristics. The length of the fibers was 12 mm. As shown in Figure 4.15b, the coconut coir fiber used in this study are matured coir fibers. Fibers were cut into pieces with length ranging from 10 to 15 mm and diameter within 0.02 to 0.25 mm.

4.2.2.1.3 Bacterial Suspension and Cementation Solution

The bacteria and growth medium and cementation media followed are those in Section 3.1.1 and Section 3.1.2.

4.2.2.1.4 Rigid Full-Contact Mold

The rigid full-contact mold used is the same with that in Section 3.3.2, as shown in Figure 3.8.

4.2.2.1.5 Sample Preparation

The MICP technique was used to treat the loose sand in this study. The MICP process was conducted in a batch tank reactor using the immersing method. The reactor shown in Figure 3.32 included a plastic box to contain samples, cementation media, sample supported shelf, and air pumps to provide oxygen for bacteria. For non-reinforced bio-bricks, the uniform mixture of 900 g sand and 250 mL bacteria solution was compacted into the rigid full-contact mold well-defined.

The average content of fibers included in a fiber-reinforced sand sample is defined as a percentage of weight of dry sand:

$$\rho = \frac{W_f}{W_s} \times 100\% \quad (4.3)$$

where W_f is the weight of the fibers, and W_s is the weight of the dry sand.

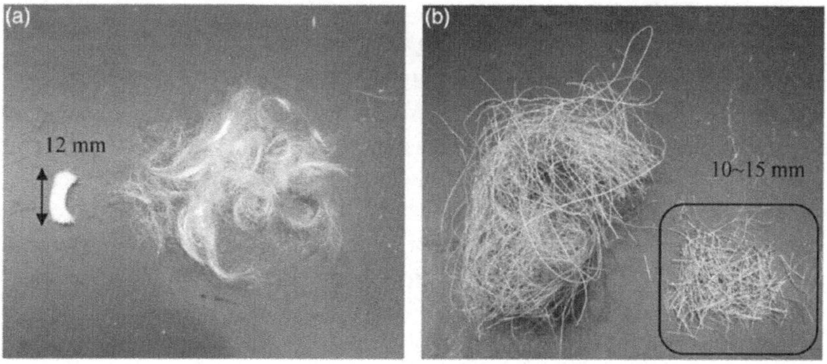

FIGURE 4.15 (a) Fibermesh 150 synthetic fiber. (b) Coconut coir fiber.

For synthetic fiber–reinforced bio-bricks, the proportion of fiber was 0.3% by weight. The synthetic fiber was uniformly mixed with sand, then placed in the rigid full-contact mold, and a 250 mL bacteria solution was added in the sand through a geotextile cover. For coconut coir fiber–reinforced bio-bricks, the well-mixed sand and coconut coir fiber were placed into a rigid full-contact mold, and then a 250 mL bacteria solution was added in the sand through the geotextile cover. Four different coconut coir fiber contents (0.3%, 0.6%, 0.9%, and 1.2% by weight of dry sand) were used for this study. At least four samples were prepared for each case to produce a representative condition.

In the single-MICP process, soil samples were submerged in a 0.5 M Ca cementation media for seven days to allow the reaction to proceed. The entire process was conducted at ambient laboratory temperature. After treatment, the samples were removed from the cementation media, and the geotextile mold in the rigid full-contact mold was carefully cut to extract the samples. The treated samples were then thoroughly washed in fresh water to remove the residual loose particles and unbonded precipitates. Following the cleaning process, the samples were oven-dried until their weight stabilized. Once dried, these samples were ready for subsequent testing or additional MICP treatments.

To enhance the properties of bio-bricks, a multiple-MICP treatment method was applied, repeating the process up to three times. Each cycle involved several steps. First, the rigid layer was demolded, and the geotextile in the rigid full-contact mold was cut to remove the MICP-treated bio-bricks. The extracted bio-bricks were then soaked in fresh water to wash off any remaining loose particles or unbonded precipitates. After getting cleaned, the bio-bricks were oven-dried for 48 h. The dried bio-bricks were then soaked in a fresh bacteria solution with an optical density of 0.6 at 600 nm until fully saturated. Once saturated, the bio-bricks were placed on a shelf and immersed in a batch reactor containing fresh cementation media, where they remained for another seven days of MICP reaction. This set of steps constituted a single-MICP treatment cycle, which could be repeated as needed to achieve the desired enhancement of the bio-brick's properties.

4.2.2.2 Effect of Fiber Reinforcement on Flexure Behavior of Bio-Bricks

The stress–strain curves obtained from four-point bending tests on MICP-treated bio-bricks without fiber, with 0.3% synthetic fiber, and 0.3% coconut coir fiber are shown in Figure 4.16. Li et al. (2015c) and Wen et al. (2018a) found that the optimum synthetic fiber content in the reinforcement of bio-specimens was 0.3%. The improvement of the ductility of bio-specimens was significant when 0.3% fiber was added to reinforce the bio-beam. The flexure stress and strain of the MICP-treated samples increased by 105% and 210%, respectively, through 0.3% synthetic fiber addition in bio-bricks. These results indicated that the ductility of bio-bricks was improved significantly with 0.3% synthetic fiber content. When 0.3% coconut coir fiber was added in the bio-bricks, flexure stress was improved from 1,185 kPa to 1,300 kPa, while flexure strain was improved only 20%. Thus, 0.3% of coconut coir fiber may have limited contribution to the improvement of the ductility of reinforced bio-bricks.

To study the effect of randomly distributed coconut coir fiber content on the flexure behavior of reinforced bio-bricks, 0.3, 0.6, 0.9, and 1.2 wt.% fiber content were used in this study. The peak flexure strength of non-reinforced bio-bricks was 1,185 kPa.

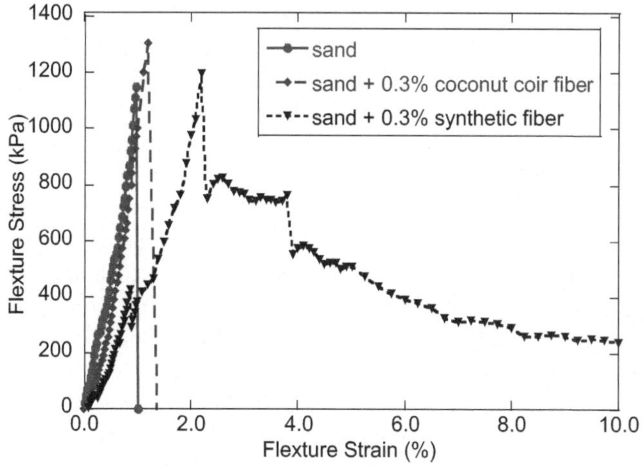

FIGURE 4.16 Flexure stress–strain curves for MICP-treated samples.

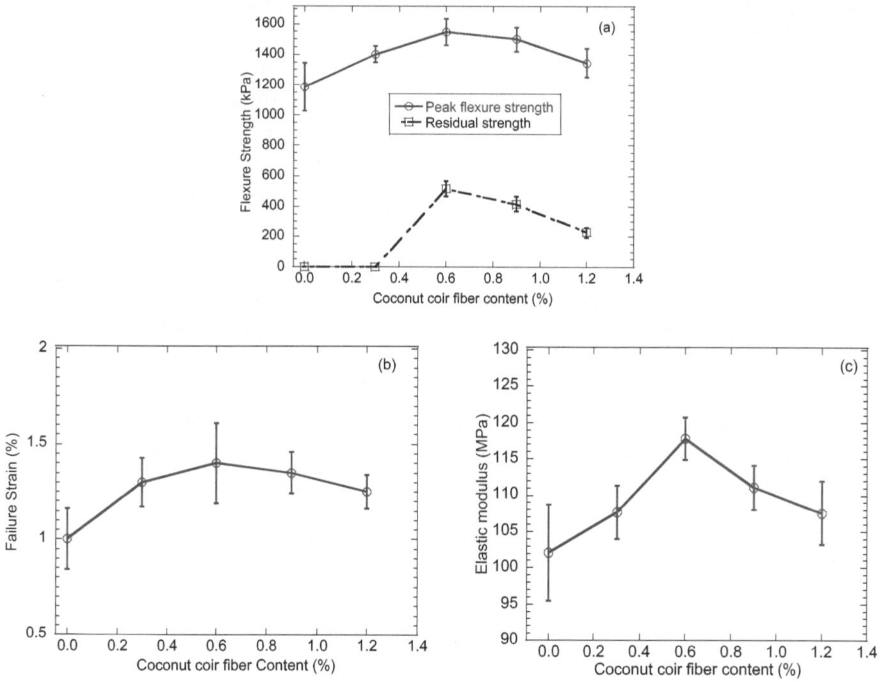

FIGURE 4.17 Effect of fiber content on coconut coir fiber–reinforced bio-bricks' (a) flexure strength, (b) flexure strain, and (c) elastic modulus.

As shown in Figure 4.17a, the peak flexure strength of coconut coir fiber–reinforced bio-bricks increased along with fiber content up to 0.6 wt.%. When 0.3%, 0.6%, 0.9%, and 1.2% coconut coir fiber were mixed with soil, peak strength increased by 118%, 131%, 127%, and 114%, respectively. Peak strength of 0.6–0.9% coconut coir fiber–reinforced bio-bricks is about 30% higher than that of non-reinforced bio-bricks. The increase in flexure strength should be linked to the increased friction between the coconut coir fiber, sand particles, and calcite crystals.

The residual strength of fiber-reinforced bio-bricks is shown in Figure 4.17a, which is the flexure strength at 15% strain. The residual strength increased with fiber content to 0.6 wt.%. The residual strength of non-reinforced bio-bricks was 0 kPa because of the brittleness of MICP-treated samples. The residual strength increased to 500–400 kPa with 0.6%–0.9% coconut coir fiber content, which was about half of their peak strength. The interaction of fibers, sand particles, and calcite crystals prevented the spread of cracks in bio-bricks, as fibers form bridges across cracks and contributed to the improved residual strength. When local cracks appeared on bio-bricks, fibers across the cracks took the tension within the soil due to the fiber-soil-calcite crystal friction, which effectively impedes further development of cracks and thus improves the residual strength of bio-bricks. However, the peak strength of coconut coir fiber–reinforced bio-bricks start to decrease when the fiber content exceeded 0.6%. It should be explained by knot and overlap of increased fibers, which would result in the reduced cohesion with sand and calcite. It was also likely that fiber clogged the pores among sand particles and weakened the calcite bond with MICP, which could lead to a reduction in strength.

Figure 4.17b showed the effect of coconut coir fiber content on the failure strain of MICP-treated bio-bricks. The failure strain of MICP-treated samples increased by 130%, 140%, 135%, and 125%, as 0.3%, 0.6%, 0.9%, and 1.2% coconut fiber was added in sand, respectively. Failure strain increased with coconut coir fiber content up to 0.6%. It indicated that coconut coir fiber reinforcement improved the ductility of bio-bricks. The failure of non-reinforced bio-bricks was sudden, and only one large crack was produced. However, the failure of coconut coir fiber–reinforced bio-bricks was characterized with multiple finer cracks. The failure of fiber-reinforced bio-bricks was more gradual and more like a ductile material with increased coconut coir fiber content. These results indicate that 0.6–0.9% fiber (by weight of sand) is the most suitable fiber content. This finding is similar to that of research on coir fiber–reinforced silty sand with a similar optimal fiber content of 0.75% (Chauhan et al. 2008).

The effect of coconut coir fiber on the elastic modulus of bio-bricks is shown in Figure 4.17. The elastic modulus can be expressed with the following equation:

$$E = \frac{q_u}{\varepsilon_f} \tag{4.4}$$

where E is the elastic modulus, q_u is the peak flexure strength, and ε_f is the failure strain.

Young's modulus at the peak flexure strength (E) increased with coconut coir fiber content up to 0.6%, which reached peak value of 118 MPa at 0.6% fiber content. This is 110% of that of bio-bricks without coconut coir fiber reinforcement. When fiber content was higher than 0.6%, E decreased with a higher fiber content. E was only slightly higher than that of bio-bricks when coconut coir fiber content was 1.2%.

A higher content of coconut coir fibers would impede the uniformity of bacteria distribution in the samples. Therefore, it affected calcite distribution in the MICP-treated samples, and thus the strength of samples. When the fiber content increases, more bacteria would be obstructed by the fiber. The calcite precipitation formed around the fibers instead of around the sand particles. This could explain why the peak strength and elastic modulus of MICP-treated samples decreased when 1.2% coconut coir fiber was added. This conclusion was similar to that in Li et al. (2015c).

4.2.2.3 Effect of Fiber Addition on the Accelerated Erosion of Bio-Bricks

Figure 4.18 shows the maximum erosion rate of non-reinforced 0.3% synthetic fiber and variable content coconut coir fiber–reinforced bio-bricks under an accelerated erosion test. The maximum erosion rate of 0.3% synthetic fiber–reinforced bio-bricks was 0.08 mm/min, which was around 51.6% reduction from an erosion rate of 0.165 mm/min of non-reinforced bio-bricks. This implied that the inclusion of 0.3% synthetic fibers in the bio-bricks would increase the resistance of bio-bricks against accelerated erosion. Liu et al. (2019) studied the effect of 0.3% synthetic fiber addition on wet–dry cycles of MICP-treated samples. They found that the failure strain of fiber-reinforced samples reached 1.6% compared with non-reinforced samples of 0.4% after five wet–dry cycles. The synthetic fibers prevented the sand particles and calcite crystals from being washed away by water in these erosion setups.

Figure 4.18 shows a rapid reduction in accelerated erosion with the coconut coir fiber content of 0.6%. The maximum erosion rate of coconut coir fiber–reinforced bio-bricks decreased slightly after 0.9% coconut fiber content. There was around a 40% reduction in the maximum erosion rate for 0.6% coconut coir fiber–reinforced bio-bricks compared to the non-reinforced bio-bricks. Danso et al. (2015) investigated the durability properties of soil building blocks reinforced with natural fibers. They found that the durability properties of blocks were generally improved, and a recommendation of 0.5% fiber content. Like their research, the increased resistance of the reinforced bio-bricks can be explained by the coconut coir fibers' ability to

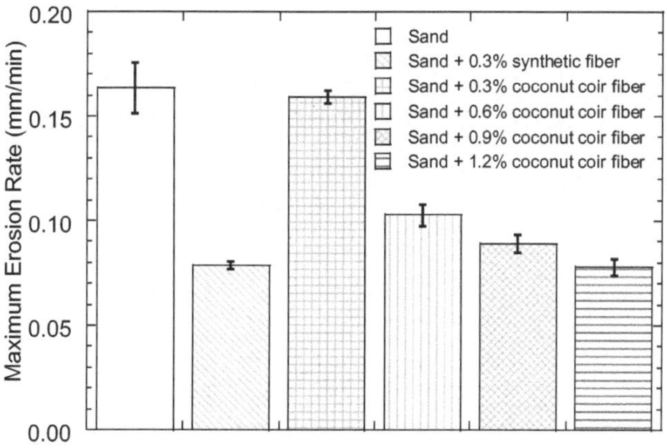

FIGURE 4.18 Maximum erosion rate for MICP-treated sample with fiber reinforcement.

protect the particles from being washed away, thereby reducing the eroding effect on the bio-bricks.

4.2.2.4 Effect of Fibers Addition on Water Absorption of Bio-Bricks

The results obtained from water absorption tests are presented in Figure 4.19. It shows that 0.3% of synthetic fiber–reinforced bio-bricks reduced water absorption slightly. The water absorption of coconut coir fiber–reinforced bio-bricks increased with coconut coir fiber content. Danso et al. (2015) reported a similar result that water absorption of blocks or bricks increases with natural fiber content. The higher water absorption is caused by the amount of water absorbed by the cellulose of the coconut coir fibers. This was due to the void volume and the number of cellulose materials presented in the coconut coir fiber–reinforced bio-bricks (Jeefferie et al. 2011). The natural absorbent properties of coconut coir fibers contribute to the free pathway in the bio-bricks, since it allows more water to be absorbed by the coconut coir fiber–reinforced bio-bricks (Ghavami et al. 1999).

4.2.2.5 Effect of Multiple Treatment on Flexure Behavior and Compressive Strength of Bio-Bricks

The variation of flexure strength as a function of flexure strain of single-, double-, and triple-MICP-treated bio-bricks is presented in Figure 4.20a. The single-MICP-treated bio-bricks obtained a peak flexure strength of 1,200 kPa, while the double- and triple-MICP-treated bio-bricks got peak flexure strengths of 1,700 kPa and 2,600 kPa, which were 41.7% and 116.7% higher than for single-treated bricks, respectively. The compressive strength of the MICP-treated bio-bricks increased with MICP treatment cycles, as presented in Figure 4.20b. This trend was consistent with flexure behavior. The increase in compressive strength from single- to double- and triple-MICP-treated bio-bricks was 145% and 254%, respectively.

Liu et al. (2019) and Wen et al. (2018a) reported that multiple MICP treatment could increase $CaCO_3$ precipitation, and these extra calcite crystals could bond the

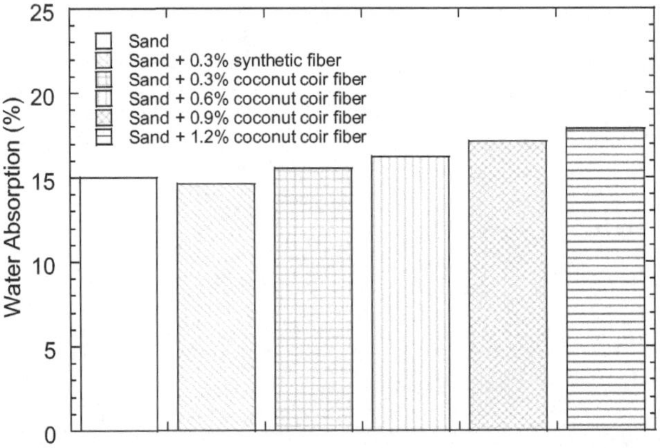

FIGURE 4.19 Water absorption for MICP-treated sample with fiber reinforcement.

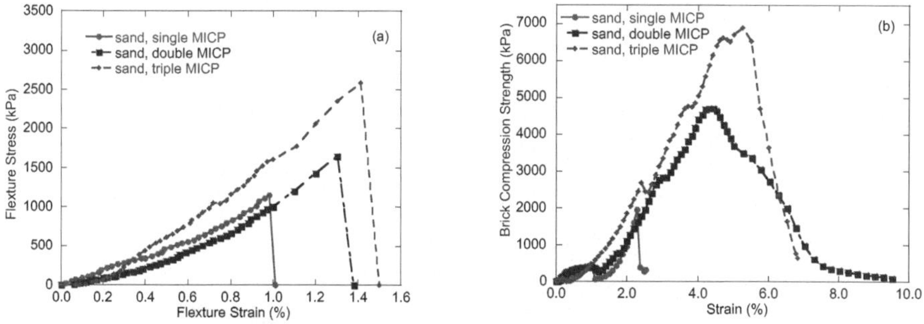

FIGURE 4.20 (a) Flexure stress–strain curves. (b) Stress–strain curves for multiple MICP-treated samples.

sand particles stronger. Like their findings, Figure 4.23a shows the $CaCO_3$ content of single-, double-, and triple-MICP-treated bio-bricks; $CaCO_3$ content was increased from 8.2% to 13.4% and 16.4% by double and triple MICP treatment, respectively. The bio-bricks got 63.4% and 100% higher $CaCO_3$ content by extra MICP treatments. This could explain how strength was increased with MICP treatment cycles.

4.2.2.6 Effect of Multiple Treatments on the Accelerated Erosion of Bio-Bricks

The maximum erosion rate of single-, double-, and triple-MICP-treated bio-bricks is shown in Figure 4.21. The single-MICP-treated samples were subjected to erosion with a maximum rate of around 0.16 mm/min in the period of 60 min. The maximum erosion rate decreased rapidly to 0 mm/min when multiple MICP treatment cycles were applied on bio-bricks. The images of bio-bricks with single, double, and triple MICP treatments at the end of the 60 min accelerated erosion test are presented in Figure 4.23. It was observed that the single-MICP-treated bio-brick was eroded, whereas the double- and triple-treated samples did not illustrate any measurable erosion over the 60 min testing period. These results indicate that multiple MICP treatment enhances bio-brick's ability to resist accelerated erosion. Liu et al. (2019) reported that multiple MICP treatments enhanced the MICP-treated samples' resistance to wet–dry durability cycles. The UCS of multiple-MICP-treated samples had no significant reduction after wet–dry cycles. The increased $CaCO_3$ content (as shown in Figure 4.22) can bond the sand particles and make a strong connection that prevents these particles from washing away by accelerated water erosion.

4.2.2.7 Effect of Multiple Treatments on Water Absorption of Bio-Bricks

Figure 4.24 shows the results of water absorption test on bio-bricks with multiple MICP treatment. All these single- and multiple-treated bio-bricks remained intact during the water absorption tests. The water absorption of single-treated bio-bricks was around 15%, while it reduced to around 11% and 8% after double- and triple-MICP treatment, respectively. Water absorption decreased continuously with more MICP treatments. Sarda et al. (2009) applied the MICP technique to improve the properties of bricks, and they found that MICP contributes a favorable effect on

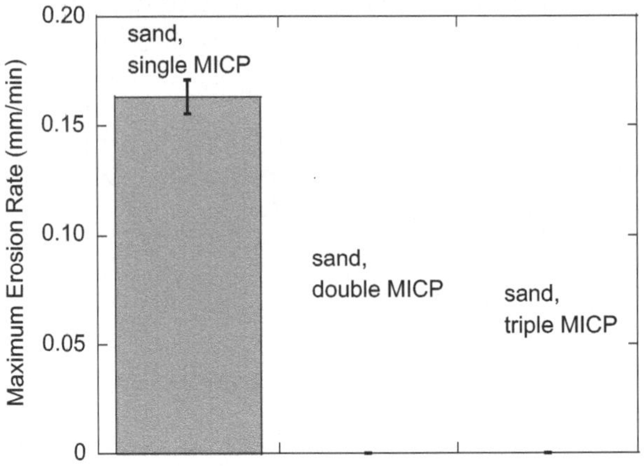

FIGURE 4.21 Maximum erosion rate for multiple-MICP-treated samples.

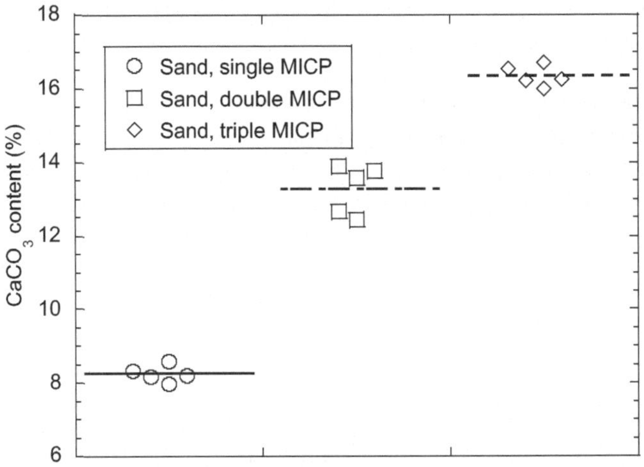

FIGURE 4.22 $CaCO_3$ content.

microbes in the improvement of the durability of bricks by reducing water absorption. They also reported that the precipitated calcite blocked the pores on the surface of bricks, thus preventing water and other pollutants from penetrating into the body of the bricks. It is found that the water absorption of bio-bricks is affected by enhancing $CaCO_3$ content. The increased calcite crystals by multiple MICP treatment occupied the pores in bio-bricks and resulted in a decrease of captivity. Therefore, multiple MICP treatment can improve the resistance of bio-bricks against water deterioration.

4.2.2.8 SEM Analysis

To investigate the interfacial interactions between fibers, calcite crystals, and sand particles, SEM analysis was conducted in this study. SEM images of MICP-treated

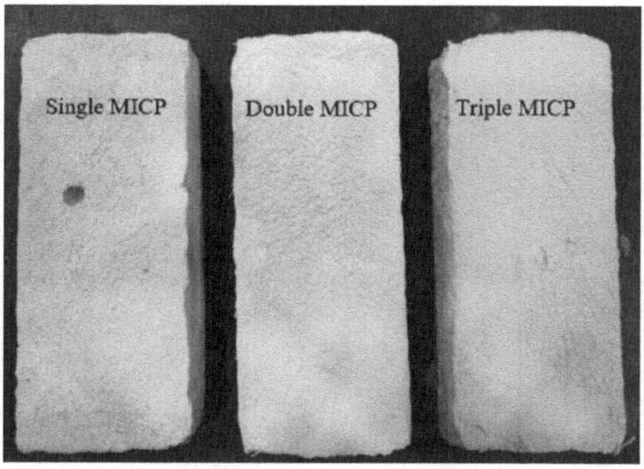

FIGURE 4.23 60 min exposure to accelerated erosion for single-, double-, and triple-MICP-treated samples.

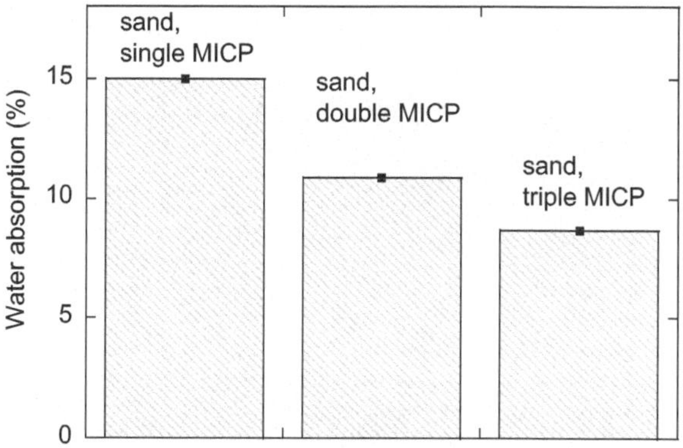

FIGURE 4.24 Water absorption for multiple-MICP-treated samples.

sand particles with synthetic and coconut coir fiber reinforcements are shown in Figure 4.25. Figure 4.25a shows that synthetic fiber acted as bridges and tension members between calcite crystals and sand particles. When local cracks, shown in Figure 4.26, appeared on synthetic fiber–reinforced bio-brick, fibers across the cracks took the tension within the soil due to fiber–soil friction. This should impede the development of cracks and improve the strength and ductility of bio-bricks. This finding is the same as in Liu et al. (2019). Figure 4.25b shows that the coconut coir fiber surface attracted calcite crystals, which contributed to the bond strength and friction between the fiber and the sand particles. The distributed discrete coconut coir fibers acted as a three-dimensional spatial network to interlock sand particles and helped sand particles form a unitary coherent matrix and restrict the displacement. That should explain why the resistance of bio-bricks against accelerated erosion and strength behavior was improved by fiber reinforcement. Tang et al. (2007) investigated the effects of discrete short polypropylene fiber on the strength and mechanical behavior of uncemented and

FIGURE 4.25 SEM images of MICP-treated sand particles with fiber reinforcement: (a) synthetic fiber; (b) coconut coir fiber.

FIGURE 4.26 Failed MICP-treated bio-breaks reinforced by synthetic fiber.

cemented clayey soil. They reported that the bond strength and friction at the interface between the fiber surface and the soil matrix seem to be the dominant mechanism controlling the reinforcement benefit. Similar to their findings, both synthetic and coconut coir fibers played an important role for the interfacial forces in this study. As the sketch drawing in Figure 4.27 shows, the interfacial forces made the fibers in the soil matrix difficult to slide, and thus to bear tensile strength.

Figure 4.28 shows the SEM images of MICP-bonded sand particles after single and triple MICP treatment. It can be observed that the precipitated calcite crystals formed a thin shell on the surface of the sand particle after single MICP treatment,

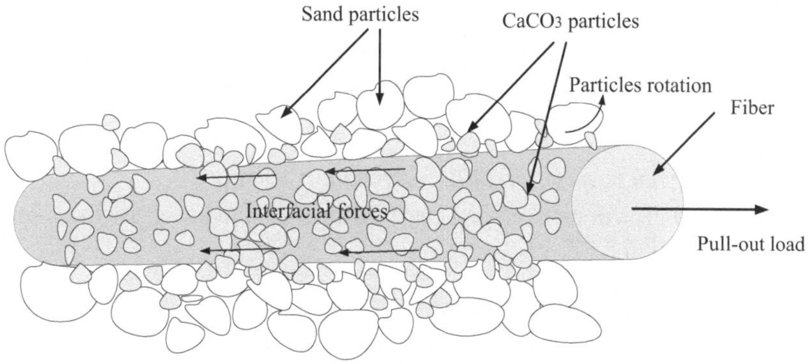

FIGURE 4.27 Sketch drawing of interfacial mechanical interactions between sand particles, carbonate particles, and fiber.

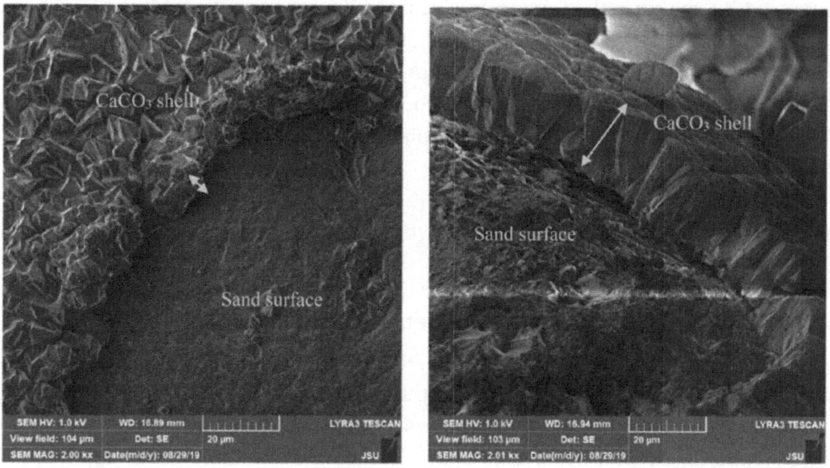

FIGURE 4.28 MICP-bonded sand particles after (a) single MICP treatment and (b) triple MICP treatment.

whereas a thick shell coated the sand surface after triple MICP treatment. The thick calcite shell on the sand surface filled in the gaps between the sand particles and made contributions to improving the resistance of multiple MICP-treated bio-bricks against accelerated erosion and their strength behavior.

4.2.3 Enhanced Erosion Resistance of Cement-Treated Bricks Using Multiple Bio-Surface Treatment

4.2.3.1 Materials and Sample Preparation

4.2.3.1.1 Sands

The Ottawa sand used in this section is the same as that in Section 3.2.1.

4.2.3.1.2 Portland Cement

Portland cement (TYPE I/II, ASTM C150) with a specific gravity of 3.15 was used as the cementing agent for the cement-treated bricks. Its early strength gain allowed for various curing times ranging from 7 to 21 days.

4.2.3.1.3 Bacterial Suspension and Cementation Solution

The bacteria and growth medium and cementation media followed are those in Section 3.1.1 and Section 3.1.2.

4.2.3.1.4 Cement Treatment Process

Cement-treated bricks were prepared by mixing dry sand with cement and water. The proportion of added cement was 10% by mass of dry sand in this study. For the preparation of bricks, 900 g dry sand, 90 g cement, and 250 mL tap water were mixed to achieve a uniform mixture, which was similar to Bu et al. (2018b). The mixture was added into a rigid mold, as shown in Figure 4.29a, in three layers and compacted by shaking, then cured in a relative humidity of 95 ± 5% and a constant temperature of 25°C for 7 days. The size of the cement-treated brick was 177.8 mm in length, 76.2 mm in width, and 38.1 mm in height. The bricks were oven-dried at 105°C to dismiss the excess moisture after demolding (Figure 4.29b). The following testing or additional treatments were conducted on the samples after oven-drying.

4.2.3.1.5 MICP Treatment Process

The MICP technique was used to treat cement-treated bricks. The treatment process is the same with that in Section 4.2.2.

4.2.3.2 Effect of Bio-Surface Treatment on the Flexure Behavior of Cement-Treated Bricks

The flexural-stress-and-strain relationship for cement-treated bricks with or without bio-surface treatments is shown in Figure 4.30. The peak flexural strength increased from 500 kPa to 750 kPa when single bio-surface MICP treatment was applied to the cement-treated bricks. The flexural strength kept increasing along with more bio-surface treatment. The cement-treated bricks achieved flexural strength of 900 kPa

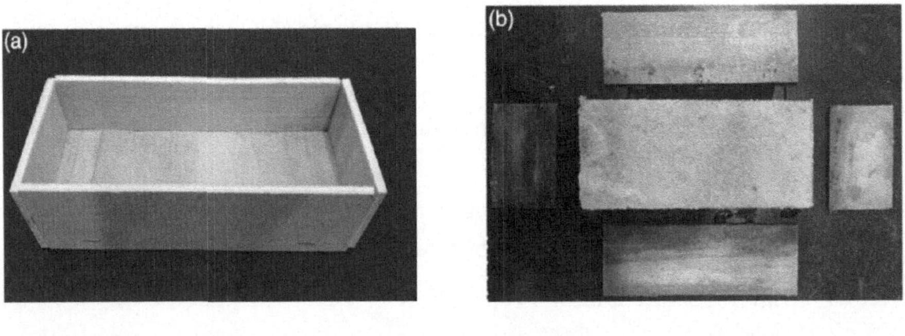

FIGURE 4.29 (a) Rigid mold for bricks of cement treatment. (b) Demolding a cement-treated brick.

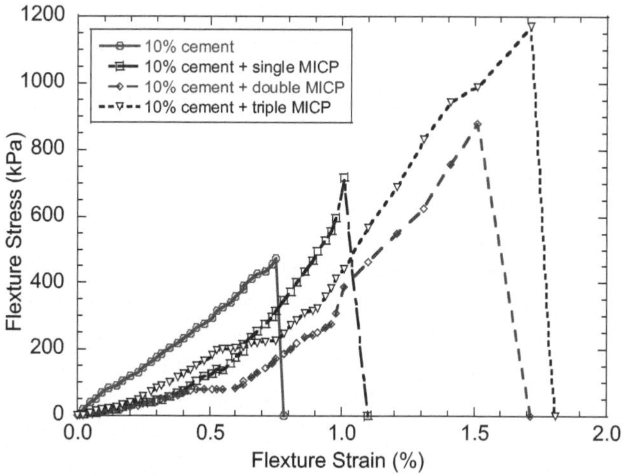

FIGURE 4.30 Flexural stress–strain curves for cement-treated sample and cement-treated samples with bio-surface treatments.

TABLE 4.7
$CaCO_3$ Content of Cement-Treated Bricks with Bio-Surface Treatments

Sample Type	Single-MICP-Treatment $CaCO_3$ (%)	Double-MICP-Treatment $CaCO_3$ (%)	Triple-MICP-Treatment $CaCO_3$ (%)
10% cement + MICP	10.5 ± 0.97	17.4 ± 1.32	23.7 ± 1.41

and 1,200 kPa with double and triple bio-surface MICP treatments, respectively. Especially for cement-treated bricks with triple bio-surface MICP treatments, nearly a 140% increase in peak flexural strength occurred compared with cement-treated bricks without any bio-surface treatment. The results indicate that bio-surface treatments significantly enhance the flexural strength of cement-treated samples. As shown in Table 4.7, the precipitated $CaCO_3$ on the cement-treated bricks also increased with MICP treatment. Multiple MICP treatment can increase $CaCO_3$ precipitation, and those $CaCO_3$ precipitates make the bricks stronger. Therefore, the strength of cement-treated bricks with multiple MICP treatment increased with treatment cycles. This finding is similar to that of Wen et al. (2018a), where only MICP-treated soil was studied (no cement). For cement-treated samples with bio-surface MICP treatment, the results demonstrated that MICP treatment could increase flexural strength. Multiple bio-surface MICP treatment resulted in the higher flexural strength of cement-treated samples, which might be caused by more $CaCO_3$ precipitation on the outer layer of the cement-treated bricks through MICP treatment.

Cracks initially appeared on the bottom and extended gradually to the top of the samples in the bending tests. The samples failed until the cracks fully extended to the top edge of the bricks. All samples failed into halves, and the fracture crack was parallel to the loading direction during bending testing, which indicated that all the cement-treated samples failed in bending mode.

4.2.3.3 Effect of Bio-Surface Treatment on Compression Strength of Cement-Treated Bricks

Figure 4.31 showed the stress and strain curves for cement-treated bricks with or without bio-surface treatment obtained from brick compression strength tests. Brick compression strength was enhanced by bio-surface treatment. The compression strength of cement-treated bricks increased from 3,000 kPa to 4,000 kPa, 5,200 kPa, and 6,300 kPa after single, double, and triple bio-surface MICP treatment, respectively. A significant increase in the compression strength was achieved for the cement-treated samples with triple bio-surface treatments, which was a nearly 100% increase. The precipitated $CaCO_3$ produced by the bio-surface treatment strengthened the cement-treated samples. Amidi et al. (2015) applied the calcium carbonate precipitation method on concrete bricks to seal surface cracks/voids. They also reported that compression strength could be significantly increased by their method. For their multiple three-day-treated bricks, the compression strength of bricks had been improved by 36.7%. Porter et al. (2018) applied the MICP method on unstabilized and cement-stabilized rammed earth blocks and reported that the addition of MICP surface treatment on both types of blocks could increase the overall compression strength. In the case of cement-stabilized rammed earth blocks, the addition of MICP increased the compression strength by 24%, while for unstabilized rammed earth blocks, the addition of MICP increased the strength by 25%.

4.2.3.4 Effect of Bio-Surface Treatment on the Erosion Resistance of Cement-Treated Bricks

Accelerated erosion tests by water jet were conducted on cement-treated samples with or without bio-surface treatment. The appearance of cement-treated samples with single, double, and triple bio-surface treatments at the end of 60 min accelerated

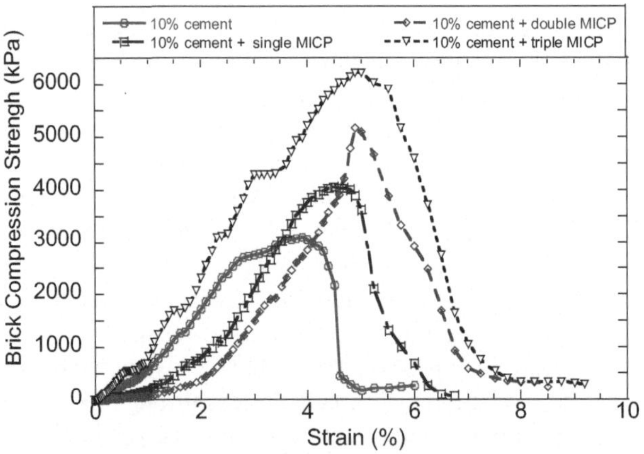

FIGURE 4.31 Stress–strain curves for cement-treated sample and cement-treated samples with bio-surface treatment.

FIGURE 4.32 Single-, double-, and triple-bio-surface-treated cement-treated samples after 60 min exposure to accelerated erosion.

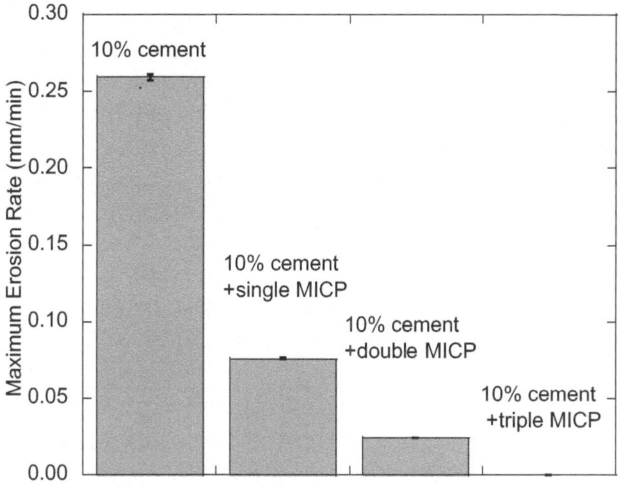

FIGURE 4.33 Maximum erosion rate for cement-treated samples with or without bio-surface treatments.

the erosion testing, as shown in Figure 4.32. The cement-treated samples with single bio-surface treatment were eroded more than the samples with double bio-surface treatments. Furthermore, the samples with triple bio-surface treatments did not experience any measurable erosion over a 60 min time.

The maximum erosion rate of samples is shown in Figure 4.33. Cement-treated sample without bio-surface treatment was subjected to rapid erosion, and the maximum erosion rate was 0.26 mm/min during the test of 60 min period. The maximum erosion rate dropped sharply when bio-surface treatments were applied to the cement-treated samples. The maximum erosion rate decreased to 0.075 mm/min after a single bio-surface treatment, which was a reduction of 71.2%. Surface treatment with MICP impeded the erosion of cement-treated samples under accelerated erosion tests. When multiple bio-surface treatments were applied to the cement-treated samples, the maximum erosion rate decreased along with more bio-surface treatments. The maximum erosion rate dropped to 0.0025 and 0 mm/min after the cement-treated sample was double- and triple-treated by the bio-surface MICP treatment method, respectively. The bio-surface treatment can enhance cement-treated bricks to resist accelerated erosion. Porter et al. (2018) also reported that the time of the collapse of unstabilized rammed earth blocks was extended from 8 min to 21 min by surface treatment with MICP; their MICP method reduced the erosion of unstabilized rammed earth blocks.

4.2.3.5 Effect of Bio-Surface Treatment on the Water Absorption of Cement-Treated Bricks

Water absorption tests were conducted on the cement-treated samples with and without bio-surface treatments, as shown in Figure 4.34. All samples remained intact during the water absorption tests. When bio-surface treatments were applied to the cement-treated samples, the precipitated $CaCO_3$ provided a lower water absorption than did cement-treated samples without bio-surface treatment. These surface-treated samples were, therefore, likely to be less water-absorbent. The cement-treated sample with single bio-surface treatment achieved a reduction of water absorption by 16.3% compared with the cement-treated sample without bio-surface treatment. As shown in Figure 4.34, water absorption kept decreasing with more surface treatment of MICP. The cement-treated sample with triple bio-surface treatments achieved a 32% lower water absorption than untreated cement samples. Achal et al. (2011) found that the deposition of calcium carbonate crystals on the surface of concrete resulted in a decrease of sorptivity. Jongvivatsakul et al. (2019) also reported that the water-tightness of cracked mortar was improved with $CaCO_3$ formation. From these results,

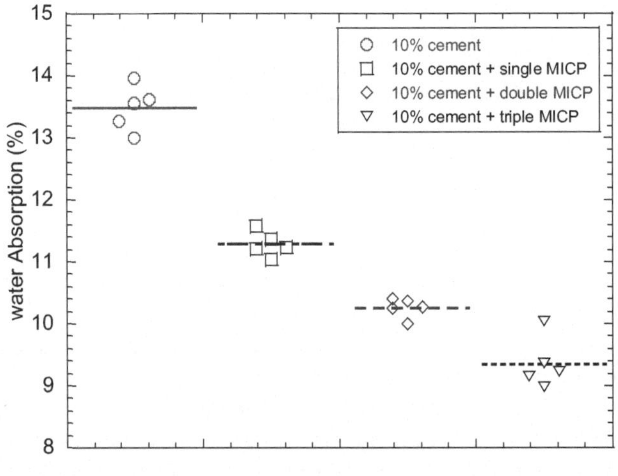

FIGURE 4.34 Water absorption for cement-treated samples and cement-treated samples with bio-surface treatment.

it can be concluded that bio-surface treatment has the potential to improve the resistance of cement-treated materials toward the deterioration process.

4.2.3.6 SEM and EDS Analysis

For the investigation of microstructures, the fractured cement-treated samples with bio-surface treatment were divided into two portions: surface zone and central zone, as shown in Figure 4.35. The SEM images of cement-treated brick are shown in Figure 4.36. As shown in Figure 4.36a, no mineral deposition could be observed in the central zone, while in Figure 4.36b, of the surface zone, a layer of mineral deposition could be clearly observed, indicating that bio-surface treatment does produce precipitated mineral deposition on the surface of cement-treated bricks. EDS analysis was carried out to determine the elements in the bricks. As shown in Figure 4.37, the EDS analysis shows that most elements in the central zone of the bricks are silicium (Si), calcium (Ca), oxygen (O), aluminum (Al), magnesium (Mg), and carbon (C), which are the elements of sand and cement particles, while in the surface zone of bricks, the main elements deposited on the surface are calcium (Ca), oxygen (O), and carbon (C), which are the elements forming calcium carbonate ($CaCO_3$). Hence, the bio-surface treatment precipitated calcium carbonate on the surface zone of cement-treated bricks, but no calcium carbonate was formed in the central area by this method. It might be caused by the precipitates blocking the surface area, which would prevent the cementation media from entering the central area to form calcium carbonate. Figures 4.36c and d show the SEM images of the surface zone of cement-treated bricks after single and triple bio-surface treatment, respectively. Individual $CaCO_3$ crystals were similar in size and were separately distributed on the surface of single-bio-surface-treated bricks. However, for triple-bio-surface-treated bricks, as shown in Figure 4.36d, the more uniform and denser $CaCO_3$ crystals were flake-like and acted as a shell to coat the surface of bricks. This will explain why multiple bio-surface-treated bricks can achieve higher flexural and compression strength, are more durable in accelerated erosion tests, and present more water absorption reduction.

FIGURE 4.35 Cross section of cement-treated brick with double bio-surface treatment.

FIGURE 4.36 SEM image of cement-treated bricks: (a) central zone, (b) surface zone, (c) surface zone with single bio-surface treatment, (d) surface zone with triple bio-surface treatments.

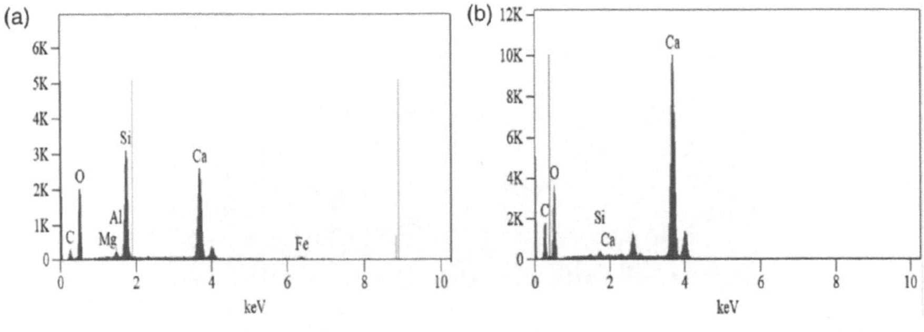

FIGURE 4.37 EDS analysis of cement-treated bricks: (a) central zone; (b) surface zone.

4.2.4 Experimental Investigation on Compression Resistance of Bio-Bricks

4.2.4.1 Materials and Sample Preparation

4.2.4.1.1 Sand

The Ottawa sand used in this section is the same as that in Section 3.2.1. The bacteria and growth medium and cementation media followed are those in Section 3.1.1 and Section 3.1.2.

4.2.4.1.2 Fiber

Fibermesh 150 synthetic fiber (Figure 4.38b) and natural fiber (Figure 4.38c) were chosen and used in this study. The synthetic fiber is 100% homopolymer polypropylene multifilament fiber with a specific gravity of 0.91 and a length of 12 mm (Shao et al. 2014). The natural fiber applied in this study was coconut fiber with a length of 15 mm.

4.2.4.1.3 Bacteria Solution and Cementation Media

The bacteria and growth medium and cementation media followed are those in Section 3.1.1 and Section 3.1.2.

4.2.4.1.4 Sample Preparation

The full-contact rigid mold (FCRM) used is the same with that in Section 3.2.2, as shown in Figure 3.8.

4.2.4.2 Effects of Fibers on the Brick Compression Strength of Bio-Bricks

Representative brick compression strengths for synthetic and natural fiber–reinforced bio-bricks are shown in Figure 4.39. For fiber-reinforced bio-bricks, the optimum fiber content is found to be 0.3%. This is in agreement with what Li et al. (2015c) and Wen et al. (2018a) reported, that the 0.3% fiber-reinforced MICP-treated samples gained the highest unconfined compression strength (UCS). It can be seen from Figure 4.39 that the increase of compression strength with treatment times for all bio-bricks seems to be intuitive—the more treatment times, the higher the compression strength. The growth trends of the compression strength reduced after double treatment. The study by Wen et al. (2018b) on UCS of MICP-treated sand also revealed that the efficiency of multiple treatments decreased after double treatment.

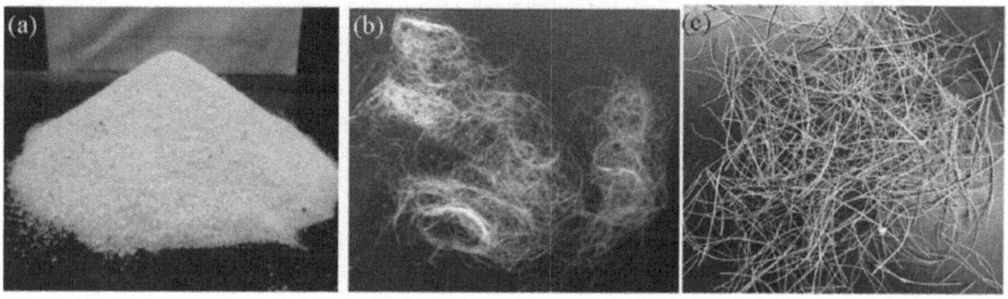

FIGURE 4.38 Images of (a) Ottawa sand, (b) polypropylene fiber, and (c) natural coconut fiber.

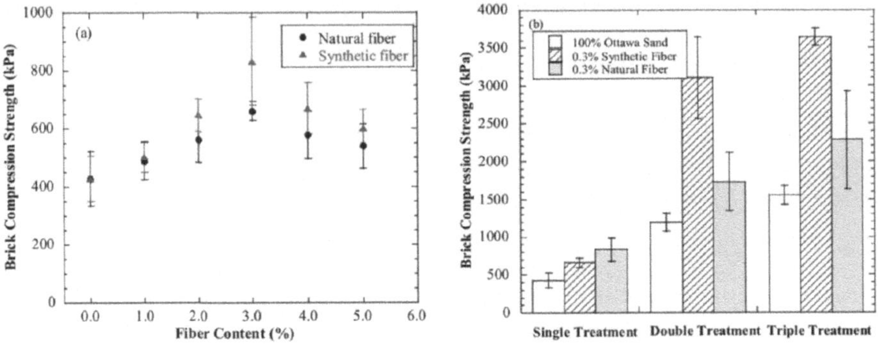

FIGURE 4.39 Brick compression strength of (a) synthetic and natural fibers–reinforced bio-bricks; (b) multiple treated bio-bricks.

Moreover, the addition of 3% synthetic and natural fibers can result in the increase of the compression strength of the bio-bricks by 50–70%, and the increased rate of synthetic fiber is greater than that of natural fiber. The difference in the increase rate could be because synthetic fiber (polypropylene fiber) is made to overcome chemical and biological degradation, while the natural fiber (coconut fiber) has low tenacity.

4.2.4.3 Effects of Clay on the Brick Compression Strength of Bio-Bricks

Figure 4.40a shows the brick compression strength of clay-reinforced bio-bricks at different clay contents. The brick compression strength increased with clay content to 2% and then decreased with clay content. Clay with negative charge in the surface could attract the bacteria to induce more $CaCO_3$ content, which results in an increase of brick compression strength at a small amount of clay content (2%). When the clay content keeps increasing, it can still attract bacteria, but the void space is occupied by clay, thereby reducing the total amount of $CaCO_3$ content. This could explain why the optimum clay content of strengthening brick compression strength is 2%. Meanwhile, brick compression strength increases with treatment times, as expected (Figure 4.40b).

4.2.4.4 Effects of Coarse Sand on the Brick Compression Strength of Bio-Bricks

Coarse sand is selected and added to samples to study the effect of void space on the brick compression strength of bio-bricks. The results of brick compression strength of bio-bricks with additional 5% and 10% coarse sand are plotted in Figure 4.41. It was found from single treatment that additional coarse sand had little or even negative impact on brick compression strength. From results of double and triple treatments, the contribution of void space from additional coarse sand on brick compression strength is insignificant. Meanwhile, the improvement through coarse sand is negligible compared to reinforcement by additional fibers (0.3%). Therefore, it can be suggested that the increase of void space is not an effective method to improve the mechanical properties of MICP-treated samples.

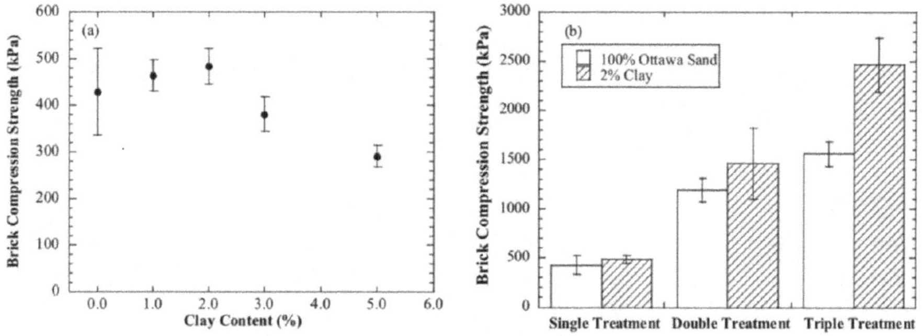

FIGURE 4.40 Brick compression strength of clay-reinforced bio-bricks.

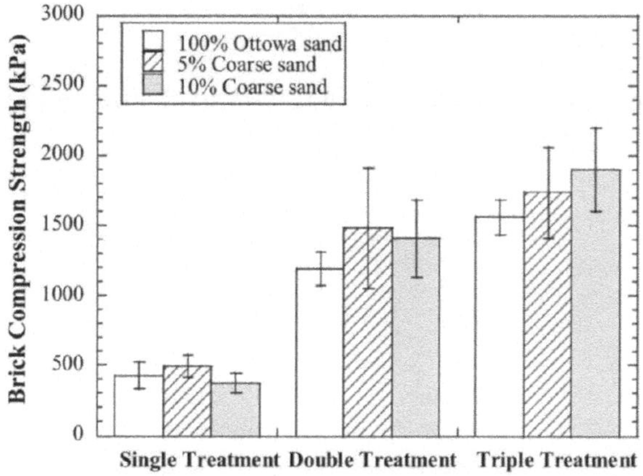

FIGURE 4.41 Brick compression strength of coarse sand–reinforced bio-bricks.

4.2.4.5 Relation to Calcium Carbonate Contents

Results presenting the $CaCO_3$ content of unreinforced and additions-reinforced bio-bricks with multiple MICP treatments are summarized in Table 4.8. The trends of $CaCO_3$ content with multiple treatments seem to be intuitive, indicating a higher compression strength gained with multiple treatments. The sharp increase of compression strength for all reinforced bio-bricks with multiple treatments coincides with the sharp increase of $CaCO_3$ content with multiple treatments. The mild increase in $CaCO_3$ content with additions-reinforced bio-bricks correlates to a slight increase in compression strength with additions-reinforced bio-bricks. The highest $CaCO_3$ content is found in the 0.3% synthetic fiber–reinforced bio-bricks, which also have the highest compression strength. The lower $CaCO_3$ content found in coarse sand–reinforced bio-bricks indicated a lower brick compression strength. This is consistent with the result concluded in Figure 4.41.

TABLE 4.8
$CaCO_3$ Content (by Weight) of Unreinforced and Additions-Reinforced Bio-Bricks with Multiple MICP Treatments

	$CaCO_3$ (%)		
Sample Type	**Single Treatment**	**Double Treatment**	**Triple Treatment**
100% Ottawa sand	8.79 ± 0.81	12.33 ± 0.44	15.96 ± 0.47
5% coarse sand	8.66 ± 0.08	13.45 ± 0.91	16.61 ± 0.35
10% coarse sand	7.57 ± 0.87	13.33 ± 0.48	16.30 ± 0.33
0.3% synthetic fiber	10.12 ± 1.51	17.23 ± 0.74	21.77 ± 0.51
0.3% natural fiber	9.55 ± 1.62	16.31 ± 0.54	19.79 ± 0.52
2% clay	9.46 ± 0.65	15.03 ± 0.85	18.83 ± 1.32

TABLE 4.9
Water Absorption of Unreinforced and Additions-Reinforced Bio-Bricks with Multiple MICP Treatments

	Water Absorption (%)		
Sample Type	**Single Treatment**	**Double Treatment**	**Triple Treatment**
100% Ottawa sand	15.5	10.01	7.86
5% coarse sand	10.16	9.78	7.13
10% coarse sand	13.27	10.46	10.51
0.3% synthetic fiber	17.82	15.03	11.26
0.3% natural fiber	17.89	15.1	12.87
2% clay	13.98	12.56	10.28

4.2.4.6 Water Absorption of Bio-Bricks

Table 4.9 summarizes the results of water absorption for unreinforced and additions-reinforced bio-bricks with multiple MICP treatments. The water absorption of bio-bricks with single treatment ranged from 10.16% to 17.89%, which is comparable with that of traditional red brick (12.5%, as reported by Samonin et al. (2004)). Water absorption decreases with additions-reinforced bio-bricks, except with the addition of fibers. This could be because fibers possess a high content of cellulose and high diffusion coefficients, resulting in high water absorption. Water absorption decreases with treatment times, as expected. An expected reduction was found in the extent of water absorption for all samples, which indicated that the void space was reduced in the bricks because of bacterial calcification and high $CaCO_3$ content precipitations (Table 4.8). Samonin et al. (2004) and De Muynck et al. (2010b) reported that microorganisms prefer to stay in the pores, which are normally five times larger than microbial cells, indicating that precipitation initially happens the pores with larger diameter. Dhami et al. (2012), however, found that precipitation processes will occur on the surface of the pores due to the alkalization of the reaction environments, which will further block the smaller pores and finally result in a decrease of water absorption. This is in agreement with what Amidi and Wang (2015) reported, that the water absorption of MICP-treated concrete bricks decreases with multiple treatments.

4.2.4.7 Crystal Morphology of Calcium Carbonate Precipitation

The XRD pattern of MICP-treated bio-bricks is presented in Figure 4.42. The vaterite and calcite phases were both identified in the XRD results, and their diffraction peaks were found at a diffraction angle of 27.1° and 30.1°, respectively. The existence of vaterite and calcite phases is confirmed by the result of FTIR (Figure 4.43). The absorption bands at 1,085 and 747 cm^{-1} are represented for vaterite and 711 cm^{-1} for

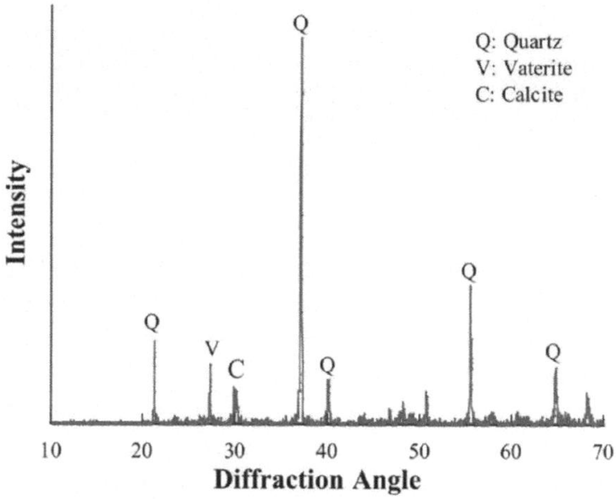

FIGURE 4.42 XRD pattern of MICP-treated bio-bricks.

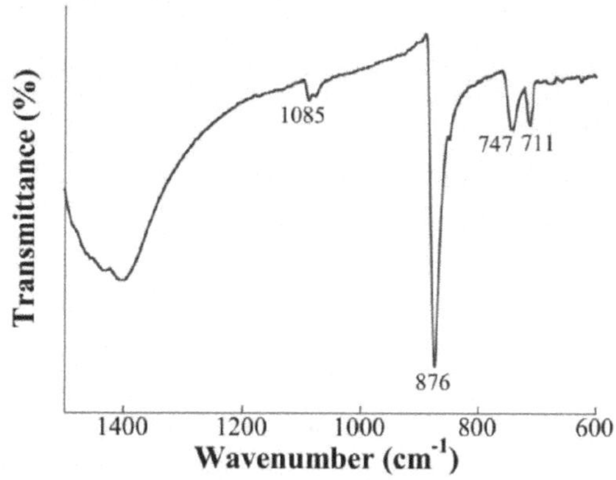

FIGURE 4.43 FTIR spectra of MICP-treated bio-bricks.

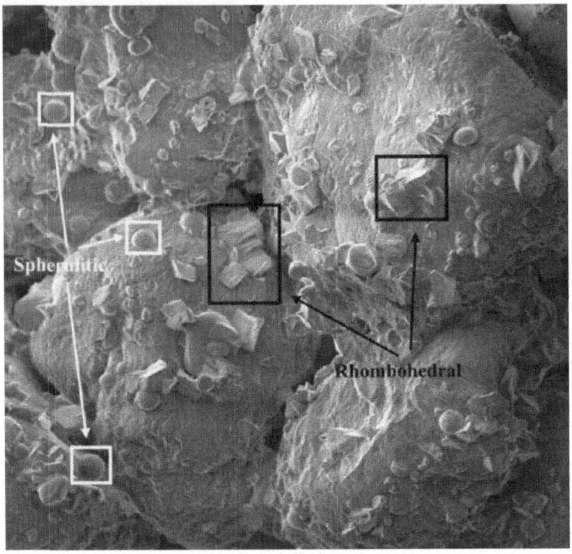

FIGURE 4.44 SEM image of MICP-treated bio-bricks

calcite. The representative bands of vaterite at 1,085/1,084 and 747 cm^{-1} and calcite at 712/711/708 cm^{-1} were also confirmed by Li et al. (2013) and Wen et al. (2019). The microstructure analysis of the MICP-treated bio-bricks is imaged in Figure 4.44. Both spherulitic and rhombohedral crystals were found in the SEM image, indicating a formation of vaterite and calcite. This is in agreement with what Wen et al. (2019) reported, that both spherulitic- and rhombohedral-shaped crystals were found in the MICP-treated sand surface after 168 h. Overall, the integrated findings of FTIR, XRD, and SEM proved the process of MICP and confirmed the formation of both vaterite and calcite in the MICP-treated bio-bricks.

4.3 APPLICATION OF BIO-BEAMS TREATED WITH MICP

4.3.1 Development of Bio-Cemented Constructional Materials through MICP

4.3.1.1 Flexure Behavior of Cement-Treated, Lime-Treated, and MICP-Treated Beam Specimens

The flexure-stress-and-strain relationship for different ratios of cement-treated beams and MICP- treated bio-specimen is shown in Figure 4.45a. The flexure strength of cement-treated beams increased approximately one order of magnitude from 150 kPa to 1,300 kPa when the cement content increased from 10% to 25%. However, the flexure strength of higher-cement-content specimens (30% and 40% of cement by weight) did not show a significant improvement compared to the 25% cement specimen. Yun (2013) found that the flexure strength of plain concrete beam (100 mm × 100 mm × 400 mm) was only around 3,000 kPa. The increment of flexure

FIGURE 4.45 Bending failure sample image of MICP-treated bio-specimen.

strength with an increase in the cement content was limited. Namikawa and Koseki (2007) also studied the tensile strength of cement-treated toyoura sand. They found that the cement-treated beam can achieve around 700 kPa flexure strength with 10% cement and 5% bentonite mixture.

As shown in Figure 4.45b, the flexure strength of lime-treated sand beams increased with the lime content until 30%. When the lime content increased from 30% to 40%, the flexure strength reduced from 280 kPa to 100 kPa. This result may be explained by the fact that 40% lime could not be fully reacted in the current treatment condition, and the unreacted lime particles with poor bond can weaken the strength performance of lime-treated samples.

The flexure behaviors of MICP-treated bio-specimen are also shown in Figure 4.55 to compare with cement and lime-treated sand beams. The result indicates that the flexure strength of the MICP-treated bio-specimen (950 kPa) was similar to that of 20–25% cement-treated beams (600–1,300 kPa). The failure strain of the bio-specimen (0.9%) was lower than that of 20–25% cement-treated beam (1.5%), which indicated that the bio-specimen were stiffer. The flexure strength of 30% lime-treated sand beam was 280 kPa, which was far below the flexure strength of MICP-treated bio-specimen (950 kPa).

The failure cracks pattern of all beams was studied. Figure 4.45 shows one example of MICP-treated bio-specimen after failure. Cracks were not shown when the load was increased initially. The cracks initially extended from the bottom of the beams to the top of the beams in the bending tests. When the cracks fully extended to the top edge of the beams, the specimens failed. All beam specimens failed into half, and the fracture crack was parallel to the loading direction, which illustrated that the beam specimens failed in the bending mode. The failure patterns of those beam samples were similar to those of plain concrete beam reported by Yun (2013).

4.3.1.2 Brick Compression Strength of Cement-Treated, Lime-Treated, and MICP-Treated Brick Specimens

Figure 4.46 shows a box plot of brick compression strength of cement-, lime-, and MICP-treated specimens as a function of cement content and lime content. Each box

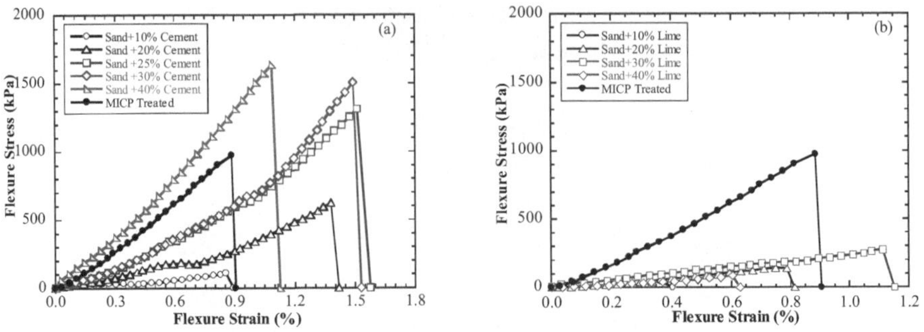

FIGURE 4.46 Flexure stress–strain curve of different mixing ratio of (a) cement-treated beam and MICP-treated bio-specimen and (b) lime-treated beam and MICP-treated bio-specimen.

shows the median value and ±25% of the brick compression strength population as the top and bottom of the box. The lines extending from the top and bottom of each box mark the minimum and maximum brick compression strength. The outlier is shown as an individual point. From Figure 4.46, the brick compression strength of cement-treated and lime-treated specimens increased with higher cement and lime content. Moreover, cement-treated bricks showed a rapid increase with cement content. The averaged brick compression strength of cement-treated bricks reached 6,000 kPa in 40% cement-treated specimen. The MICP-treated bio-specimen only achieved 420 kPa of brick compression strength after 7 days' treatment, roughly equivalent to 30% lime-treated bricks. In terms of brick compression strength, cement-treated specimen showed a better performance than MICP-treated bio-specimen. Agbede and Manasseh (2008) indicated that the compression strength increased with cement content up to 9%. Fang et al. (2011) also found that the compression strength of lime-treated bricks increased by increasing lime content up to 13.3%.

4.3.1.3 Unconfined Compression Strength of Cement-Treated, Lime-Treated, and MICP-Treated Sand Specimens

The effect of cement-treated and lime-treated sample on UCS is shown in Figure 4.47. Results showed that a higher cement content could increase the UCS and failure strain of cement-treated specimens. The UCS of cement-treated specimen was around 9,500 kPa at 30% cement.

Bernardi et al. (2014) found that with 15.3% cement and 28 days' treatment, the UCS was 2,500 kPa, which is similar to the results shown in Figure 4.47a (15% cement, 2,400 kPa after 7 days' treatment). Figure 4.47b showed that the UCS of lime-treated sample increased until 15% of lime content. After the leveling off of UCS from 15% to 30% of lime content, the UCS of the lime-treated sample decreased with more lime content. The optimum lime content reduced to 15% in the cylinder-shaped sample was influenced by the different air contact area in the sample preparation. A lime-treated beam sample was prepared in a rectangular mold with a dimension of 177.8 mm × 76.2 mm × 38.1 mm (L × W × H). The exposed area of the beam-shaped

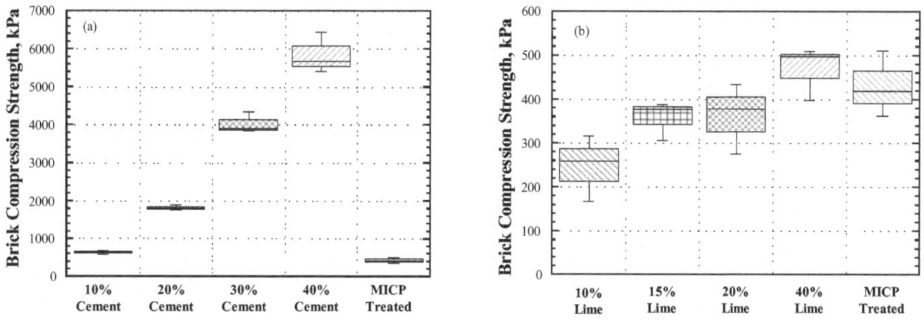

FIGURE 4.47 Brick compression strength with different ratio of (a) cement-treated brick and MICP-treated bio-specimen, and (b) lime-treated brick and MICP-treated bio-specimen. The cement-treated brick and lime-treated brick were cured in 7 days.

sample was $1.35 \times 104\ mm^2$, which occupied 41.2% of the superficial area of the beam-shaped sample, while the cylinder sample for the UCS test prepared by a cylinder mold had a diameter of 38.1 mm and a 76.2 mm height. The proportion of the exposed area reduced to 5.6% of the superficial area of the cylinder-shaped sample. The lime reacted with CO_2 to form $CaCO_3$ during the treatment period. Therefore, the beam-shaped sample with a higher exposed area achieved better performance with higher lime content. Kumar et al. (2007) also found that the expansive soil-lime mixtures containing more than 8% lime showed a decrease in unconfined compression strength.

The decrease of UCS may be attributed to the platy shapes of the unreacted lime particles in the soil mass. In terms of failure strain, the lime-treated samples showed ductile failure when lime content was less than 20%, whereas the sample changed to brittle failure after 20% lime content. Bernardi et al. (2014) found a similar trend, where bricks exhibited ductile failure with 20.9% lime. In Figure 4.47, the UCS of MICP-treated sample was similar to that of the 10% cement-treated sample, which was around 1,300 kPa, whereas the UCS of lime-treated sample (maximum of 140 kPa) was far lower than that of MICP-treated specimens.

Figure 4.48 shows the image of MICP-treated bio-specimen failure after unconfined compression strength test. The fracture pattern of MICP-treated bio-specimen showed a columnar vertical, which was similar to that of the lime-treated sample. In cement-treated specimen, fracture cracked through both ends a well-formed cone on the top, and the vertical cracks ran though the sample from top to bottom. From the SEM image of a MICP-treated bio-specimen in Figure 4.48a, the sand particles were bonded by $CaCO_3$ crystals. Meanwhile, there was a lot of void space between sand particles. When the MICP-treated sample carried the load during the unconfined compression test, the sample failed along with the weakest path. However, for cement-treated sand, the bond material almost filled the void between the sand particles with a strong bond and uniform strength. The cement-treated sample failed in the cone-shaped sample, which was similar as concrete. Tariq and Maki (2014) also observed cone and split failure modes in cement-treated sand. While these failure patterns both indicated that the cement-treated, lime-treated, and MICP-treated

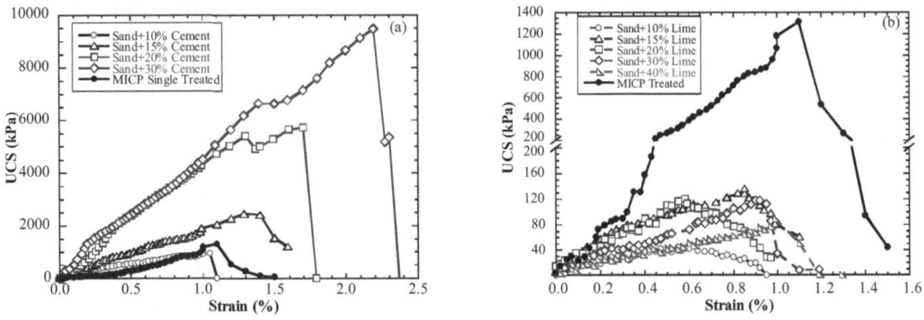

FIGURE 4.48 Unconfined compression stress–strain curve of different ratio of (a) cement-treated sample and MICP-treated bio-specimen, and (b) lime-treated sample and MICP-treated bio-specimen.

samples were well-bonded, it should be noted that curing time is one of the important factors for cement-treated and lime-treated specimens (Bell 1996; Kim et al. 1998). The UCS of 10% cement and 15% lime-treated specimen increased with treatment time. When the treatment time increased from 7 days to 14 days, and then to 21 days, the UCS of cement-treated specimens increased from 950 kPa to 1,100 kPa and to 1,500 kPa, respectively. The UCS of lime-treated specimen increased up to 220 kPa after 21 days' treatment. Bernardi et al. (2014) found that with 20.9% lime treatment for 28 days, the UCS of the lime-treated sample could achieve around 550 kPa. The difference may be caused by a different type of lime and a different curing temperature. Fang et al. (2011) showed that a higher curing temperature can increase compression strength. However, the UCS of MICP-treated bio-specimen (7 days' treatment) was still much higher than the UCS of 15% lime-treated sample with 21 days' treatment.

4.3.1.4 $CaCO_3$ Distribution

The precipitated $CaCO_3$ content was measured in the MICP-treated bio-specimen. As shown in Figure 4.49, the MICP-treated bio-specimen was divided into four different areas. In each area, the $CaCO_3$ content was measured by acid washing method. The $CaCO_3$ content was indicated as a percentage of sample weight. The $CaCO_3$ content in the rectangular-shaped MICP-treated bio-specimen prepared from the immersing method ranged from 8.4% ± 1.5% to 9.4% ± 1.2% by weight, and the $CaCO_3$ content in the cylinder UCS sample ranged from 6.6% ± 0.5% to 8.0% ± 0.9% by weight. The $CaCO_3$ distribution results indicated that the precipitated $CaCO_3$ was distributed uniformly in the soil sample.

4.3.1.5 SEM Analysis

The SEM images of MICP-treated, cement-treated, and lime-treated specimen are shown in Figure 4.50a, b, and c, respectively. According to the SEM images, there were slight differences among the bonding patterns for MICP-treated, cement-treated, and lime-treated specimens. For MICP-treated bio-specimen, the $CaCO_3$ precipitation in the inter-particle contact point was used to bond the sand particles.

FIGURE 4.49 Failure image of MICP-treated bio-specimen after UCS test.

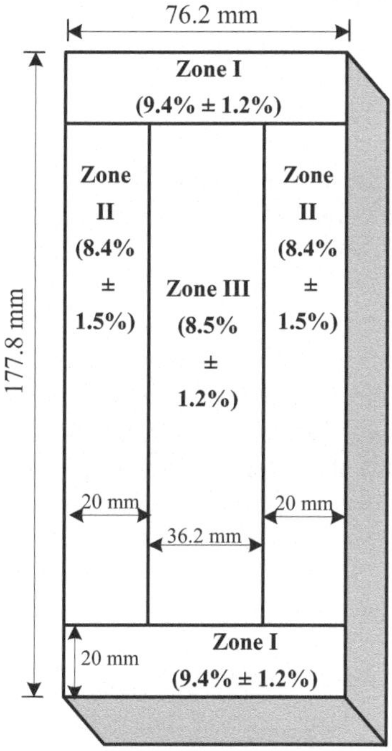

FIGURE 4.50 Four different areas of MICP-treated soil specimens for $CaCO_3$ uniformity evaluation.

However, there were also numerous $CaCO_3$ crystals deposited on the surface away from the inter-particle contact point in the MICP-treated bio-specimen. In contrast, the sand particles were fully surrounded by cement and lime, since the cementation materials just directly mixed with sand uniformly. Meanwhile, it is apparent that clear cracks (indicated by a white arrow) took place between sand particles and the cementation bond, which indicated that the failure patterns of MICP-treated, cement-treated, and lime-treated specimens were bond particle failure.

4.3.2 Experimental Investigation of Flexure Resistance Performance of Bio-Beams Reinforced with Discrete Randomly Distributed Fiber and Bamboo

4.3.2.1 Materials and Methods

Bamboo has been proved to be a flexible material to sustain tension and flexure strength (Rahman et al. 2011). Laminated bamboo strips, produced from China, with a density of 0.8 g/cm3, tensile strength of 150 MPa, and elasticity modulus of 10 GPa, were used to investigate reinforcement of the bio-beam. Five different arrangements with various cross-sectional sizes and number of bamboo strips are shown in Figure 4.51. The five arrangements were thus: bamboo arrangement (BA) BA-1, two bamboo strips with cross-sectional dimension of 3 mm × 6 mm (H × W); bamboo arrangement BA-2, three bamboo strips with cross-sectional dimension of 1.5 mm × 6 mm (H × W); bamboo arrangement BA-3, three bamboo strips with cross-sectional dimension of 3 mm × 3 mm (H × W); bamboo arrangement BA-4, four bamboo strips with cross-sectional dimension of 2 mm × 3 mm (H × W); and bamboo arrangement BA-5, six bamboo strips with cross-sectional dimension of 2 mm × 2 mm (H × W). The five bamboo arrangements were selected based on the width of the bio-beams (38.1 mm) and their own dimension. The bamboo strips function as rebar reinforcement, so the arrangement considered that the reinforcement ratio (~1%) and minimum spacing should be larger than two times the width of the bamboo strip.

As shown in Figure 4.52a, bamboo arrangement BA1 strips were placed in the rigid full-contact mold, and the two bamboo strips were fixed by three small bamboo pieces. The position of bamboo arrangement BA-1 bamboo strips in the bio-beam is at the bottom, with a 1.5 mm gap, and the spacing of bamboo arrangement BA-1 is 24 mm, as shown in Figure 4.52c. Details for the other bamboo strips are shown in Table 4.10. A different ratio of reinforcement and spacing of bamboo may affect the behavior of bamboo-reinforced bio-beams. The ratio of reinforcement of the bamboo strips was calculated by the cross-sectional area of bamboo strips divided by the cross section of bio-beams, and the spacing of bamboo strips was the clear distance between each bamboo strip (Figure 4.52 (c)).

4.3.2.2 Effect of Fiber Addition on the Flexure Behavior of Bio-Beam

The stress–strain curve obtained from a four-point bending test with 0.3% fiber-reinforced bio-beam and unreinforced bio-beam is shown in Figure 4.53. Both flexure strength and flexure strain improved at 0.3% fiber addition in the bio-beam, which indicated that fiber could contribute to the improvement of ductility.

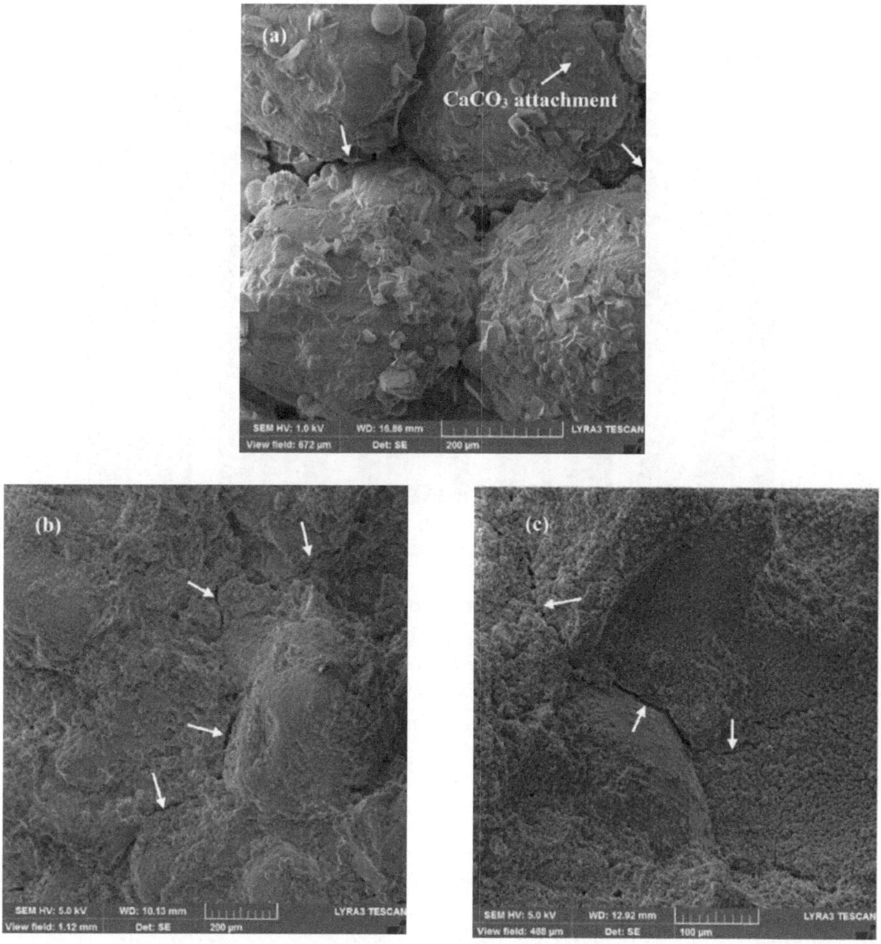

FIGURE 4.51 SEM image of (a) MICP-treated bio-specimen, (b) 10% cement-treated specimen, and (c) 15% lime-treated specimen. The cement-treated and lime-treated samples were cured for 7 days.

The effect of randomly distributed fiber content on the flexure strength of the fiber-reinforced bio-beam is shown in Figure 4.54a. The peak flexure strength of unreinforced bio-beam was around 890 kPa, whereas the peak flexure strength increased with fiber content up to 0.3%. The peak strength of the 0.3% fiber-reinforced bio-beam was 40% higher than that of the unreinforced bio-beam. When the fiber content exceeded 0.3%, the peak strength of the fiber-reinforced bio-beam decreased. This is caused by the fiber clogging the pores between soil particles and weakening the MICP bond, which decreased flexure strength at a higher fiber content (0.4% or more).

The residual strength of the fiber-reinforced bio-beam is also shown in Figure 4.54a. *Residual strength* is defined as the flexure strength at 15% strain. The residual strength

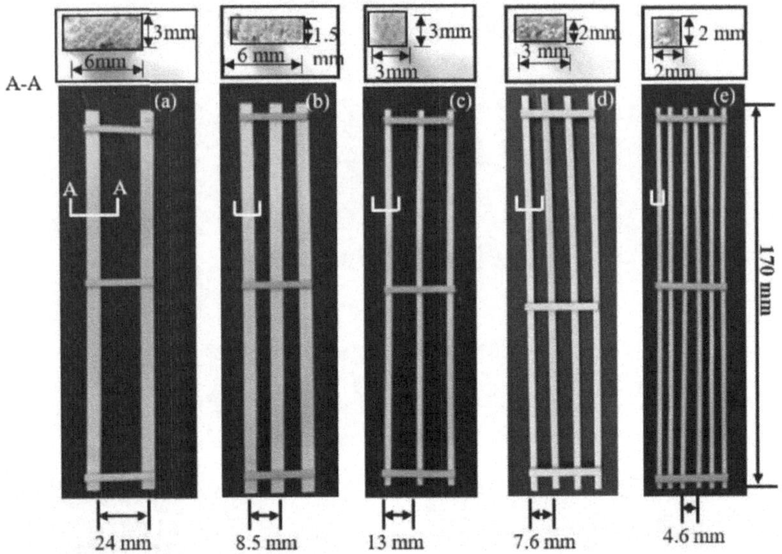

FIGURE 4.52 Different arrangement with various cross-sectional sizes and numbers of bamboo strips: (a) BA-1, two bamboo strips with cross-sectional dimension of 3 mm × 6 mm (H × W); (b) BA-2, three bamboo strips with cross-sectional dimension of 1.5 mm × 6 mm (H × W); (c) BA-3, three bamboo strips with cross-sectional dimension of 3 mm × 3 mm (H × W); (d) BA-4, four bamboo strips with cross-sectional dimension of 2 mm × 3 mm (H × W); (e) BA-5, six bamboo strips with cross-sectional dimension of 2 mm × 2 mm (H × W). The length of each bamboo strip was 170 mm.

TABLE 4.10
Detailed Parameters for Different Bamboo Reinforcement in the Bio-Beam

Bamboo Arrangements	No. of Bamboo Strips	Cross-Section Area of Bamboo Strips (mm^2)	Ratio of Bamboo Reinforcement* (%)	Spacing between Bamboo Strips (mm)	Bending Stiffness† (E × I, kN•mm^2)
BA1	2	(3 mm × 6 mm) × 2	1.24	24.0	135.0
BA2	3	(1.5 mm × 6 mm) × 3	0.93	8.5	16.9
BA3	3	(3 mm × 3 mm) × 3	0.93	13.0	67.5
BA4	4	(2 mm × 3 mm) × 4	0.83	7.6	20.0
BA5	6	(2 mm × 2 mm) × 6	0.83	4.6	13.3

Note:

* Reinforcement ratio = cross-sectional area of bamboo strips divided by cross-sectional area of the bio-beam. The cross-sectional area of the bio-beam was 38.1 mm × 76.2 mm (2,903 mm^2).

† Bending stiffness = E × I, where *E* is the elastic modulus of bamboo strips (10 GPa), and *I* is the moment of inertia of the beam cross section.

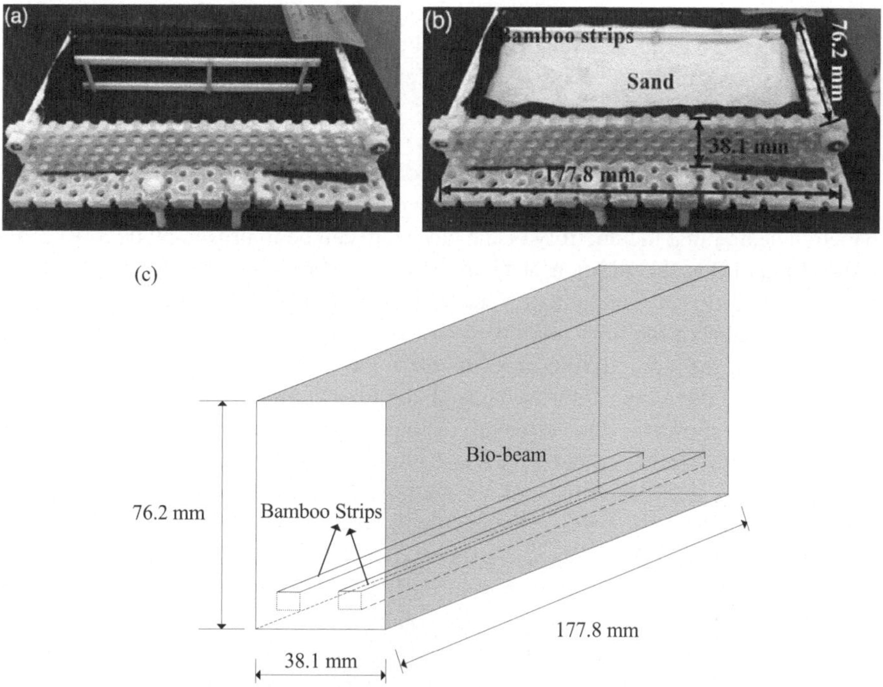

FIGURE 4.53 Sample preparation of bio-beam reinforced by bamboo strips: (a) bamboo strips placed inside the geotextile at the desired position; (b) sand added into the rigid mold; and (c) sketch of bio-beam reinforced by bamboo strips.

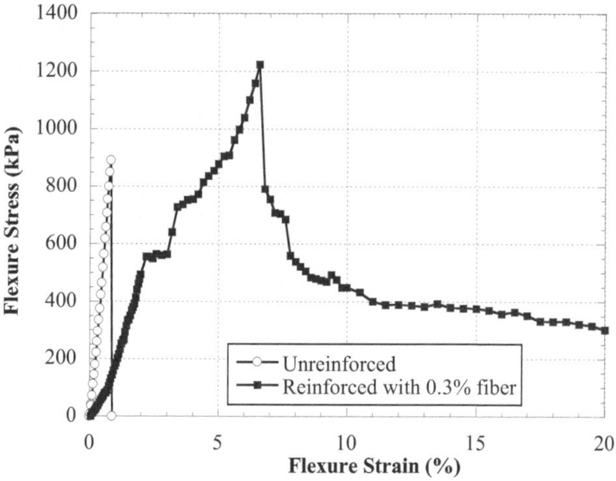

FIGURE 4.54 Flexure stress as a function of flexure strain for unreinforced bio-beam and (0.3%) fiber-reinforced bio-beam in single MICP treatment.

of unreinforced bio-beam was 0 kPa due to the brittleness of the MICP-treated sample. However, the residual strength of the fiber-reinforced bio-beam increased with the fiber content until 3%. The residual strength of the fiber-reinforced bio-beam achieved 400 kPa with 0.3% fiber.

The effect of fiber content on the failure stain of the bio-beam is shown in Figure 4.54b. The strain increased with fiber content up to 0.3%. The failure strain of the fiber-reinforced bio-beam reached the highest at 6.0% with 0.3% fiber addition, which indicated that the ductility of the bio-beam can be improved by fiber. Li et al. (2015b) found that the addition of fiber in the MICP process can enhance the ductility of MICP-treated sand and prevent the loss of post-peak strength. However, the flexural modulus of the fiber-reinforced bio-beam decreased when fiber was added, indicating that the strength reinforced by fiber was mainly caused by extending the elastic strain, rather than improvement of the elastic modulus (see Figure 4.54c). Table 4.11 also shows the peak strength, residual strength, failure strain, and flexural modulus of fiber-reinforced bio-beams and unreinforced bio-beams.

Figure 4.55a shows the image of a 0.3% fiber-reinforced bio-beams after failure in the four-point bending test. The specimen was still bonded by the fiber when cracks appeared, and the inclined cracks clearly indicate that the crack was formed in shear failure near the support points at the bottom. Figure 4.55b shows the image of unreinforced bio-beams in the failure condition. No cracks were shown on the unreinforced bio-beam when load increased initially, and the specimen failed into two parts directly after the first crack appeared in the bending test. This result shows that the ductility of the bio-beam can be improved with the addition of fiber.

4.3.2.3 Effect of Bamboo Strip on the Flexure Behavior of the Bio-Beam

The variations of flexure stress as a function of flexure stain of unreinforced bio-beam and bamboo-reinforced bio-beam are shown in Figure 4.56a. Bamboo

TABLE 4.11
Effect of Fiber Content on the Flexure Behavior of Fiber-Reinforced Bio-Beam

	Flexure Strength			
Fiber Content (By Weight)	**Peak Strength (kPa)**	**Residual Strength (kPa)**	**Failure Strain (%)**	**Flexural Modulus (MPa)**
0	890 ± 58	0	0.92 ± 0.06	4.26 ± 0.15
0.1%	910 ± 79	297 ± 25	2.99 ± 0.21	1.34 ± 0.20
0.2%	970 ± 80	318 ± 26	3.29 ± 0.20	1.30 ± 0.13
0.3%	1,220 ± 157	400 ± 53	6.10 ± 0.42	0.88 ± 0.17
0.4%	960 ± 178	332 ± 38	2.99 ± 0.63	1.42 ± 0.27

Note: Flexural modulus = $0.21\ mL^{3}/bd^{3}$ for four-point bending test (ASTM D6272), where, m = slope of the tangent to the initial straight line of the stress–strain curve, L = span of the bio-beam (152.4 mm), b = width of the bio-beam (38.1 mm) and d = depth of the bio-beam (76.2 mm).

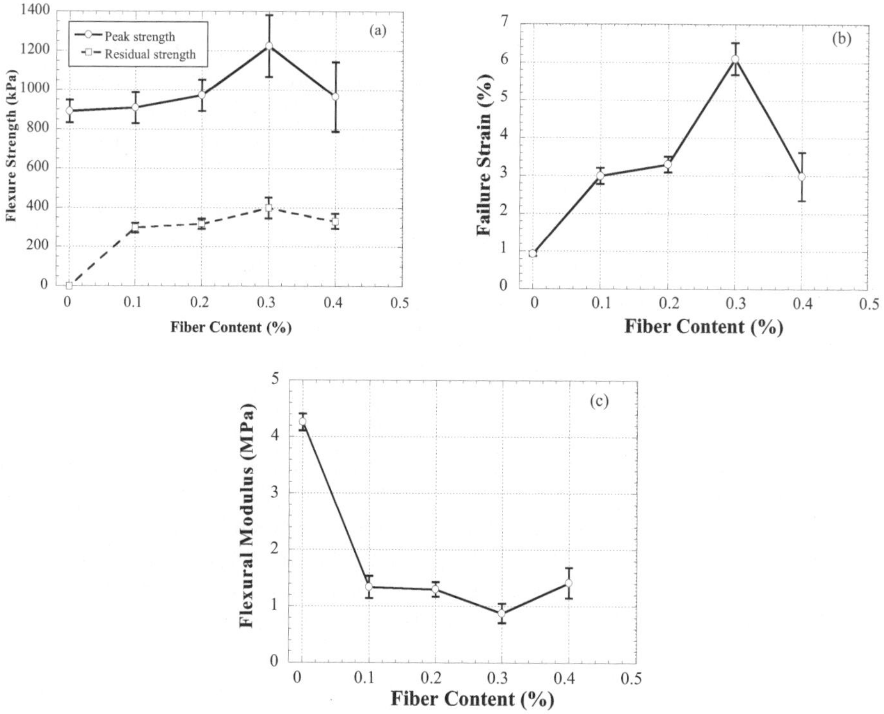

FIGURE 4.55 Effect of fiber content on fiber-reinforced bio-beam: (a) flexure strength, (b) failure strain, (c) flexure modulus.

FIGURE 4.56 (a) Image of failed sample of fiber-reinforced bio-beam. (b) Image of failed sample of unreinforced bio-beam.

arrangements BA-2, BA-3, and BA-4 enhanced the flexure strength of the bamboo-reinforced bio-beam. With three 1.5 mm × 6 mm cross-sectional bamboo strips, BA-2 bamboo-reinforced bio-beam achieved peak flexure strength of 1,590 kPa, which is around a 78% improvement compared to unreinforced bio-beam (890 kPa). With three 3 mm × 3 mm cross-sectional bamboo strips, BA-3 bamboo-reinforced bio-beam caused a nearly 100% enhancement on peak flexure strength (1,750 kPa) compared with the unreinforced bio-beam (890 kPa). The BA-3 bamboo-reinforced bio-beam shows shear failure in the bending test (as shown in Figure 4.56b). While BA-4 bamboo strips were used to reinforce the bio-beam, the peak flexure strength of the bamboo-reinforced sample reached 1,410 kPa. However, there is no remarkable change after BA-1 bamboo strips were used to reinforce the bio-beam. BA-1 has the largest reinforcement ratio (1.24%) in these five bamboo arrangements, but the bio-beams failed before bamboo strips could play a role on resistant in the bending test (as shown in Figure 4.56c). Sandy soil fell from the bottom of the bio-beam and separated from the bamboo strips. Figure 4.56a shows that the BA-5 bamboo arrangement in the bio-beam reinforcement had a negative effect on the peak flexure strength. This is caused by the six bamboo strips and the narrow spacing, which may weaken the MICP bond of sand particles in the zone subjected to tensile force.

A comparison of flexure stress as a function of flexure strain for 0.3% fiber, bamboo arrangement BA-3, and both 0.3% fiber and BA-3-reinforced bio-beam is shown in Figure 4.57. The flexure strength of both BA-3-reinforced bio-beam and 0.3% fiber + BA-3-reinforced bio-beam was higher than 0.3% fiber-reinforced bio-beam. The failure strain of 0.3% fiber + BA-3-reinforced bio-beam was improved compared with the BA-3-reinforced bio-beam, indicating that the 0.3% fiber + BA-3 reinforced bio-beam possessed the advantage of both fiber and BA-3 reinforcement.

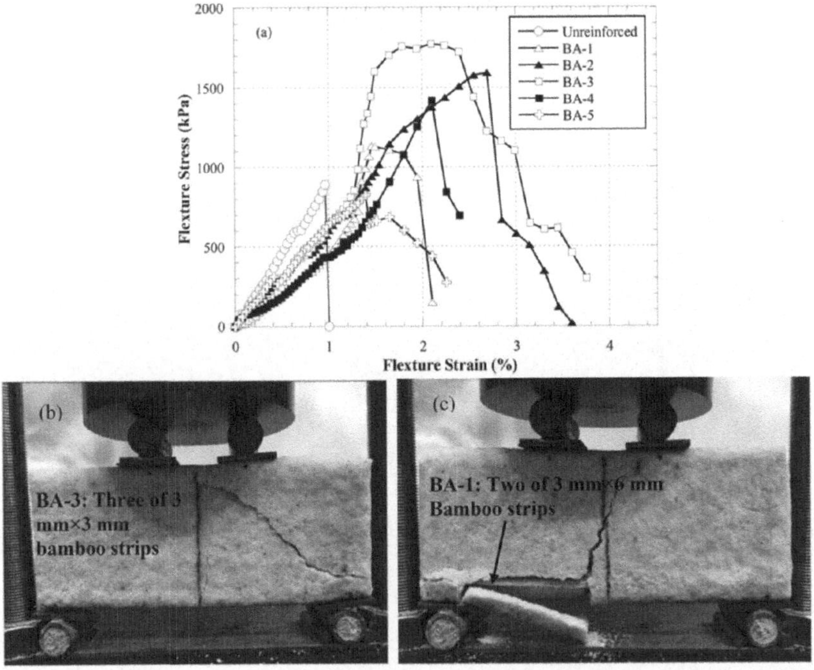

FIGURE 4.57 (a) Flexure stress as a function of flexure strain for unreinforced bio-beam and bamboo-reinforced bio-beam in single-MICP treatment. (b) Image of failure of BA-3-reinforced bio-beam. (c) Image of failure of BA-1-reinforced bio-beam.

4.3.2.4 Effect of Multiple MICP Treatments on the Flexure Behavior of Bio-Beam

The effects of multiple MICP treatments on the flexure behavior of unreinforced bio-beams and with 0.3% fiber–reinforced bio-beam are shown in Figure 4.58. The single-treated unreinforced bio-beam had peak flexure strength around 890 kPa, and the peak flexure strength increased to 1,550 kPa and 1,950 kPa after double MICP treatments and triple MICP treatments, respectively. Meanwhile, the precipitated $CaCO_3$ content also increased with MICP treatment cycles, as shown in Table 4.3. Multiple MICP treatment can increase more $CaCO_3$ precipitation, and those $CaCO_3$ precipitates can bond the sand particles stronger. Therefore, the strength of multiple-MICP-treated soil increased with the treatment cycles.

As shown in Figure 4.58b, the bio-beams reinforced by 0.3% fiber also had a higher peak flexure strength after multiple MICP treatments. The peak flexure strength of the double-treated fiber-reinforced bio-beams increased from 1,250 kPa of a single MICP treatment to 2,400 kPa of the double MICP treatment. However, the ductility improvement by the added fiber disappeared at double or triple MICP treatments. Double-treated fiber-reinforced bio-beams performed as a brittle material again, the same as unreinforced bio-beam. Triple-treated fiber-reinforced bio-beams had a similar pattern with double-treated fiber-reinforced bio-beams. The reason could be explained by the fact that excessive $CaCO_3$ precipitation almost encapsulated fibers and hindered the fibers from providing the desired ductility improvement. Figure 4.58c shows the failure specimen of triple-treated fiber-reinforced bio-beam. The fiber-reinforced multiple-treatment bio-beam showed a different failure pattern compared with the failure pattern of fiber-reinforced single treatment in Figure 4.55a. This may be caused by the multiple treatment enhancing the $CaCO_3$ bond between

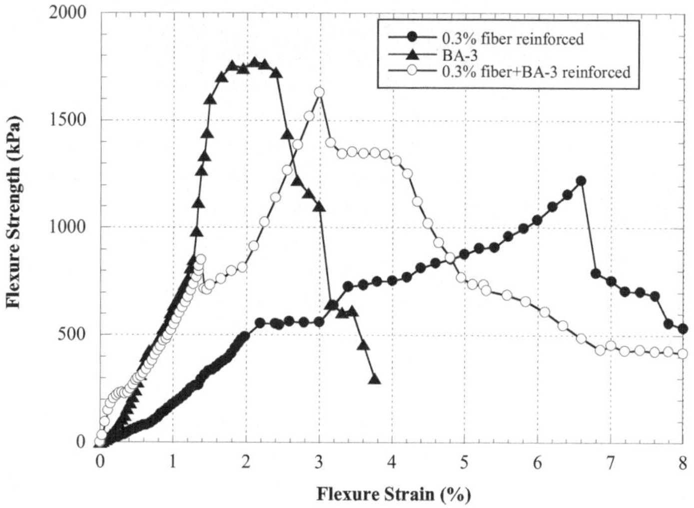

FIGURE 4.58 Flexure stress as a function of flexure strain for 3% fiber–, BA-3-, and 3% fiber + BA-3-reinforced bio-beam.

sand particles and fiber. Fiber can barely prevent post-strength loss in this situation. The strength loss was over 70% immediately after peak strength for double-treated and triple-treated fiber-reinforced bio-beams.

4.3.2.5 SEM Analysis

A comparison of $CaCO_3$ precipitation for the fiber-reinforced bio-beam in multiple treatments is shown in Figure 4.59. In Figure 4.59a, $CaCO_3$ precipitation with fiber-reinforced bio-beams (single treatment) had an average radius of 15.44 μm,

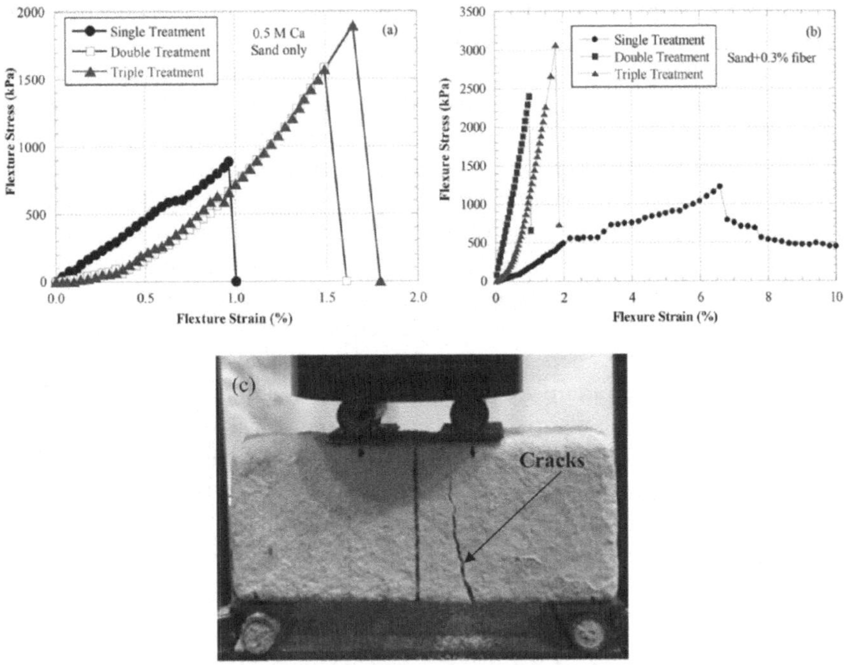

FIGURE 4.59 Flexure stress as a function of flexure strain of bio-beams in the multiple MICP treatments: (a) unreinforced bio-beams, (b) 0.3% fiber-reinforced bio-beams, (c) image of triple-treated fiber-reinforced bio-beams after failure.

TABLE 4.12
$CaCO_3$ Content (by Weight) of Unreinforced and Fiber-Reinforced Bio-Beams with Multiple MICP Treatments

Sample Type	Single Treatment $CaCO_3$ (%)	Double Treatment $CaCO_3$ (%)	Triple Treatment $CaCO_3$ (%)
100% Ottawa sand	8.23 ± 0.85	13.89 ± 0.87	16.09 ± 0.55
100% Ottawa sand with 0.3% fiber by weight of dry sand	8.55 ± 1.70	15.01 ± 0.40	19.55 ± 0.52

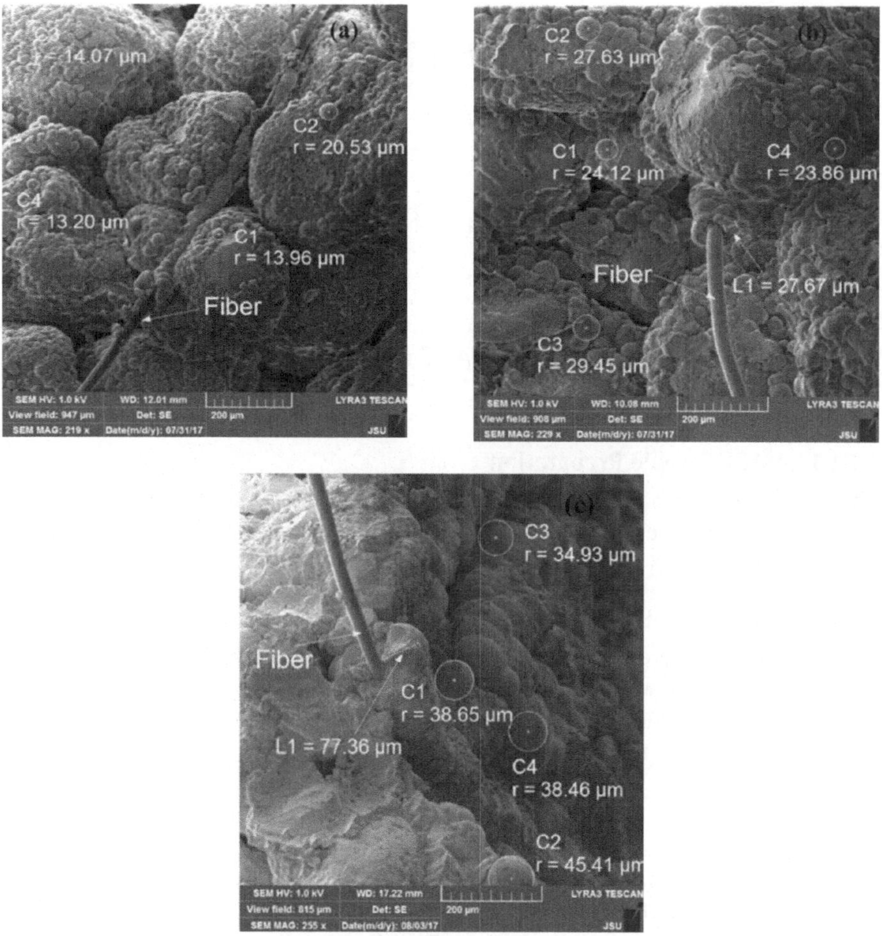

FIGURE 4.60 SEM image of fiber-reinforced bio-beam: (a) single-treated sample, (b) double-treated sample, and (c) triple-treated sample.

widely deposited on the sand particle surfaces and inter-particle contact points. The average radius increased to 25.82 µm after double treatment, and 39.36 µm after triple treatment (Figure 4.59b and c). The diameter of fiber was around 30 µm. The thickness of the $CaCO_3$ precipitation, wrapped around the fiber, increased from 27.67 µm after double MICP treatment of fiber-reinforced bio-beams to 77.36 µm after triple MICP treatments of fiber-reinforced bio-beams. The observation confirms that $CaCO_3$ precipitation increased with more MICP treatment cycles. More $CaCO_3$ precipitation bond surrounding the fiber may weaken the ductility improvement through fiber reinforcement.

5 Application of MICP in Erosion Control and Durability

5.1 EFFECT OF MICP ON SURFACE EROSION

5.1.1 Experimental Methods

5.1.1.1 MICP Sample Preparation

The specimens were prepared in full-contact flexible molds (FCFMs) made of geotextile, as described in Section 3.1.3, and MICP sample preparation for details described in Section 3.1.

5.1.1.2 Erosion Tests

Three flume tests were conducted to investigate the enhancement in the erodibility of MICP-treated Ottawa graded sand. All tests were carried out in a Plexiglas-sided recirculating flume with a length of 3 m, a width of 0.3 m, and a depth of 0.45 m. The flume was supported by a steel framework and was kept horizontal throughout the experiments. The setup included a tank, two pumps, and a network of pipes. A carriage that could be moved in both longitudinal and transverse directions was mounted on the flume top to carry the scour depth-measuring point gauge. The point gauge could also be moved vertically with an accuracy of 0.1 mm. A straw screen filter was installed at the flume inlet to encourage a uniform flow distribution and to dampen fluctuations at the water surface. The discharging rate Q could be adjusted by the control valve of the pipe connected to water pumps, and the flow depth h could be adjusted using a tailgate fitted at the end of the flume. A sand recess (0.9 m × 0.3 m × 0.08 m) was positioned 1.2 m downstream of the flume inlet to retain the standard Ottawa sand used as bed materials. The top of the sand bed in the recess was constructed at the same elevation as the upstream planer bed. This was to ensure that the fully developed turbulent flow entered the sediment bed smoothly. Eroded sand was collected by a sediment trap at the downstream end of the flume. Five test cases were conducted; the configuration of the tests is illustrated in Figure 5.1.

Test case no. 1 was designed to investigate the erodibility of MICP-treated and untreated sand materials under a surface erosion scenario. For this reason, the MICP-treated sample was placed at one corner of the flume, leaving the remaining section untreated (Figure 5.1a). The treated zone had a width of 15 cm, which was just half of the total width of the flume. Therefore, both the treated and untreated zones were under the same hydraulic conditions at the beginning of the test. The depth-averaged inlet flow velocity was controlled at 0.26 m/s, which is slightly higher than the critical velocity (~0.23 m/s) for the Ottawa graded sand used in this study.

DOI: 10.1201/9781003519638-5

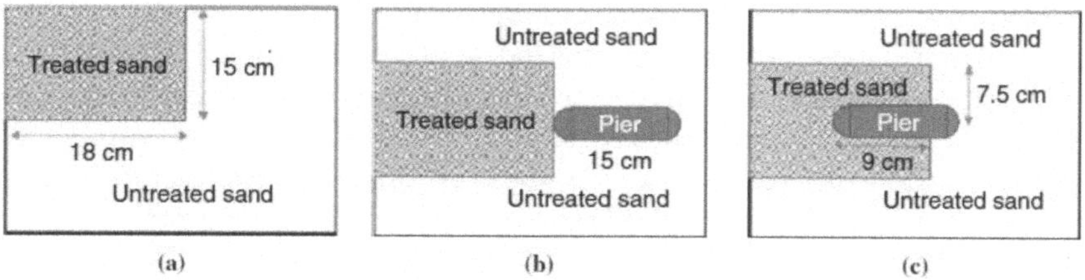

FIGURE 5.1 Top view of the flume bed. The flow direction is from left to right. Only a portion of the flume is illustrated to highlight the treated sand: (a) case no. 1, (b) case no. 3, and (c) case nos. 4 and 5.

Test case nos. 2, 3, and 4 were conducted to investigate the benefit of the MICP technique under a bridge local-scour scenario (Figure 5.1b and c). An oblong model pier with rounded noses was used in these three tests. The aspect ratio of pier width to length was 1:4. The pier width was 30 mm, which resulted in a blockage ratio of 10% at the flume transverse cross section. Such a magnitude was below the value (16%) suggested by Raudkivi and Ettema to avoid the wall friction effect in the flume (Landers 1992). The pier was vertically mounted at the centerline of a location that is about 2.5 m downstream of the flume inlet, with its long axis aligned with the flow direction. By adjusting the flume tailgate, the approaching flow depth (*h*) was kept as 10 cm in test case nos. 2, 3, and 4. This was to ensure that there was no significant influence of flow depth on the scour depth, as suggested by Melville and Sutherland as $h/b \geq 3.5$ for a uniform cylindrical pier (b is pier width) (Lagasse et al. 2001). To satisfy clear-water conditions, the depth-averaged inlet velocity was set at about 0.15 m/s for test case nos. 2, 3, and 4. Such a magnitude was determined as 70% of the critical velocity for the Ottawa sand ($D_{50} = 0.35$ mm) based on the classic diagram of Shields (1936). Test case no. 5 was conducted to model the live-bed scour condition, for which the inlet velocity was set as the limit of the flume (about 0.5 m/s), and the flow depth was decreased to about 5 cm.

Test case no. 2 was a control scour test in which only untreated sand was involved. In test case no. 3, the treated sand sample was placed in front of the pier, while the centerline of the sample aligned with that of the pier (Figure 5.1b). In test case no. 4 and test case no. 5, a slot was created in the treated sample so that the pier could be placed inside, as shown in Figure 5.1c. After the sample was trimmed, there was a small gap between the pier surface and the slot. To prevent any jet impingement in this gap, the gap was filled with waterproof sealant. Previous studies suggested that the scour hole mainly develops within 2 pier widths (6 cm in this case) from the pier. In test case nos. 4 and 5, the treated sample covers this region, as shown in Figure 5.1c.

Each test case was run for a period nominally expected to allow reaching equilibrium scour stage. The equilibrium condition was achieved when negligible (<0.5 mm) variation in the maximum scour depth was observed at an interval of 2 h after running the test for 15 h. This time was considered sufficient based on the calibration run by Ataie-Ashtiani and Beheshti, who found that 86% and 92% of the equilibrium scour depth were attained after 5.5 h and 8 h, respectively (Ataie-Ashtiani and

Beheshti 2006). When the equilibrium stage of scour was reached, water was gently drained from the flume to protect the final topography of the scoured bed. The topography of the flume bed after erosion was obtained using the structure from motion (SFM) technique, which can reconstruct a three-dimensional scene from overlapping photos taken from different angles (Westoby et al. 2012). Two open-source software applications, VisualSfM (Wu 2013; Wu et al. 2011) and CloudCompare, were used to generate the digital elevation model from overlapping photos of the scoured bed.

5.1.1.3 Erodibility Test with a Pocket Erodometer

The erodibility of soil can be determined with other devices and tests, such as an erosion function apparatus, the hole erosion test, the jet erosion test, the cohesive strength meter, the rotating cylinder test, and the pocket erodometer test (PET) (Briaud et al. 2012). In this study, PET was conducted to categorize the erodibility of the MICP-treated sand. PET is a simple test using a mini jet generator, such as a water gun. The pocket erodometer directs a horizontal jet of water 20 times to the same spot on the vertical face of a soil sample. The depth of the induced erosion hole is recorded and compared with an erosion function apparatus erosion chart to reach a preliminary determination of the erodibility category of the soil.

The treated sample was taken out of the reactor after 4 days of MICP treatment. The entire sample was hardened. As shown in Figure 5.2a, on the top of the sample is a thin layer of white precipitation, which was identified by XRD to be pure $CaCO_3$. The fabric mold was then removed to expose the sample, as shown in Figure 5.2b, where the sample was turned upside down. The sample was intact, except for the very limited damage on the edges. The red substances in Figure 5.2b are fiber residuals from the fabric mold. They were left on the sample surface to serve as a visual indicator in the erosion tests.

5.1.2 Characterization

The SEM images of untreated and MICP-treated Ottawa graded sand are shown in Figure 5.3. The untreated Ottawa graded sand has sub-round smooth particles, and

FIGURE 5.2 MICP-treated sand sample: (a) in the mold and (b) upside-down view after the mold was removed.

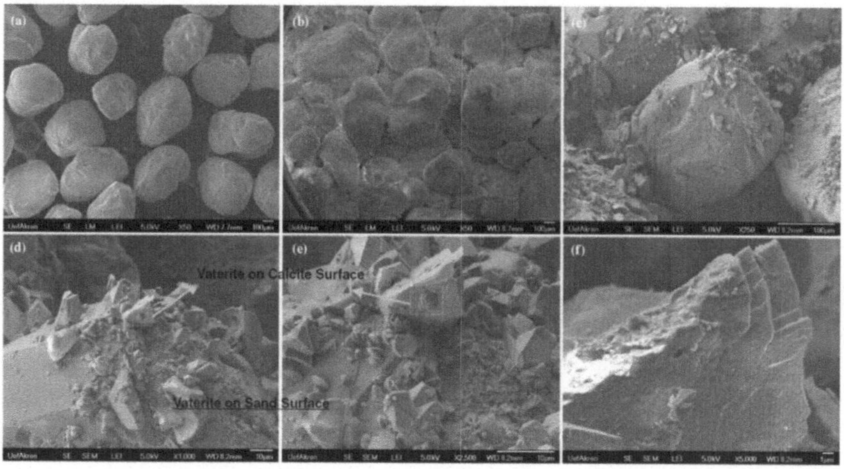

FIGURE 5.3 SEM images of (a) untreated Ottawa graded sand and (b–f) MICP-treated sand, with magnifications of 50, 250, 1,000, 2,500, and 5,000, respectively.

the contacts between particles are not bonded (Figure 5.3a). As shown in Figure 5.3b, however, the particles in the treated specimen are bonded together, and the pores are partially filled with precipitation; the distribution of the precipitation is, in general, uniform. Figure 5.3c shows that precipitated crystals exist both on the open surface of the particles and at the contact areas. Two polymorphic forms of $CaCO_3$, namely, calcite and vaterite, can be seen in Figure 5.3d and e. Calcite shows a rhombohedron surrounded with smooth surfaces; vaterite shows a spherical shape with crossing planes. Vaterite can be deposited on the surface of either sand or calcite. Figure 5.3f shows the multiple-layer planes of calcite and a small hole, which is likely a site for the bacteria cell. Figure 5.3f may indicate that calcite precipitates layer by layer in the proximity of the bacterial cells. The SEM images prove that $CaCO_3$ can both coat and bridge particles. The precipitated $CaCO_3$ modified the morphology of the sand particles and provided bonding at the contacts.

The existence of calcium and carbon in the MICP-treated sample was confirmed by EDX (Figure 5.4a). The XRD results, compared with typical diffraction spectra for silicon dioxide (SiO_2) and $CaCO_3$, confirmed the abundance of calcite and proved that the MICP process was successful (Figure 5.4b). However, with the two samples prepared for the XRD test, there was only one tiny peak, corresponding to vaterite. This was perhaps because of the very low concentration of vaterite in the prepared samples.

The average $CaCO_3$ content for the upper half of the sample was 7%, while that for the lower half, it was 3.86%. The upper half has a higher calcite concentration, which is likely because the top surface of the sample was left open, and the bottom surface was in contact with the fabric and seated on the shelf. It was easier for the cementation solution to enter the sample from the top surface than from the bottom surface.

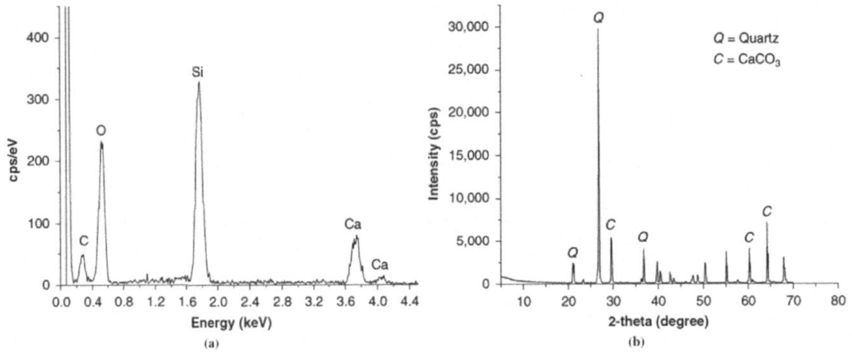

FIGURE 5.4 Results of MICP-treated samples from (a) EDX analysis and (b) XRD analysis.

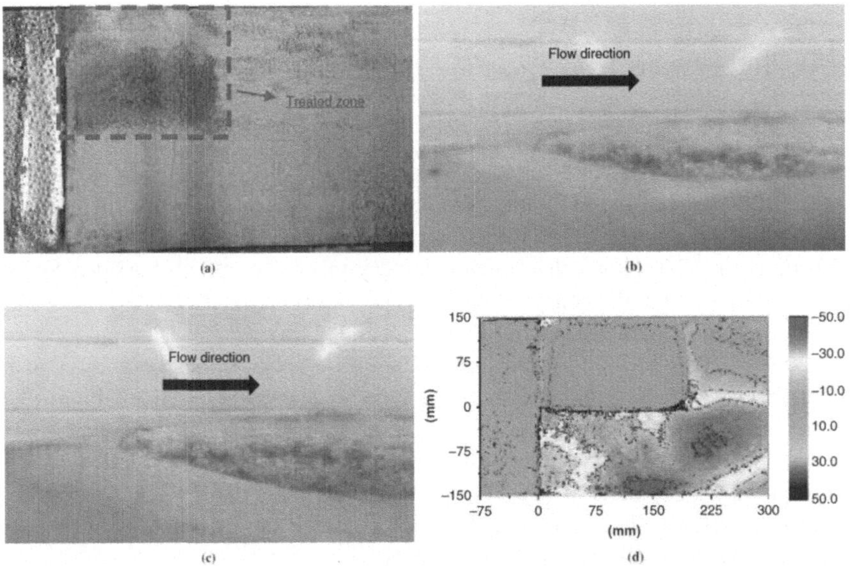

FIGURE 5.5 Surface erosion test case no. 1 with inlet velocity V_{in} = 0.26 m/s: (a) top view of the initial flume bed before erosion—the upper half (red and wet) is the MICP-treated sand, and the lower half is untreated; (b) side view of the flume bed after 10 min erosion; (c) side view of the flume bed after 50 min erosion; and (d) bed elevation contour after 15 h.

5.1.3 Erosion Tests

The surface erosion test case no. 1 was run for 15 h. Before erosion, the flume bed was prepared according to Figure 5.1a and Figure 5.5a. Because the inlet velocity was set close to the critical velocity of sand with a D_{50} of 0.35 mm, smaller particles were ready to be eroded. Figures 5.5b and c show the side view of the flume bed after 10 min and 50 min, respectively. Clearly, the treated zone was barely eroded, while the untreated zone kept eroding with time. After 15 h, the flow velocity gradually decreased to zero, and the remaining water was allowed to drain freely.

Overlapped pictures were taken on the flume bed, and the bed elevation contour was obtained with the SFM method (Figure 5.5d). The maximum erosion depth was about 45 mm on the untreated side; the treated side was intact. Note that 45 mm is larger than the thickness of the sample (40 mm), and erosion started to occur under the edge of the treated sample. Slight erosion was also observed near the downstream edge of the sample (Figure 5.5d).

Because of the development of vortices in the vicinity of piers, the scour hole is nonuniform. The maximum scour depth at different time instances was recorded to track the evolution of the scour hole. In the control scour test case no. 2, the maximum scour depth increased fast in the first 2 h and then increased gradually until an equilibrium depth of about 25 mm was reached after a total of 20 h (Figures 5.6a and b). The scour hole has the shape of a horseshoe; the scour depth reaches its maximum at the upstream nose of the pier and gradually decreases away from the pier. Deposition occurs around the downstream nose. In test case no. 3, while the treated region stayed intact, a scour hole started to develop immediately downstream of the treated sample, surrounding the upstream nose of the pier (Figure 5.6c).

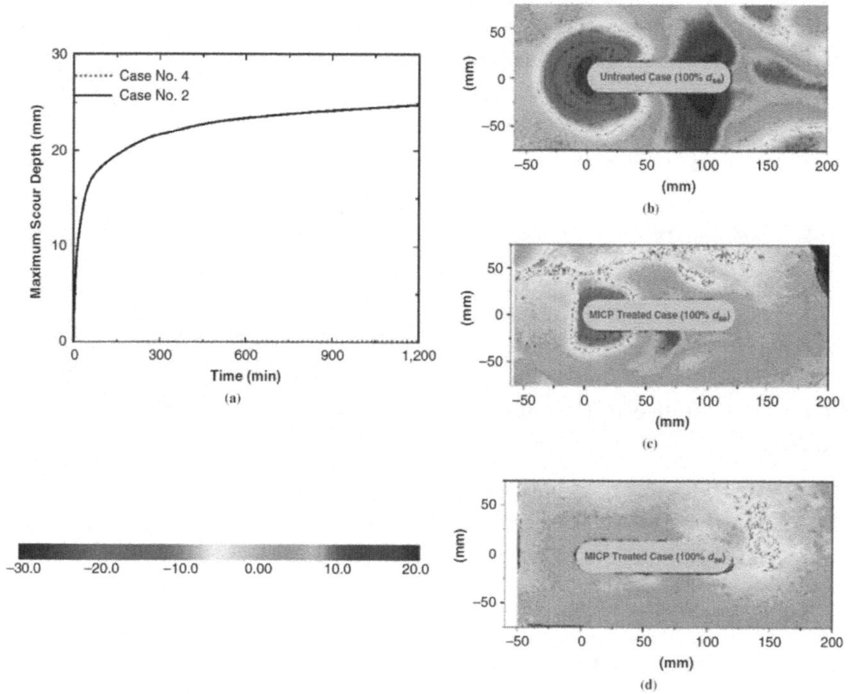

FIGURE 5.6 Results for test case nos. 2, 3, and 4 with inlet velocity of V_{in} = 0.15 m/s: (a) evolution of the maximum scour depth for test case nos. 2 and 4, and (b–d) illustration of the equilibrium scour bed topography for test case nos. 2, 3, and 4, respectively.

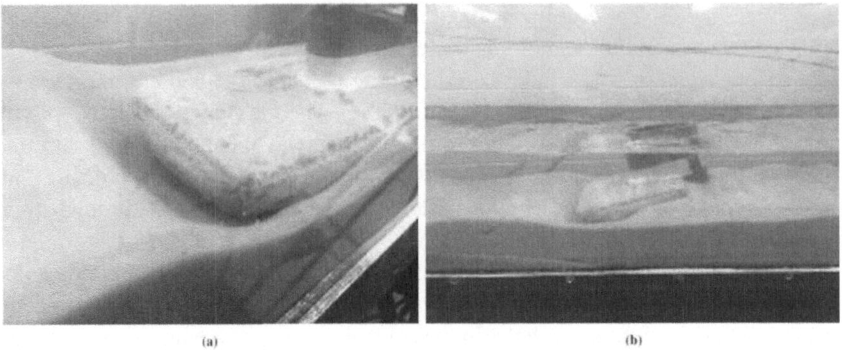

FIGURE 5.7 Undermining erosion in test case no. 5 with inlet velocity $V_{in} = 0.5$ m/s: (a) after 2 h and (b) after 4 h, tilting occurs.

For test case no. 4, however, the erosion was negligible (Figures 5.6a and d). Test case no. 5 modeled the live-bed scour condition, where the inlet velocity was much larger than the clear-water condition. Such conditions represent the extreme flow conditions caused by severe precipitation or flooding events. It was found that the sediment particles began to be eroded at the boundaries of the treated sample. This erosion model is like undermining erosion or contact erosion. When the erosion depth kept increasing and became larger than the thickness of the treated sample, the treated sample was entirely undermined and lost support (Figure 5.7a). As a result, differential settlement was observed, and eventually, the pier and the treated sample failed by tilting (Figure 5.7b).

5.1.4 Erodibility

Following the procedures detailed in Wu, the erodibility of the MICP-treated soil was determined using PET (Wu 2013). After 20 applications of calibrated jets from a water gun, there was no noticeable erosion on the vertical face of the MICP-treated sample. According to the PET erosion categories developed in Wu, the MICP-treated sample in this study falls into the "very low erodibility" category (Wu 2013).

5.2 EFFECT OF RAINFALL-INDUCED EROSION ON THE STABILITY OF SANDY SLOPES TREATED BY MICP

5.2.1 Experimental Methods

5.2.1.1 MICP Sample Preparation

The specimens were prepared in full-contact flexible molds (FCFMs) made of geotextile, as described in Section 3.1.3, and MICP sample preparation for details, as described in Section 3.1.

5.2.1.2 Rainfall-Induced Erosion System

Rainfall simulators are basic equipment to duplicate the physical characteristics of natural rainfall as closely as possible. It can be separated into two main types based on the way in which raindrops are produced: (1) non-pressurized nozzle simulators; (2) pressurized nozzle simulators (Aksoy et al. 2012). In the non-pressurized nozzle simulators, waterdrops fall under the effect of gravity. These simulators are unrealistic for field use, where a huge height (10 m) is needed for waterdrops to achieve terminal velocity. The waterdrops strike the ground at a velocity much lower than the terminal velocity and with lower kinetic energy only if the simulator is hoisted very high. That is why pressurized nozzle simulators are extensively preferred for studies in large-area field. In these simulators, raindrops were produced through single or multiple nozzles, while the drop intensities and velocities are usually exaggerated as the water is released under pressure. Since, a pressurized nozzle rainfall simulator was designed in this study.

5.2.1.3 Design of Rainfall Simulator and Erosion Flume

The experimental design to perform erosion experiments consists of a rainfall simulator and an erosion flume (Figure 5.8). A rainfall simulator was made of a PVC frame attached to a raindrop plate. The raindrop plate with multiple nozzles was installed on the PVC frame at a height of 1 m from the flume bed to ensure the terminal velocity of raindrops. Water was supplied from the water supply system. Lying under the rainfall simulator at a height of 5 cm from the ground was the erosion

FIGURE 5.8 Photo of rainfall simulator and erosion flume.

flume, which was fixed at a slope angle of 35°. The slope angle was considered to represent the inclination of natural soil slopes at which shallow slope failures are common (Rahardjo 1999).

5.2.1.4 Rainfall Uniformity

The coefficient of uniformity (CuC) defined by Christiansen (Christiansen 1941) is the most widely used measure of spatial uniformity, which is in percentage as

$$\text{CuC} = \left(1 - \frac{\sum_{1}^{N} \left|X_i - \bar{X}\right|}{N\bar{X}}\right)100 \tag{5.1}$$

where X_i is rainfall amount at location i, $\bar{X}$ is the average amount of rainfall, and N is the number of points where measurement cups are placed over the flume to collect rainfall. The *CuC* is a useful index of spatial uniformity of rainfall. The more uniform the pattern of rainfall, the closer the *CuC* approaches 100%. Rainfall can be considered uniform when *CuC* is higher than 80% (Moazed et al. 2010).

Table 5.1 shows the uniformity results. *CuC* was found to be higher than 80% for the four rainfall intensities. But *CuC* does not give any indication on the spatial pattern, which means that it is possible for different patterns to get the same *CuC* value. So the spatial patterns of rainfall were examined, and spatial relative rainfall intensity distribution maps were created with the Surfer software, shown in Figure 5.9. The difference between the average rainfall intensity among the repetitions was less than 5%; thus, the average rainfall intensity can be considered as constant during the experiments with each flow rate applied in this study (Lassu et al. 2015). The spatial distribution of rainfall can be considered uniform enough over the flume.

5.2.1.5 Raindrop Size

The raindrop size was determined by the flour pellet method (Kohl 1974), supported with an image processing technique. A 12.5 cm diameter circular pan filled with wheat flour (Figure 5.10a) was exposed to rainfall for each rainfall intensity (Figure 5.10b). Exposure time was restricted to 2 s to minimize the coalescence of

TABLE 5.1
Rainfall Uniformity Test

Rainfall Intensity (mm/h)	Coefficient of Uniformity (CuC) (%)
105	93.1
85	95.6
65	86.1
45	81.0

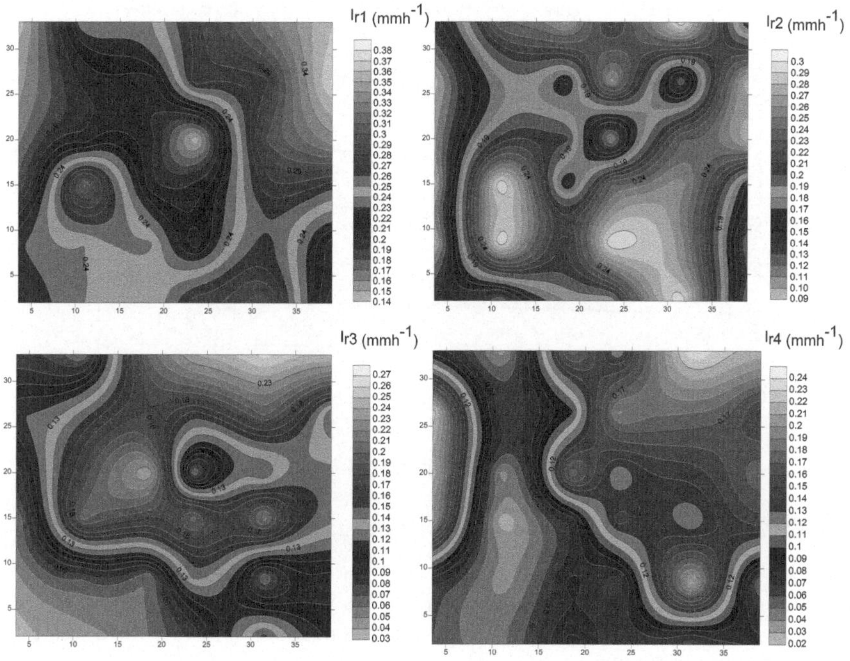

FIGURE 5.9 Spatial rainfall intensity distribution in the rainfall simulator with the four different rainfall intensities (I_r = relative rainfall intensity).

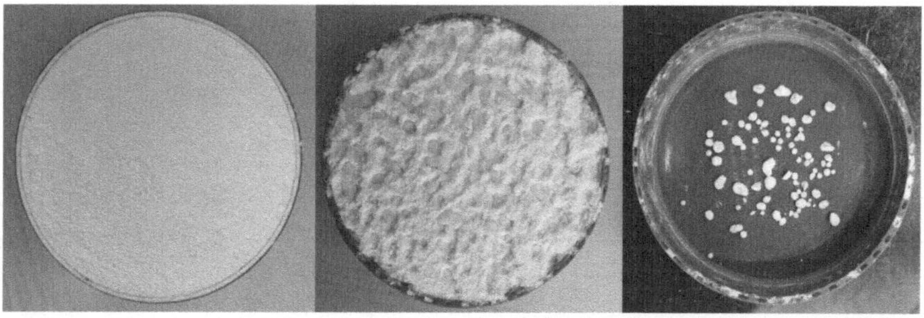

FIGURE 5.10 (a) Flour cup. (b) Flour cup exposed to rainfall. (c) Flour pellets after oven-drying.

the pellets in the flour. Flour was dried for 12 h at 105°C, and then the pellets were sieved and photographed (Figure 5.10c). The photograph was then processed by an image processing software (Image-Pro Plus 6.0) to automatically distinguish and classify the raindrops based on the size of their surface area. Each raindrop was characterized by a circle area with a diameter to get the median diameter of the raindrops.

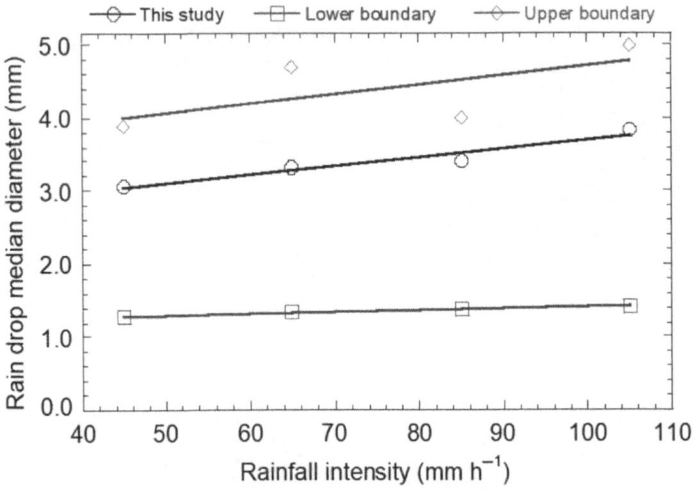

FIGURE 5.11 Changes in median diameter with rainfall intensity.

Raindrop size determined as detailed previously was compared to those found in literature. An empirical equation fitted to the data available in literature was given by Van Dijk et al. (2002) as

$$D_{50} = \alpha R^{\beta} \tag{5.2}$$

where R is rainfall intensity, given in mmh^{-1}. The parameters in equation 5.2 have ranges of $0.8 \leq \alpha \leq 1.28$ and $0.123 \leq \beta \leq 0.292$, with which envelope curves in Figure 5.11 were graphed. The determined median raindrop sizes of this study were found within this range, which shows the performance of the rainfall simulator in terms of drop size.

Mississippi local sand was used in this study to perform the rainfall-induced erodibility of MICP-treated soil. Before the formal experiments, rainfall with four different intensities was examined by eroding the MICP-treated soil that was treated by 0.25 M Ca cementation media concentration for 24 h. No noticeable erodibility could be observed on the MICP-treated soil under the rainfall with the intensity of 45 mm/h, 65 mm/h, and 85 mm/h. As shown in Figure 5.12, the 0.25 M Ca MICP-treated soil was eroded seriously by rainfall with the intensity of 105 mm/h, since, to obtain noticeable erodibility, rainfall with the intensity of 105 mm/h was selected to perform rainfall-induced erosion of MICP-treated soil in this study.

5.2.1.6 Roughness Testing

The LS-40 Portable 3D Surface Analyzer, as shown in Figure 5.13a, is a 3D surface measurement and analysis device which scans a 10.8 cm × 15.2 cm or 25.4 cm area and produces a high-resolution (0.01 mm) digital surface structure with an intensity image and a surface depth (height) related range image. LS-40 provides the data

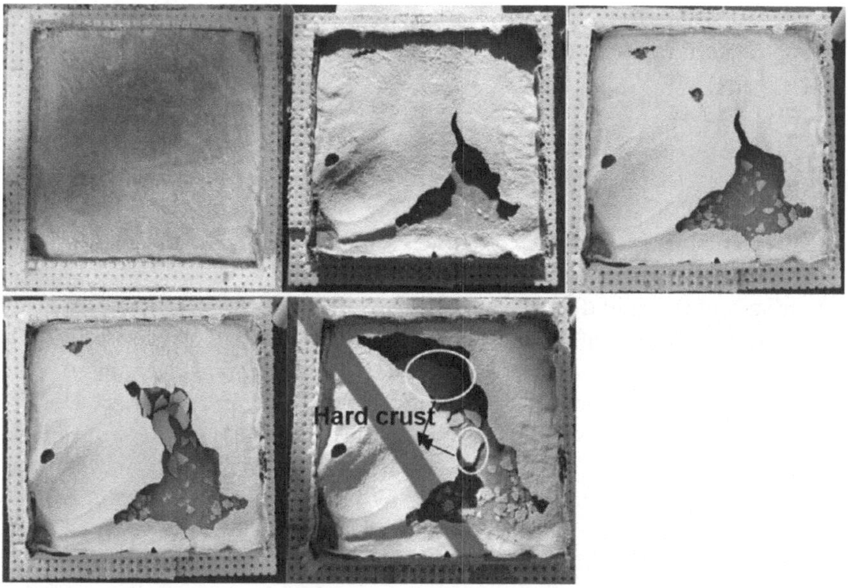

FIGURE 5.12 Photograph of 0.25 M Ca MICP-treated soil after 105 mm/h rainfall-induced erosion under different eroded times.

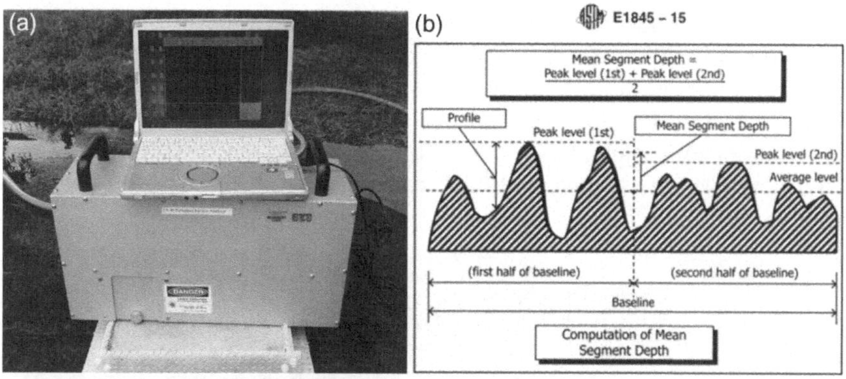

FIGURE 5.13 Surface analysis: (a) photograph of LS-40 Portable 3D Surface Analyzer; (b) computation procedure of mean segment depth (ASTM E1845).

of mean profile depth (MPD) and 3D views by processing profiles over the entire scanned surface according to ASTM E1845 specifications.

As shown in Figure 5.13b, the measured profile was divided for analysis purpose into segments each having a base length of 100 mm. The segment was further divided into half, and the height peak in each half-segment was determined. The difference between that height and the average level of the segment was calculated. The average value of these differences for all segments making up the measured profile was reported as the MPD. The MPD values were used to describe the roughness of the samples' surface in this study.

5.2.2 Visual Observation

The observed surface erosion pattern of 0.25 M Ca MICP-treated slopes under simulated rainfall at 0, 4, 6, 12, 24 h are shown in Figure 5.12. A hard surface calcite crust was induced on the entire surface of the soil sample by MICP treatment.

For the 0.25 M Ca MICP-treated sample, surface erosion was found to occur soon after the start of the rainfall erosion test (Figure 5.12). At 4 h, eroded holes appeared on the soil surface, and soil fluidization and loss were observed at the slope toe. This gradually destabilized the slope, the hole came to a greater range, and more eroded holes were observed on the soil surface, along with longer rainfall erosion. At 6 h, the soil inside the slope was washed away, and the hard surface calcite crust was suspended above the surface. Finally, at 24 h, almost a third of the slope was damaged, and the suspended surface crust was broken by rainfall erosion. The stability of soil slope was deteriorated by exposure to simulated rainfall.

For the 0.5 M Ca MICP-treated sample, no discernible surface erosion could be identified after 12 h of rainfall impact. Even after exposure to the rainfall for 24 h, only loose particles on the surface were washed away, while the slope surface remained intact.

After the rainfall-induced erosion test, the loose particles were washed away, the average thickness of the hard crust formed on the 0.25 M Ca, and the 0.5 M Ca MICP-treated surface of slopes was determined using a caliper. The 0.5 M Ca MICP-treated slope was artificially broken before the thickness measurement. For the 0.25 M Ca MICP-treated slope, the average thickness of the hard crust was 4.33 mm, while for the 0.5 M Ca MICP-treated slope, the average thickness of the hard crust was 9.82 mm, which was more than two times that of the 0.25 M Ca MICP-treated slope. And the average $CaCO_3$ content of the hard crust was 8.3% for 0.5 M Ca and 5.6%

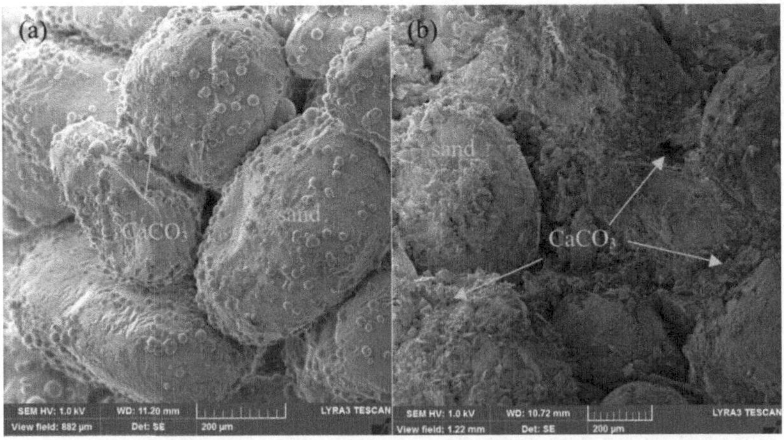

FIGURE 5.14 SEM images of (a) 0.25 M Ca, and (b) 0.5 M Ca MICP-treated sand.

for 0.25 M Ca MICP-treated specimens. The thicker hard crust with a higher $CaCO_3$ content of the 0.5 M Ca MICP-treated slope was primarily responsible for its better surface erosion resistance than the 0.25 M Ca MICP-treated slope.

Figure 5.14 shows the SEM images of soil samples from the surface of the 0.25 and 0.5 M Ca MICP-treated slopes before the rainfall erosion tests. From these SEM images, the morphology of $CaCO_3$ crystal can be clearly illustrated among these two cases. For the 0.25 M Ca MICP-treated soil, sporadic $CaCO_3$ crystals were found on the surface of sand particles, and lots of large void network remained open, while for the 0.5 M Ca MICP-treated soil, $CaCO_3$ crystals were found between particle contacts, even though not all adjacent particles were bridged. But most of the large pores were partially obstructed by $CaCO_3$ crystals. These phenomena observed from SEM images could also explain why the 0.5 M Ca MICP-treated slopes were better to resist rainfall-induced erosion.

5.2.3 UCS of MICP- and Cement-Treated Samples

Figure 5.15 shows a box plot of the UCS of MICP- and cement-treated samples as a function of the concentration of cementation media and cement content. Each box shows the median value and ±25% of the UCS population as the top and bottom of the box. The lines extending from the top and the bottom of each box mark the minimum and maximum UCS. The outlier was shown as an individual point. As shown in Figure 5.15, the UCS of MICP-treated and cement-treated samples increased with a higher concentration of the cementation media and cement content. The 0.5 M Ca MICP-treated samples achieved 1,300 kPa of UCS, which was roughly equivalent to 10% cement-treated samples. This result was like that of Bu et al. (2018b). The 0.25 M Ca MICP-treated samples only got about 200 kPa UCS, which was much lower than the 0.5 M Ca MICP-treated samples. That should be because of the difference in $CaCO_3$ content. The $CaCO_3$ content of 0.5 M Ca MICP-treated UCS samples was 7.4%, while it was only 4.3% for 0.25 M Ca MICP-treated UCS samples.

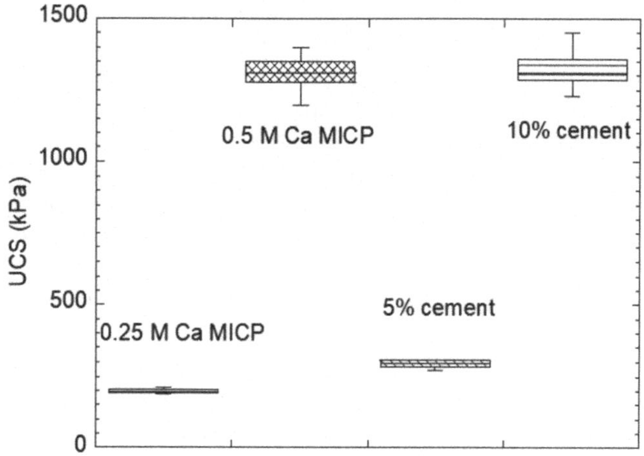

FIGURE 5.15 UCS of MICP-treated and cement-treated samples.

5.2.4 Effect of Rainfall Erosion on the Roughness of MICP and Cement-Treated Slopes

To obtain the properties of MICP-treated soil under rainfall-induced erosion, the properties of MICP-treated samples were compared with those of soil samples made through cement modification. The proportion of cement in this part was 10% by weight of dry sand, and the concentration of cementation media for the MICP-treated samples was 0.5 M Ca. These two samples had similar UCS, as shown in Figure 5.15.

Figures a, b, and c are for 0.5 M Ca MICP-treated sample, while d, e, and f are for 10% cement-treated sample. No visible erodibility could be observed on the surface of MICP-treated and cement-treated slopes after 24 h rainfall-induced erosion, since roughness testing was conducted on samples to investigate the subtle erodibility caused by rainfall-induced erosion. Soil surface roughness is an important parameter in understanding the mechanisms of soil erosion by water. Three-dimensional (3D) measurements were applied by a laser-scanner device in this study. The roughness of samples was tested after the eroded time of 0 h, 1 h, 2 h, 4 h, 6 h, 12 h, and 24 h, respectively. To investigate the roughness of soil surface, 3D views and mean profile depth were analyzed.

For the 0.5 M Ca MICP-treated slope, Figures 5.16a and b show the 3D surface views before and after 24 h rainfall-induced erosion. The eroded surface looks smoother than the uneroded surface, maybe because the powdery bonds, unbonded minerals, or loosely bonded calcite were washed away by the simulated rainfall.

Similar results could be obtained from Figure 5.16c, which shows the MPD values of the MICP-treated sample as a function of rainfall-induced eroded time. A higher MPD value represents a textured surface, which means a rougher surface. The MPD values on the surface of the MICP-treated slope decreased gradually along with a longer rainfall-induced eroded time. Around a 16.7% decrease of MPD value happened after 24 h rainfall-induced erosion. The smoother surface of the MICP-treated sample was caused by the simulated rainfall. When the soil slope was treated by MICP, a part of $CaCO_3$ precipitation was only attached to the surface of sand particles that may not bond the sand particles strongly, thus getting easily eroded by rainfall and resulting in a smoother surface.

For the cement-treated samples, the opposite results were induced by simulated rainfall. As shown in Figures 5.16d and e, the 3D views of the cement-treated sample became rougher after 24 h rainfall-induced erosion.

The same finding was found in Figure 5.16f for the cement-treated samples. The MPD values on the surface of cement-treated samples kept increasing as the rainfall-induced eroded time was growing. A 75% increase in MPD value happened because of 24 h rainfall-induced erosion. A rougher surface was induced by rainfall. This phenomenon could be caused by the hydraulic strength of cement during the rainfall-induced erosion process. The early strength of cement in this study gain allowed the various curing times to range from 7 to 21 days. The cement-treated samples were cured for 7 days in a constant humidity of 100% before the rainfall-induced erosion test. Loosely bonded sand particles were washed away by rainfall at the early stage of erosion; meanwhile, the strength of the cement-treated sample kept increasing during the erosion process to resist erosion. Afterward, a rougher surface was formed at the later stage of rainfall-induced erosion.

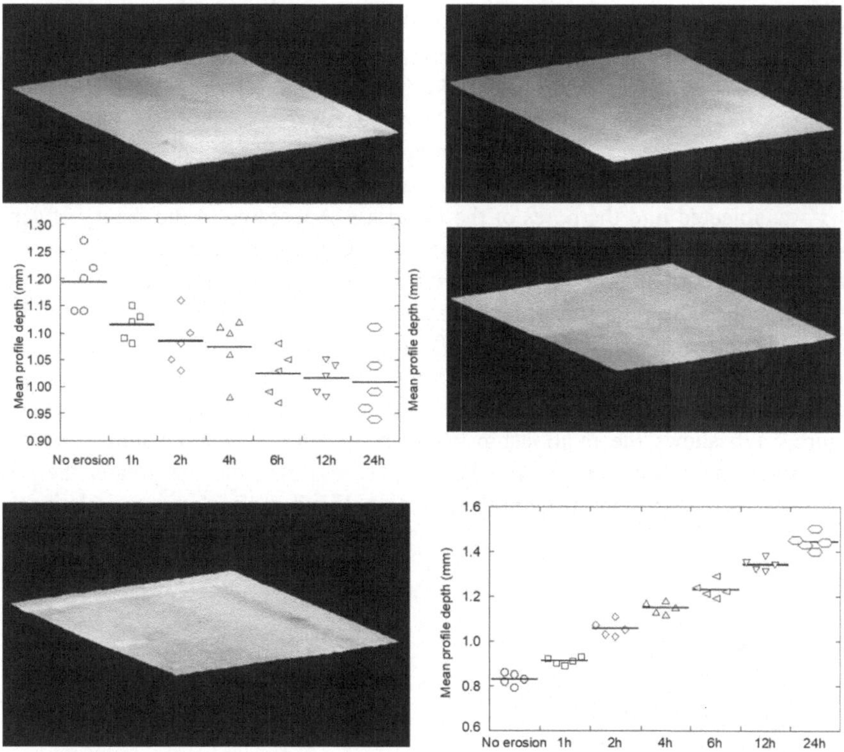

FIGURE 5.16 (a, d) 3D surface views of sand surface before rainfall induced erosion. (b, e) 3D surface views of sand surface after 24 h rainfall-induced erosion. (c, f) Mean profile depth on the surface as a function of rainfall-induced eroded time.

However, it has been reported that rough surfaces absorb raindrop impact on the soil surface and modify significantly water flow and soil surface interactions (Dunkerley 2004). Moreover, Liu et al. (2019) concluded that MICP-treated materials had a short service life during wet–dry cycles. Especially in the wetting period, water could penetrate and inundate the MICP-treated soil; the powdery bonds, unbonded minerals, or loosely bonded calcite acting as the cementing agent could be eroded by water during this process. This means the MICP-treated slope was weak to resist rainfall-induced erosion compared to the cement-treated slope. The ability of MICP-treated slopes against rainfall-induced erosion must be improved in the future.

5.3 EFFECT OF MICP ON SURFACE ACID CORROSION

5.3.1 Experimental Methods

5.3.1.1 MICP Sample Preparation

The specimens were prepared in full-contact flexible molds (FCFMs) made of geotextile, as described in Section 3.1.3, and MICP sample preparation for details, as described in Section 3.1.

5.3.1.2 Microscopic Testing of Acid-Corroded Specimens

The porosity of the acid-corroded specimens was measured to analyze corrosion depth and severity using the Autopore IV 9510 testing system (Micrometrics, Norcross, Georgia) based on a mercury porosimetry analyzer (Version 5.0) (Figure 5.17a). Because of the unwettability of mercury on the solid surface, pressure was applied to overcome the resistance caused by the surface tension the mercury created. Mercury was injected into the pores of the materials. According to the theory of capillary pressure, the smaller the capillary pore size, the greater the pressure needed. The pressure required to inject the mercury into the specimens' pores was inversely proportional to the pore size. The mercury pressure ranged from 0 to 413.8 MPa (60,000 psi), and the pore size ranged from 50×10^{-4} to 1,000 μm. Porosity, pore characteristics, and main pore distribution of the specimens were measured based on the different levels of pressure needed for injecting mercury into the specimens. Figure 5.17b shows the relationship between mercury injection volume and pore size in the uncorroded specimen. The smaller the pore size, the greater the pressure needed and the more difficult the injection. Figure 5.17c shows the relationship

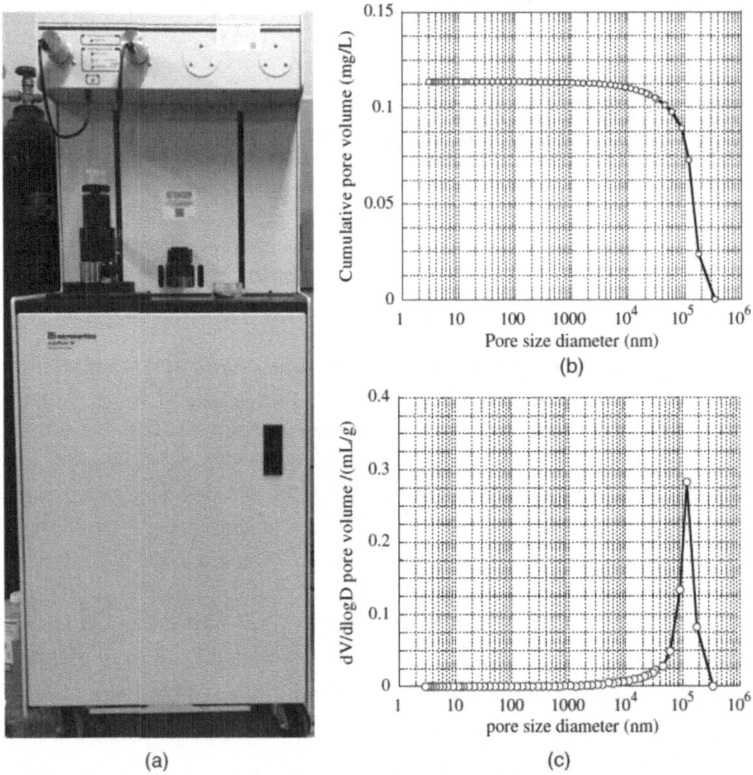

FIGURE 5.17 Microscopic porosity testing: (a) Autopore IV 9510 testing system; (b) cumulative pore volume versus pore diameter for uncorroded specimen; and (c) relationship between log-differential intrusion and pore size.

between log-differential intrusion (dV = d$logD$) and pore size. This figure also indicates internal pore volume distribution. From Figure 5.17c, it is seen that macropores take up most of the distribution, leaving mesopores and micropores to take up the rest. Pore distribution was more concentrated, indicating that the internal uniformity of the specimen was relatively good.

Philips vertical X-ray diffractometer (PW3050/60, MPSS; Philips, Amsterdam, the Netherlands) using Cu Kα (I ¼ 1.54056 A) radiation with 2α in the range of 5°–80° and a step size of 0.02° was used. Next, the cell parameters were calculated using the X'Pert HighScore Plus diffraction software. A field-emission scanning electron microscope (Quanta FEG650; Thermo Fisher, Waltham, Massachusetts) was used to obtain the specimens' fracture-surface microstructure before and after corrosion.

5.3.1.3 Fractal Analysis of Microscopic Porosity Testing

Self-similarity refers to the characteristics of a structure or process that are similar at different spatial and temporal scales. *Scale invariance* refers to the characteristics of the localized fractal. If the scale invariance is amplified, the enlarged fractal will manifest the morphological characteristics of the original.

The application of fractal theory in the specimen pore simulated the characteristics of pore structure based on the structure of Menger's sponge. The characteristic parameter of the pore volume fractal dimension is Db from Menger's sponge theory (van Paassen et al. 2010a; Vita et al. 2012). According to the famous Washburn formula, as shown in equation (5.3), the relationship between mercury pressure and pore diameter is built when the pores are assumed to be cylindrical:

$$P(r) = (-4\sigma cos\theta) / r \times 10^2 \tag{5.3}$$

where $P(r)$ = pressure of mercury (MPa), corresponding to pore size; r = pore size (nm); σ = coefficient of surface tension of mercury (generally σ ¼ 480 nN = cm); and θ = contact angle between mercury and a solid dielectric surface (140°). With all known parameters put into equations (5.3)–(5.5):

$$P(r) \times r = 1.5 \times 10^4 \tag{5.4}$$

$$1 / P(r) = 6.7 \times 10^{-5} r \tag{5.5}$$

In the experiment, total pore volume was equal to the volume of mercury injected into the pore (equation (5.6)). Equation (5.3) was used to derive equation (5.7). Equation (5.7) was transferred to equation (5.8):

$$dV_k = dV_{P(r)} \tag{5.6}$$

$$P(r)dr + rdP(r) = 0 \tag{5.7}$$

$$dr = -\left[\frac{r}{P(r)}\right] dP(r) \tag{5.8}$$

Then, equation (5.7) was taken to $dV_k/dr \propto r/P(r)^{2-Dl}$, where D_b is the fractal dimension, and equation (5.9) was generated (Maria et al. 2012; Vafai 2005):

$$dVP(r)/dP(r) \propto r/P(r)^{2-D_b} \tag{5.9}$$

Equation (5.3) was taken to equation (5.9), and equation (5.10) was generated:

$$dV_{P(r)}/dP(r) \propto r^2 r^{2-D_b} \tag{5.10}$$

Then, taking the logarithm with respect to equation (5.10), equation (5.11) was generated:

$$lg\left[dV_{P(r)}/dP(r)\right] \propto (4-D_b)\,lgr \propto (D_b-4)\,lgP(r) \tag{5.11}$$

Taking $lgP(r)$ as the horizontal coordinate and $lg[dV_{P(r)}/dP(r)]$ as the vertical coordinate, a graph was drawn, and the slope K was acquired.

In addition, $D_b - 4 = K$ and $y = Kx + b$ (b is constant), in which $y = lg[dV_{P(r)}/dP(r)]$, $x = lgP(r)$, and $K = D_b - 4$. Finally, D_b was obtained through the beeline fitting method.

5.3.2 Change in Mechanical Properties of Acid-Corroded Specimens

To explore the change in the mechanical properties of an acid corroded specimen, UCS tests were carried out on samples soaked in various concentrations of acid solution for 3, 7, 14, 28, and 60 days, respectively. Comparative analysis was performed with non-corroded specimens.

The unconfined compression tests were conducted under strain-controlled conditions at a uniform loading rate of 1.5%/min in accordance with ASTM D2166. The tests were halted when the axial strain reached 15%. The testing results for UCS at failure are summarized in Table 5.2. Later, the relationship between UCS and corrosion time in different concentrations of acid solution was analyzed, with the results shown in Figure 5.18a–e.

The surface of the specimens turned white after acid corrosion. This indicates that the chemical reaction produced a new substance with a certain strength after immersion in acid. The new substance was speculated to be calcium sulfate, which needed to be verified with additional tests. The change in specimen UCS also indicated that a new substance had been generated.

In Figure 5.18f and Table 5.2, the UCS of specimens after corrosion by sulfuric acid solution is much lower than that of uncorroded ones. As the soaking period progressed, UCS underwent a grousing-dropping process corresponding to the corrosion period (dropping at <7 days; rising at 7–28 days; dropping again at >28 days). When the corrosion time was 7 days, specimen strength went down to a minimum level and then increased slightly. At 7 to 28 days, a new substance was produced, and supplementary strength was achieved, but the substance did not return to the original

TABLE 5.2
UCS and Corrosion Time for Various Acid Solution Concentrations

Corrosion Time	2 g/L	4 g/L	8 g/L	16 g/L	20 g/L
(Days)	UCS (MPa)	UCS (MPa)	UCS (MPa)	UCS (MPa)	UCS (MPa)
0	2.0	2.0	2.0	2.0	2.0
3	1.55	1.31	0.78	1.03	1.03
7	0.96	0.47	0.49	0.96	0.71
	(A = 47.8%)	(A = 23.5%)	(A = 23.5%)	(A = 47.8%)	(A = 35.7%)
Average of A			35.7%		
14	1.01	0.64	0.63	0.97	1.23
28	1.50	0.68	0.96	1.18	1.42
	(B = 26.9%)	(B = 10.4%)	(B = 23.4%)	(B = 11.3%)	(B = 35.2%)
	(C = 74.8%)	(C = 33.9%)	(C = 47.8%)	(C = 59.1%)	(C = 70.8%)
Average of B			21.4%		
Average of C			57.3%		
60	0.84	0.52	0.78	0.68	0.54
	(D = 42.1%)	(D = 26.1%)	(D = 39.1%)	(D = 33.9%)	(D = 26.9%)
Average of D			33.6%		

Note: A = strength of specimen at 7-day corrosion time in acid corrosion solution relative to noncorrosion; B = percentage of strength complement, the ratio between the difference in the value of strength (7–28 days) and strength of noncorrosion; C = strength at 28-day corrosion time in acid corrosion solution relative to noncorrosion; and D = strength at 60-day corrosion time relative to noncorrosion.

strength of the non-corroded specimens. When the corrosion time was extended to more than 28 days, the new substance dissolved gradually, owing to slight dissolution. Pore size increased constantly, and the strength of the specimens decreased dramatically. When the bio-geomaterial was soaked in acid solution for 7 days, the average strength stayed at 35.7%, relative to that of the non-corroded specimens. Average strength was 57.3%, relative to that of the non-corroded specimens after 28 days of corrosion. The average strength complement was 21.4% from 7 to 28 days of corrosion. At 28 up to 60 days, average strength was only 33.6% relative to that of the non-corroded specimens.

Acid corrosion tests showed that when this material encountered sulfuric acid, its anti-acid-corrosion ability was not as weak as expected because of the formation of a slightly soluble new substance. The microbially induced calcium carbonate was used as cement to fill in the soil pores. Owing to the acid corrosion of calcareous cement, the internal porosity of this material increased, while its strength decreased. In the process of chemical corrosion, the new substance was produced, its volume increased gradually with corrosion time, and it filled the pores of the material, serving to supplement the material's strength. Because the strength of the new substance was much lower than that of calcium carbonate and the bio-geomaterial was not able to keep its strength after slight dissolution of the new substance, the strength of the

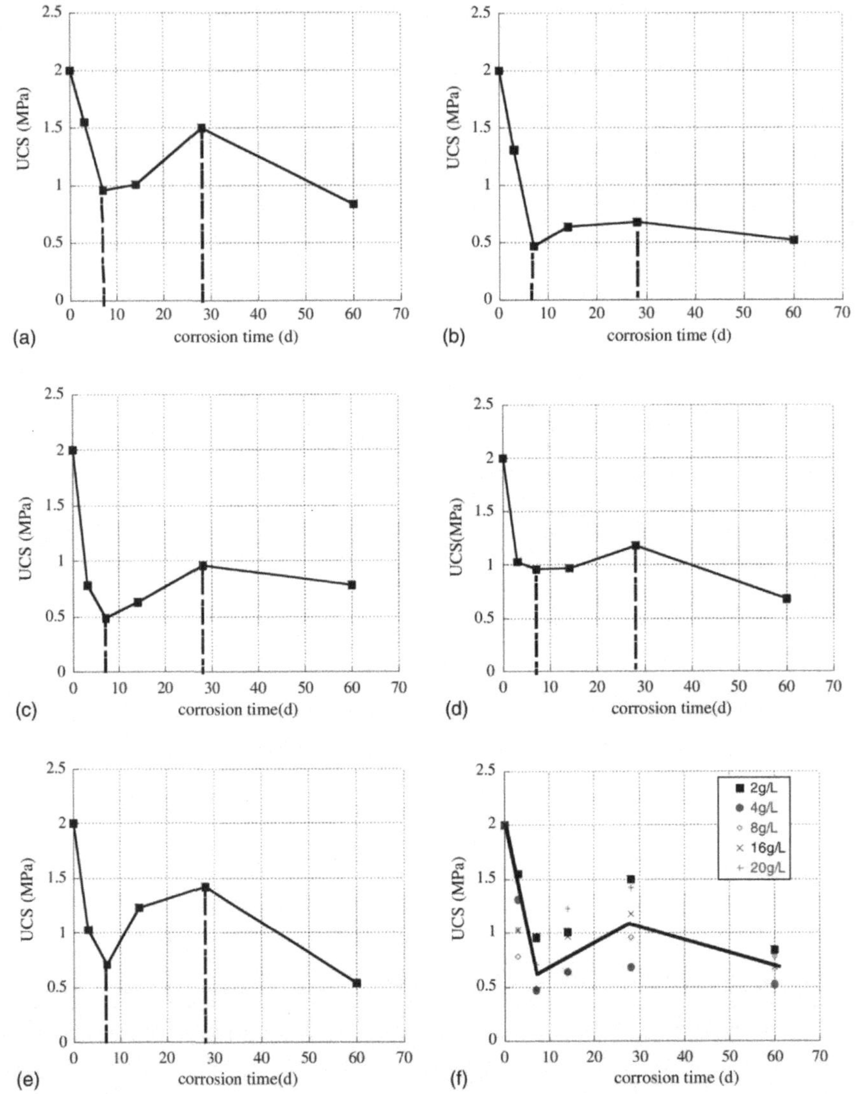

FIGURE 5.18 Changes in specimen mechanical properties with various acid solution concentrations and corrosion times: (a) 2 g/L; (b) 4 g/L; (c) 8 g/L; (d) 16 g/L; (e) 20 g/L; and (f) 2–20 g/L.

samples decreased rapidly with corrosion time. From this it could be concluded that the concentration of acid in the environment, the density and strength of the calcium carbonate crystals, and the stability of the newly produced materials determined the corrosion properties of the bio-geomaterial in an acid solution. The higher the concentration of the acid solution, the higher the dissolution rate of the calcium carbonate. In addition, the higher the concentration of calcium in the solution, the faster the new generations appear.

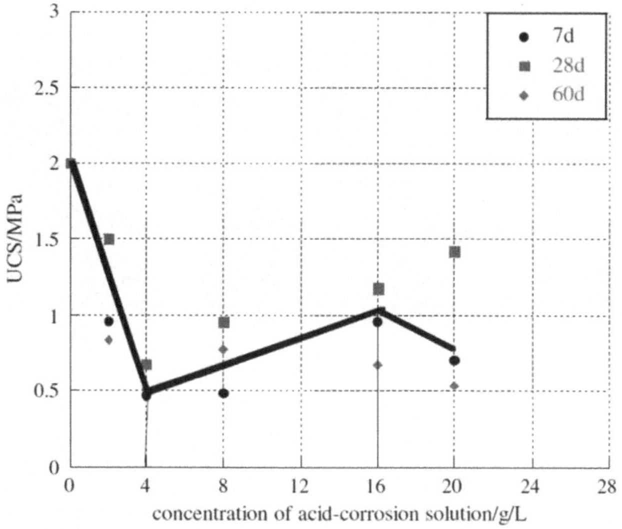

FIGURE 5.19 Relationship between UCS and acid solution concentration with corrosion time.

Based on Figure 5.19, the relationship between UCS and acid solution concentration for various corrosion times was summarized. When the concentration reached approximately 16 g/L, corrosion was feeble. When the concentration was greater than 16 g/L, the dissolution rate of calcium carbonate was higher than the speed of new substance formation, reducing the strength of the samples. The worst mechanical properties of the specimens were found in 4g/L sulfuric acid solution. The production of the new substance was not adequate to complement the strength. Thus, the dissolution of calcium carbonate in 4 g/L sulfuric acid solution was shown to inhibit the production of the new substance and so caused the worst corrosion of the bio-geomaterial.

5.3.3 Morphology Observation and Analysis

To describe the properties and crystal forms of the newly produced white material on the specimens in acid solution, the specimens were tested using scanning electron microscopy (SEM) and X-ray diffraction (XRD) before and after acid corrosion (Figure 5.20 and Figure 5.21).

From Figure 5.20a, it is seen that calcite and vaterite are the dominant calcium carbonate crystal types in the specimens before acid corrosion, but calcium carbonate and calcium sulfate coexist in the specimens after corrosion (Figure 5.20b). From Figure 5.20b, it is seen that calcium carbonate crystals, which are microbially induced cement, present as spherical clusters. These crystals are a source of strength for the specimens created using MICP technology. Similar observations were also made regarding $CaCO_3$ crystals by Al Qabany et al. (2012) and Zhao et al. (2014a). In addition, the reduction in void space was a secondary strengthening factor due to

FIGURE 5.20 SEM results: (a) before acid corrosion, and (b) after acid corrosion (28 days; 20-g/L concentration).

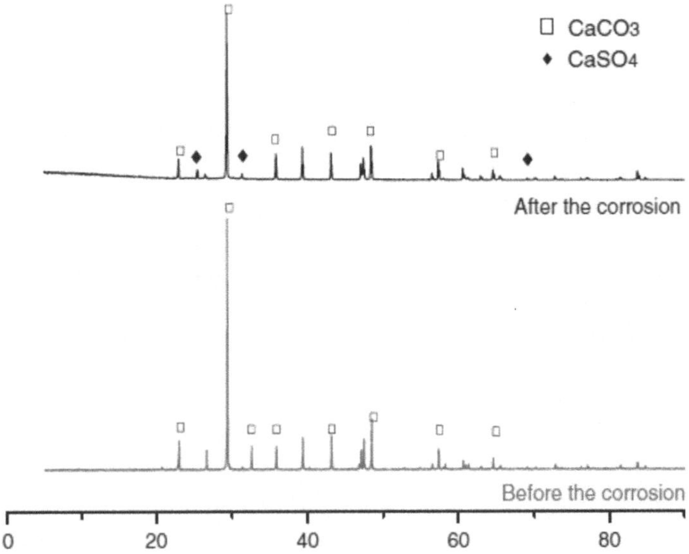

FIGURE 5.21 XRD measurement before and after sulfuric acid corrosion after soaking for 3 days in 20 g/L acid solution concentration.

the filling of $CaCO_3$ precipitation (DeJong et al. 2010). The other factor was calcium sulfate dihydrate with a monoclinic system, which is produced by the dissolution of calcium carbonate crystal in acid corrosion, presented as monoclinic prismatic crystals. The volume of calcium sulfate crystals was larger than that of calcium carbonate crystals, but their strength was lower.

XRD measurements were done to confirm the composition of the mineral before and after sulfuric acid corrosion. The results are shown in Figure 5.21. From Figure 5.21, it is seen that the white material induced by MICP is calcium carbonate before corrosion, and a mixture of calcium carbonate and calcium sulfate after corrosion. The chemical formula of the corrosion process refers to equation (5.10). As

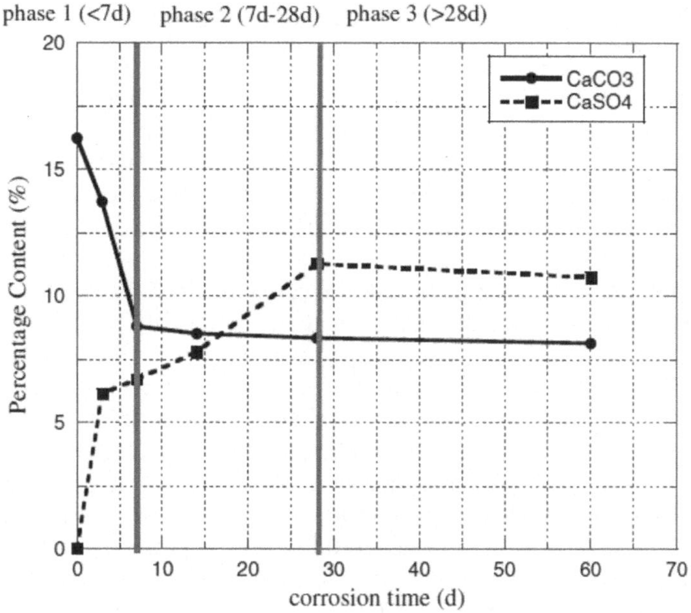

FIGURE 5.22 Percentage content of two crystals at various corrosion stages (20 g/L acid solution concentration).

the binder in the bio-geomaterial, calcium carbonate was dissolved constantly in the sulfuric acid solution. Microsoluble calcium sulfate precipitation was produced, and carbon dioxide bubbles appeared:

$$CaCO_3 + H_2SO_4 = CO_2 + H_2O \tag{5.10}$$

The percentages for the two crystals at different corrosion stages were measured and are shown in Figure 5.22. The weights of the acid-corroded samples, which were first dried and ground, were measured. Then, diluted acid was added, and the samples were cleaned using purified water. The reduced solid substance was calcium carbonate. It was dispersed in 100 mL of purified water and added to the excessive sodium carbonate solution and continuously stirred for no less than 30 min to ensure that all sulfates were in the form of free ions in the water. Then, excessive hydrochloric acid was added, dried, and cleaned. The second reduced solid substance was calcium sulfate in the sample.

Based on the crystal form revealed by SEM, the composition of the mineral revealed by XRD, and the percentage contents of the two crystals before and after acid corrosion, it was concluded that the white material in the samples induced by MICP was calcium carbonate before acid corrosion and a mixture of calcium carbonate and calcium sulfate dihydrate after corrosion. The main factor in determining the mechanical properties and microscopic pore characteristics of the bio-geomaterial lay in the dynamic equilibrium of the mixed crystals in different phases of corrosion.

5.3.4 Change in Porosity of Acid-Corroded Specimens

Specimen porosity at different corrosion times was tested with AutoPore IV 9500. The results are provided in Table 5.3. Taking the 2 g/L sulfuric acid solution as an example, with immersion time prolonged in the acid solution, the change in the porosity was divided into three phases. In the first phase (0–7 days), both strength and porosity decreased. In the second phase (7–28 days), strength increased but porosity continued to decrease. In the third phase (>28 days), strength decreased rapidly but porosity increased.

In the first phase (0–7 days), the calcium carbonate crystals near the surface of the specimens dissolved rapidly, and a small number of calcium sulfate dihydrate crystals were produced. Meanwhile, the specimen pores were filled with mixed crystals consisting of a small amount of calcium sulfate dihydrate and a large amount of calcium carbonate. Compared with non-corroded specimens, in the corroded specimens, the calcium sulfate dihydrate crystals replaced the existing calcium carbonate crystals in the pores, causing a dramatic decrease in specimen strength after soaking in the acid solution. In addition, specimen porosity decreased because the volume of mixed crystals turned out to be larger. In the second phase (7–28 days), the dissolution of the calcium carbonate crystals gradually developed from the surface to the core of the specimens. More and more calcium sulfate dihydrate crystals were generated and gradually filled the specimens. Specimen porosity decreased; overall density decreased as well because the volume of mixed crystals became larger. It was concluded that overall specimen strength was slightly improved. $CaSO_4$-$CaCO_3$ filled the specimen pores. The strength of the specimens with large amounts of calcium sulfate dihydrate crystals was obviously lower than that of specimens with calcium carbonate crystals. In the third phase of acid corrosion (after 28 days), because of the instability of the newly produced calcium sulfate dihydrate crystals, corrosion gradually increased; because of micro-dissolution of the calcium sulfate dihydrate crystals, porosity increased also.

The relationship between porosity and corrosion time is summarized in Figure 5.23. The porosity of the uncorroded specimens was 23.0%, whereas that of the corroded specimens reduced to 20.6% on the 7th day and 17.6% on the 28th day.

TABLE 5.3
Porosity and Pore Characteristics of Acid-Corroded Specimens (2 g/L Acid Solution Concentration)

Corrosion Time (Days)	Porosity (%)	Total Intrusion Volume (mL/g)	Gradient	Principal Pore Size (mm)	Fractal Dimension
0	23.0176	0.1136	0.047	0.12	2.988
3	22.7086	0.1127	0.046	0.12	2.989
7	20.596	0.0998	0.033	0.09	2.985
14	19.1455	0.0921	0.037	0.12	2.988
28	17.5877	0.0845	0.031	0.12	2.969
60	19.2774	0.0948	0.028	0.09	2.964

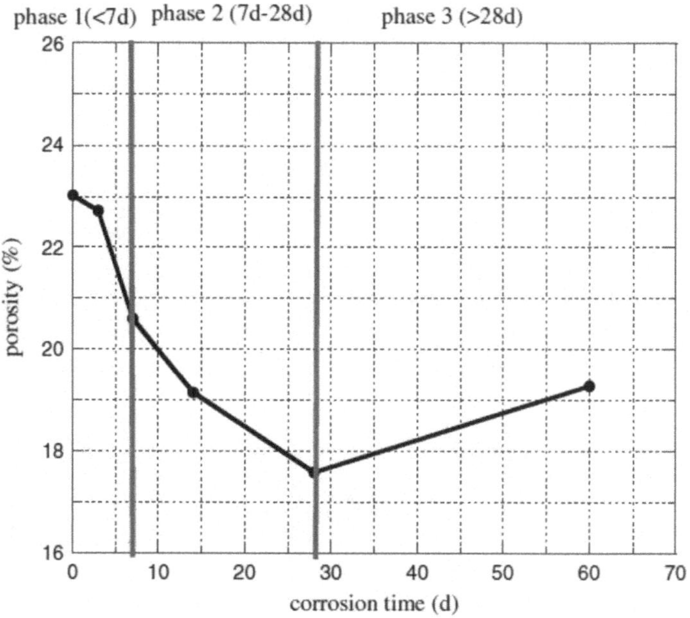

FIGURE 5.23 Change in porosity with increasing corrosion time.

This decrease in porosity indicated that the calcium sulfate dihydrate crystals were increasingly replacing the calcium carbonate crystals in the specimens' pores. When the corrosion time was more than 28 days, porosity increased to 19.3% because of the dissolution of the calcium sulfate dihydrate crystals.

Total intrusion volume was measured and is shown in Table 5.3. The results for total intrusion volume versus corrosion time corresponded with the change in porosity versus corrosion time. The fact that total intrusion volume decreased shows that the corrosion was not simply dissolving the calcium carbonate crystals. Cementation ability was also lost, causing pores to widen during corrosion, but a new substance was produced and filled the pores. The new substance had greater volume and less density than that of calcium carbonate crystals of the same quantity. Therefore, poverty decreased rather than increased.

5.3.5 Change in Pore Characteristics of Acid-Corroded Specimens

Taking 2 g/L sulfuric acid solution as an example with prolonged corrosion in acid solution, the change in the pore characteristics is summarized based on the relationship between cumulative pore volume and mercury pressure (Figure 5.24). From Figure 5.24, it is seen that pore volume increases with mercury pressure, which goes through three phases. At the beginning, the cumulative intrusion volume increases linearly as mercury pressure develops. Cumulative intrusion volume gently increases as mercury pressure develops as well. Finally, the pore volume stays constant when the pores are filled with liquid mercury. The steeper the gradient in the linear increase stage, the smaller the diversity of pore size. The gradients in the linear increase stage

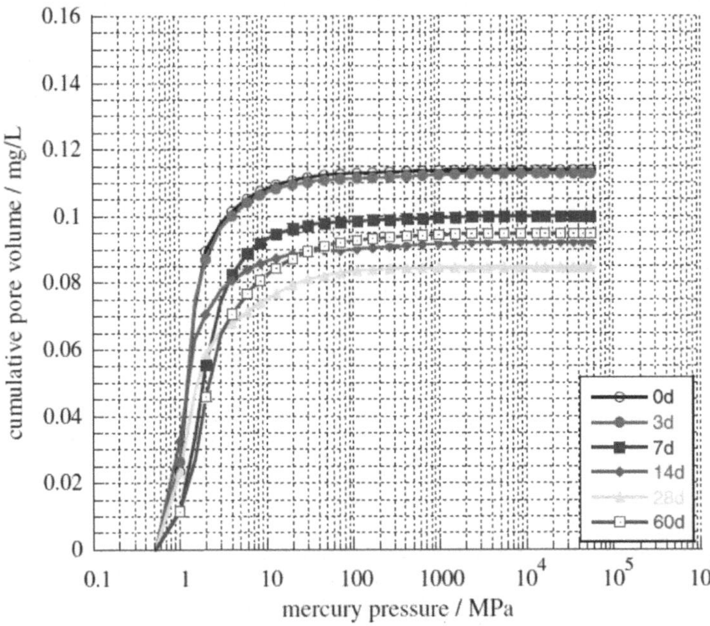

FIGURE 5.24 Relationship between cumulative intrusion volume and mercury pressure for various corrosion times.

(mercury pressure 0–3 psi) are summarized in Table 5.3. As stated in the previous discussion, uniformity in the bio-geomaterial produced using MICP technology was relatively good. The gradient in the linear increase stage of the uncorroded specimens was 0.047. When acid corrosion started, the gradient value decreased, and pore size diversity increased. With the deepening of the degree of acid corrosion, pore size diversity increased, and uneven corrosion appeared in the specimens. With the corrosion time extended from 7 to 14 days, an acid-corrosion reaction occurred in the complete sample. Pore size tended to be more centralized in the whole specimen, and diversity decreased. After 14 days, the acid-corrosion reaction continued, and along with the dissolution of calcium sulfate dihydrate crystals, pore size diversity increased.

Figure 5.25 shows the relationship between log-differential intrusion (dV=d $logD$) and pore size at various corrosion times. The distribution of pore size is different at different corrosion times, but principal pore size shows little change (approximately 0.12 mm) (Table 5.3). When corrosion time extended to 7 days, some calcium sulfate dihydrate crystals were produced and filled the pores from the specimen's surface area to its core. Porosity and principal pore size decreased. When corrosion time reached 60 days, the filled pores were released again. and porosity increased, owing to the dissolution of the calcium sulfate crystals. However, calcium sulfate dihydrate crystals are slightly soluble, so their dissolution was progressive. The principal pore size decreased to 0.09 mm.

Based on fractal theory, the pore characteristics of the specimens at various corrosion times were analyzed and fitted and are shown in Figure 5.27. The fractal dimension Db was obtained, and its values are provided in Table 5.3 for different

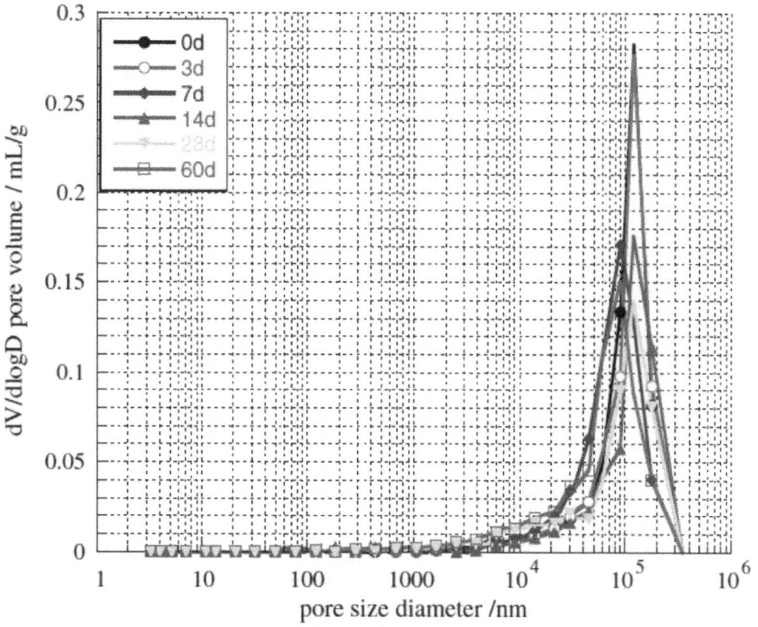

FIGURE 5.25 Relationship between log-differential intrusion (dV = dD) and pore size.

corrosion times. From Table 5.3 and Figure 5.26, it is seen that the value of Db ranged from 2 to 3. The distribution of pores for different corrosion times had self-similarity and followed fractal law. With increasing corrosion time, D_b presented a descending trend, showing that the self-similarity of pore distribution improved, the complexity of internal pores decreased, and the uniformity of specimens increased.

Changes in porosity and pore characteristics were noticeably different in different corrosion stages, but the fractal dimension D_b changed negligibly. The influence of acid corrosion did not cause any fundamental change in the structure of the specimens. The specimen skeletons were not disturbed significantly with the extension of corrosion time.

5.4 EFFECT OF MICP ON AEOLIAN SAND

5.4.1 Experimental Materials and Methods

5.4.1.1 Materials

5.4.1.1.1 Aeolian Sand

Several samples of Aeolian sand were collected from the Kubuqi Desert, which is located on the north of the Ordos plateau in the Inner Mongolia Autonomous Region, China. Tests were conducted on these samples to determine index properties, soil classification, and unit weight. A summary of the properties of the Aeolian sand is shown in Table 5.4. Particle size distribution curves for the Aeolian sand are shown in Figure 5.27. The proportion of fine particles which were smaller than 0.25 mm

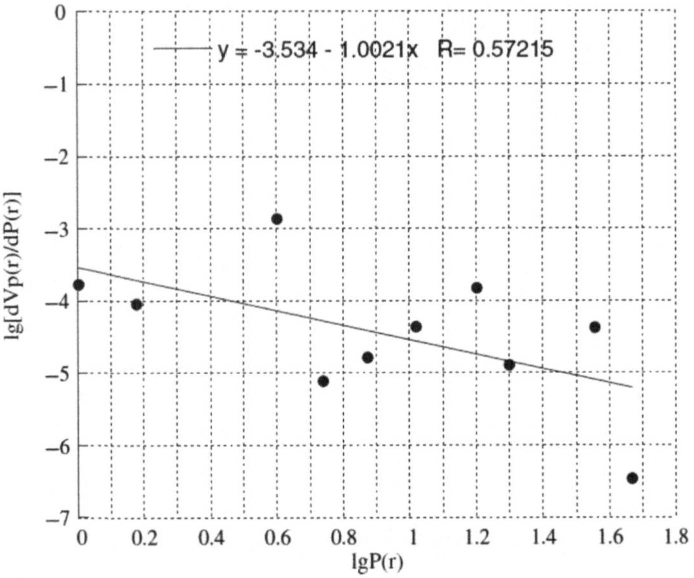

FIGURE 5.26 Beeline fitting method to obtain D_b ($K = D_b - 4$, where K = gradient of fitting curve).

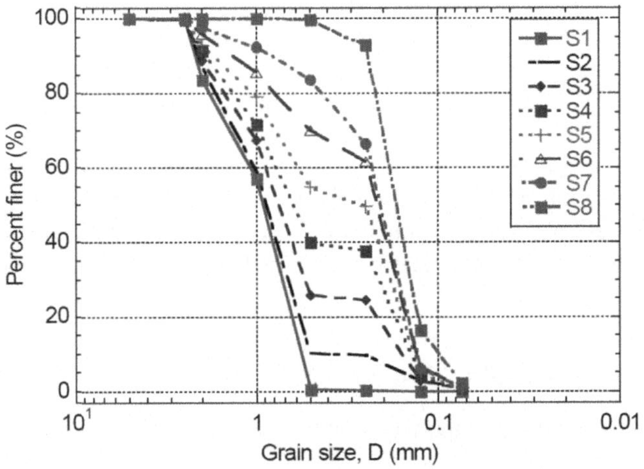

FIGURE 5.27 Particle size distribution curves: (S1) sample of standard sand, (S8) sample of Aeolian sand, and (S2-S7) samples of mixtures of standard and Aeolian sand.

TABLE 5.4
Particle Characteristics and Grade of Soil

Sample No.	W_{SS}(g)	w_{aero} (g)	Weight Ratio	Unit Weight (kN.′m³)	D_{50} (mm)	c_u	C_c	Volume of Solid v_s (cm³)	Void Space $V_{v'} = V\text{-}v_s$ (cm³)	Porosity V_v/V (%)	USCS Classification	Particle Grade
SI	155.0	–	–	15.50	0.90	1.96	0.91	59.16	39.01	39.76	SP	Poorly graded
S2	139.7	15.5	9/1	15.52	0.87	2.38	1.04	59.19	38.98	39.73	SP	Poorly graded
S3	116.9	38.9	3/1	15.58	0.74	5.80	2.07	59.35	38.82	39.56	SW	Well graded
S4	94.0	62.4	1.50/1	15.64	0.61	4.67	2.03	59.51	38.66	39.40	SW	Poorly graded
S5	70.7	86.2	0.82/1	15.69	0.25	4.14	2.25	59.64	38.54	39.27	SW	Poorly graded
S6	47.3	110.0	0.43/1	15.73	0.22	1.84	0.93	59.72	38.45	39.19	SP	Poorly graded
S7	23.2	134.7	0.17/1	15.79	0.20	1.77	0.86	59.88	38.30	39.03	SP	Poorly graded
S8	–	158.6	–	15.85	0.17	1.80	1.09	60.08	38.10	38.83	SP	Poorly graded

Note: W_{ss} is dry weight of standard sand. W_{aero} is dry weight of Aeolian sand. *Weight ratio* is the ratio of dry weight of standard sand (W_{ss}) and dry weight of Aeolian sand (W_{aero}); for mixed sand samples (S2~S7), the ratio is taken into account. D_{50} is the median particle size, that is, 50% of the sample is finer than this size. C_u is the coefficient of uniformity, and $C_u = D_{60}/D_{10}$. C_c is the coefficient of curvature, and $C_c = (D_{30})^2/(D_{10} \times D_{60})$. D_{60} and D_{10} define that 60 and 10% of the sample are finer than this size. V is the total volume of specimens (i.e., volume of experimental mold; a cylindrical mold with 50 mm diameter and 50 mm height was used in this study). The total volume (V) of each specimen is 98.17 cm³. V_s is the volume of solid particles; it can be estimated from $_{Vs} = (W_s)/(G_s \times \gamma_w)$, where W_s is the weight of dry soil, γ_w is the unit weight of water and equal to 1.0. G_s is specific gravity (G_s), measured as 2.64 for Aeolian sand and 2.62 for standard sand.

accounted for 93%. The median particle size was 0.17 mm. The sand was uniformly graded, with a coefficient of uniformity (C_u) of 1.8 and a coefficient of curvature (C_c) of 1.09. The unit weight of Aeolian sand was 15.85 kN/m^3, and its specific gravity was 2.64. The soil was classified as poorly graded sand (SP) according to the Unified Soil Classification System (USCS).

5.4.1.1.2 Standard Sand

Standard silica sand (99.7% quartz) was used in the experiments. The unit weight of the standard sand was 15.5 kN/m^3, and its specific gravity was 2.62. The sand was uniformly graded, with a median particle size of 0.9 mm. Fine particles which were smaller than 0.25 mm in diameter were excluded in the experiments. The coefficient of uniformity (C_u) of the standard silica sand was 1.96, and its coefficient of curvature (C_c) was 0.91. It was classified as poorly graded sand according to the USCS.

5.4.1.1.3 Mixed Sand

Mixtures of Aeolian sand (*As*) and standard sand (*ss*) are prepared according to weight ratios in Table 5.4. Before mixing, the standard sand and Aeolian sand were soaked in 0.1 M NaOH solution for 12 h and then washed and soaked in 0.1 M HCl solution for 12 h. The two sands were soaked in deionized water until the pH of each reached 7. After drying, samples of standard sand (sample 1), Aeolian sand (sample 8), and mixed sand samples (samples 2–7) were prepared. All the sand mixtures were classified according to USCS. The particle grade was evaluated. The volume of solids, void space, and porosity of each mixture were calculated (Table 5.4).

5.4.1.2 MICP Sample Preparation

The specimens were prepared in full-contact flexible molds (FCFMs) made of geotextile, as described in Section 3.1.3.1, and MICP sample preparation for details, as described in Section 3.1. Table 5.5 shows the mechanical and hydraulic properties of the geotextile.

TABLE 5.5
Properties of the Full-Contact Flexible Mold (FCFM)

	Mechanical Parameters					Hydraulics Parameters		Geometric Parameters	
Items	Grab Tensile Strength	Grab Elongation	Trapezoidal Tear Strength	CBR Burst Strength	Apparent Opening Size	Dielectric Constant	Water Flow Rate	Thickness	Unit Mass
ASTM testing method	D-4632	D-4632	D-4533	D-6241	D-475I	D-4491	D-4491	D-5199	D-5199
Average	172.4 kg	50%	68 kg	464.9 kg	0.15 mm	0.7 sec^1	34 mm/s	1.5 mm	200 g/m^2

5.4.2 The Effect of Optimal Reaction Time

MICP is a complex bacterial and biochemical transient process. Reaction time is an important factor in the MICP process. Standard sand MICP-treated samples after 1 day, 2 days, 3 days, 4 days, 5 days, 6 days, 7 days, and 8 days were prepared at a bacterial concentration (OD_{600}) of 0.6 and calcium source solution concentration of 0.5 M Ca. Zhao et al. (2014b) showed that standard sand samples were subjected to MICP treatment for 1, 2, 3, 4, 5, 6, 7, and 8 days at a bacterial concentration of 0.6 (OD_{600}) and a calcium source solution concentration of 0.5 M Ca, because urea and calcium ions were fully utilized at a calcium source solution concentration of 0.5 M Ca. After 1 or 2 days of reaction, the MICP samples were not strong enough for removal from the molds, but after 3 or more days of reaction, they could be tested in unconfined compression tests. As shown in Figure 5.28, the unconfined compression strength (UCS) of the MICP-treated specimens increased with an increase in reaction time. However, the rate of increase slowed after more than 5 days. The growth rates from days 3 to 5 were 0.61 MPa/d; from days 5 to 7, they were 0.26 MPa/d, and after day 7, they were 0.05 MPa/d. Thus, the most rapid growth of UCS occurred between days 3 and 5. Zhao et al. (2014b) explained that the bacteria were enclosed by crystalline calcite and that the enzyme activity producing the calcite had slowed after 5 days. The precipitated $CaCO_3$ reduced the pore volume in the samples and even blocked the pores, lowering the effective contact between the calcium source solution and the urease, thereby lowering the MICP bonding efficiency. The differences between UCS values in this study and those in the work of Zhao et al. (2014b) can be explained using different types of standard sand and the corresponding D_{50} values. In this study, the D_{50} value was 0.9 mm, whereas it was 0.46 mm in the study by Zhao et al. (2014b).

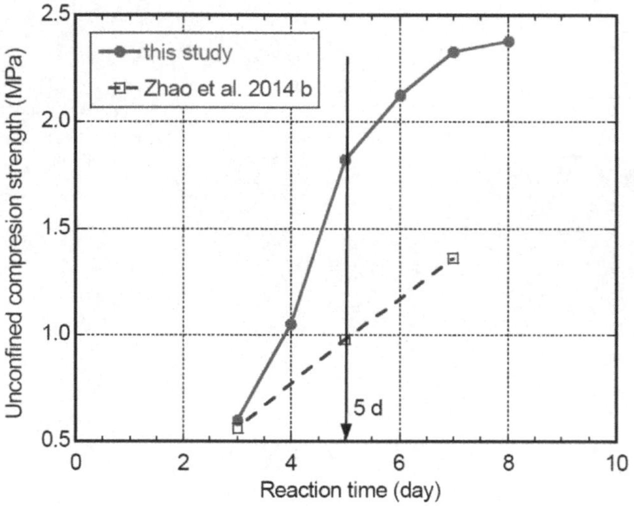

FIGURE 5.28 Unconfined compression strength of MICP-treated samples as a function of reaction time.

5.4.3 $CaCO_3$ Precipitation and Porosity Distribution in MICP-Treated Specimens

The SEM images of untreated Aeolian sand and of MICP-treated soil are shown in Figure 5.29a and b. The $CaCO_3$ crystals were mainly irregular and bulky, like the observations made by Al Qabany et al. (2012) and Zhao et al. (2014b). MICP binding was generated between the particles, which enhanced the connections within the loose sand. Figure 5.29c shows precipitated calcium carbonate induced by microorganism, and Figure 5.29d shows precipitated calcium carbonate adhering to the cell. In Figure 5.29c and d, no soil particles are noticeable. Precipitated calcites can be an adhesive material additive to sandy soil. As a result, the strength of MICP-treated sand was improved (Burbank et al. 2013; DeJong et al. 2010). The reduction in void space due to the $CaCO_3$ precipitation can be a secondary strengthening factor (DeJong et al. 2010). The notion that a negatively charged bacterial cell adsorbs calcium ions at its surface and thereby provides a nucleation site for $CaCO_3$ crystallization is also supported (Figure 5.29d).

To evaluate the uniformity of the MICP process in the sand specimen, the porosity in the core, the side, and the top and bottom of MICP-treated specimens was measured. The four areas corresponded to the top (20 mm thick), bottom (20 mm thick), core (20 mm in diameter), and side (18 mm thick). Table 5.6 shows the

FIGURE 5.29 SEM images: (a) untreated Aeolian sand, (b) MICP binding between Aeolian sand particles, (c) precipitated calcium carbonate induced by microorganism, and (d) precipitated calcium carbonate adhering to a bacterial cell.

TABLE 5.6
Measured Porosity of MICP-Treated Standard Sand

OD_{600}	Measuring Positions	Porosity (%)	Average Porosity (%)	Standard Deviation (%)
0.5	The average of top and bottom	17.88	20.14	3.29
	The side	17.75		
	The core	24.78		
0.6	The average of top and bottom	19.87	22.51	1.99
	The side	23.02		
	The core	24.65		

Note: Autopore IV 9510 testing system was used to measure the porosity of the MICP-treated standard sand (S1).

porosity of four areas of MICP-treated specimens at different bacterial concentrations (OD_{600} values). Porosity in the core was 24.8%, and porosity in the top/bottom/side was around 17.8% for the MICP-treated standard sands at OD_{600} of 0.5. A higher porosity is attributable to a decrease in $CaCO_3$ precipitation in the soil specimens. There was more $CaCO_3$ precipitated in the top, bottom, and side than in the core, which illustrates the non-uniformity of the MICP-treated sample. The average porosity was 20.1% for the MICP-treated specimens at a bacterial concentration of OD_{600} of 0.5, which was lower than the porosity of 22.5% for the MICP-treated specimens at a bacterial concentration OD_{600} of 0.6. However, the porosity of the core exhibited no obvious change with an increase in bacterial concentration at OD_{600} values.

5.4.4 UCS at Different Cell Concentrations and Porosities of Sand Specimens

MICP-treated specimens of standard sand and mixed sand at a calcium source solution concentration of 0.5 M were prepared at cell concentrations corresponding to OD_{600} of 0.3, 0.4, 0.5, 0.6, 0.7, and 0.8. The reaction time for all specimens was 7 days. MCIP-treated specimens formed at a bacterial concentration corresponding to OD_{600} of 0.3 were not strong enough to be removed from the mold. Although the MICP-treated specimens can reach a uniform state in the full-contact flexible mold, there was still a slightly different distribution of $CaCO_3$ within each sample (Zhao et al. 2014b), which may cause UCS to fluctuate during the test. At least three samples were prepared for each test condition to produce a representative result. At OD_{600} of 0.4 to 0.8, the MICP-treated soil specimens were strong enough to be removed from the batch reactor and to be tested in unconfined compression tests.

Table 5.7 shows the unconfined compression strength (UCS) as a function of OD_{600} and the type of soil. The stress–strain curves obtained from unconfined compression tests for sample 3 at the different OD_{600} values are shown in Figure 5.30. Figure 5.31 shows the effect of different soil porosities and cell concentrations on the UCS of MICP-treated soil specimens. It is seen that the UCS of MICP-treated soil specimens

TABLE 5.7
Unconfined Compression Strength (UCS) Test Results in MPa

Sample No.	Porosity (%)	OD_{600} = 0.4	OD_{600} = 0.5	OD_{600} = 0.6	OD_{600} = 0.7	OD_{600} = 0.8	Avg. UCS
S1 (standard sand only)	39.76	1.14	1.74	1.13	1.16	1.24	1.32
S2	39.73	1.27	2.61	1.90	2.34	2.01	2.22
S3	39.56	1.42	2.69	2.26	2.59	2.63	2.54
S4	39.40	1.03	2.06	1.48	1.96	2.02	1.88
S5	39.27	0.74	1.23	1.20	1.18	1.09	1.18
S6	39.19	0.65	1.35	1.10	1.22	1.04	1.18
S7	39.03	0.32	0.66	0.66	0.65	0.67	0.66
S8 (Aeolian sand only)	38.83	0.29	0.78	0.64	0.59	0.62	0.66

Note: Reaction time was 7 d. *Avg. UCS* is average UCS between OD_{600} of 0.5 and 0.8.

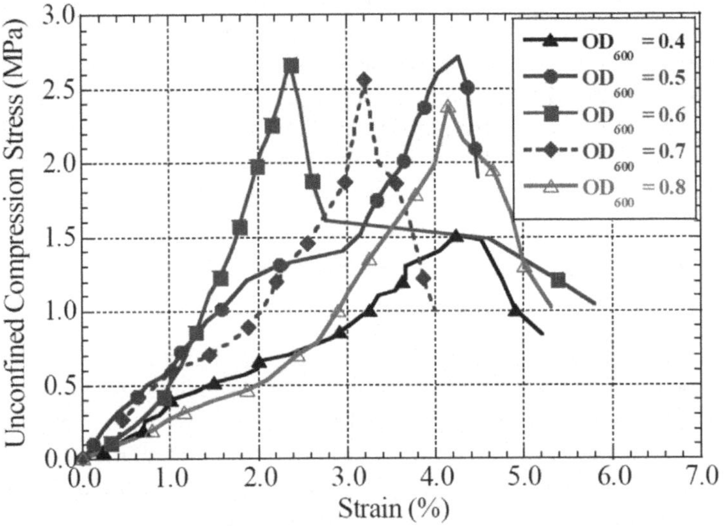

FIGURE 5.30 Unconfined compression stress of MICP-treated soil sample S3 at different bacterial cell concentrations.

increased with bacterial concentration (OD_{600} value). The testing results are consistent with previous findings at OD_{600} values of 0.3–0.8 (DeJong et al. 2010; van Paassen et al. 2010a; Zhao et al. 2014b). But the UCS of MICP-treated soil specimen did not increase linearly with OD_{600}. In this study, the UCS of MICP-treated specimen reached a maximum at OD_{600} of 0.5. The bacterial cells had an optimal effect on the MICP at an OD_{600} of 0.5. A greater concentration of bacteria would result in

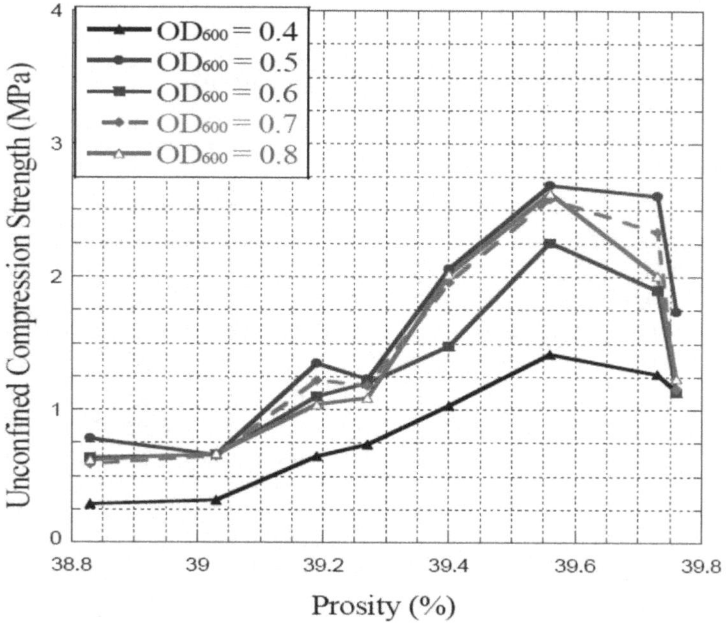

FIGURE 5.31 Unconfined compression strength of MICP-treated samples as a function of porosity and different bacterial concentrations.

greater enzyme activity. It would provide more nucleation sites for MICP, but only if sufficient pore space was available. The cell surface is assumed to be saturated with adsorbed Ca ions. Thus, at the same bacterial concentration (OD_{600}), soil porosity plays a key role in MICP. With an increase of soil porosity, the MICP process can be adequately developed. The precipitated calcite induced by urea hydrolysis will fill the pore spaces of loose sands. The MICP-treated specimens had higher UCS in soil, with greater porosity. An interesting observation was that the UCS of MICP-treated soil specimens was higher with well-graded soil than with poorly graded soil. With poorly graded sample 1 and well-graded sample 3, the difference in their average UCS was 1.22 MPa, while the difference in porosity was only 0.2%. So the compatibility between microbes and soil porosity with grain grading should be considered.

5.4.5 TSS at Different Cell Concentrations and Porosities of Sand Specimens

MICP-treated specimens of standard sand and mixed sand at cell concentrations corresponding to OD_{600} of 0.5, 0.6, 0.7, and 0.8 and at a Ca concentration of 0.5 M were prepared. At least three soil samples were prepared in triaxial tests at three different confining pressures, 800, 900, and 1,000 kPa. The deviator stresses were recorded upon completion of a test. Table 5.8 summarizes the cohesion and inner friction angle of the MICP-treated soils at different OD_{600} values. Figure 5.32 shows the inner friction angle of the MICP-treated specimen as a function of porosity at different

TABLE 5.8
Triaxial Shear Compression Strength (TSS) Test Results

	Porosity	OD$_{600}$ = 0–5		OD$_{600}$ = 0–6		OD$_{600}$ = 0–7		OD$_{600}$ = 0–8		Avg.	Avg.
Sample No.	(%)	c	0	c	0	c	0	c	0	c	θ
SI (standard sand only)	39.76	0.39	38.2°	0.56	37.3°	0.79	37.7°	0.66	38.1°	0.60	37.8°
S2	39.73	1.01	40.8°	0.93	39.1°	0.40	39.3°	0.98	39.9°	0.83	39.8°
S3	39.56	0.50	44.6°	1.01	43.2°	1.00	43.8°	1.02	44.2°	0.88	44.0°
S4	39.40	0.44	40.5°	0.48	39.8°	0.50	40.2°	0.68	40.3°	0.53	40.2°
S5	39.27	1.03	39.1°	0.80	37.9°	0.62	38.2°	0.34	38.7°	0.70	38.5°
S6	39.19	0.60	38.6°	0.56	37.1°	0.79	37.9°	0.48	38.3°	0.61	38.0°
S7	39.03	0.79	37.2°	0.39	35.9°	0.56	36.2°	0.57	36.8°	0.58	36.5°
S8 (Aeolian sand only)	38.83	0.26	36.7°	0.49	35.2°	0.47	35.8°	0.21	36.1°	0.36	36.0°

Note: Reaction time was 7 days, *c* is cohesion (MPa), *θ* is inner friction angle (°), *Avg. c* is averaged cohesion between OD$_{600}$ of 0.5 to 0.8, and *Avg. θ* is averaged inner friction angle between OD$_{600}$ of 0.5 to 0.8.

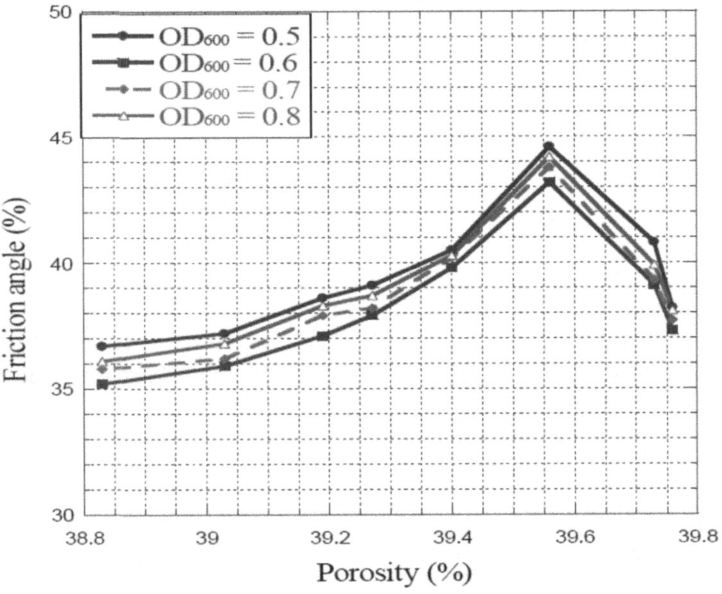

FIGURE 5.32 Friction angle of MICP-treated samples as a function of porosity and different bacterial concentrations.

bacterial concentrations. The inner friction angle of the MICP-treated specimen increased with porosity of the soil until the porosity reached 39.56%. After that, the inner friction angle of the MICP-treated specimen decreased. The porosity of the soil had limited impact on the cohesion of the MICP-treated specimen (Table 5.8).

The precipitated $CaCO_3$ induced by urea hydrolysis filled the pore spaces of loose sand. The sand was reinforced by the interlocking binding among the particles. The inner friction angle of the MICP-treated specimen reached the maximum at OD_{600} of 0.5. The bacterial cells had an optimal effect on the MICP process, with the interlocking binding being the strongest at OD_{600} of 0.5. The inner friction angle was higher for the well-graded soil than the poorly graded one. For example, the difference of the inner friction angle was 17%, while the difference in porosity was only 0.2% between the poorly graded sample 1 and the well-graded sample 3 at an OD_{600} of 0.5.

5.5 EFFECT OF MICP-TREATED SANDY SOILS AGAINST ENVIRONMENTAL DETERIORATION

5.5.1 Experimental Methods

The specimens were prepared in full-contact flexible molds (FCFMs) made of geotextile, as described in Section 3.1.3, and MICP sample preparation for details, as described in Section 3.1.

5.5.1.1 Wet–Dry Durability

There is no standard testing method to perform wet–dry (WD) cycles for MICP-treated samples. Consequently, ASTM D 559-15 was adopted for WD cycles of MICP-treated samples. These samples were subjected from one to five WD cycles. In each WD cycle, the samples were submerged in a tank with distilled water for 24 h and then were oven-dried for 48 h to permit the dissipation of excess moisture. The samples were then tested for their unconfined compression strength and compared with un-weathered samples to determine the impact of WD cycles on the MICP-treated samples.

5.5.1.2 Freeze–Thaw Durability

There is no standard testing method to perform freeze–thaw (FT) cycles for MICP-treated samples. Consequently, ASTM D 560-16 was adopted for freeze–thaw cycles of MICP-treated samples. Each procedure of FT cycle was subjected to water-saturated samples for 24 h frozen at -18°C, followed by 24 h thaw at 23°C. Five, ten, and fifteen FT cycles were operated to simulate the effect of temperature changes on the performance of the MICP-treated samples. The effect of FT cycles on the MICP-treated samples was evaluated by comparing the UCS values before and after the FT cycles.

5.5.1.3 Acid Rain Durability

Accelerated test was applied in acid rain simulation experiments. Submerging and spraying are two major methods to accelerate the deteriorations caused by acid rain in the laboratory. Based on the study by Xie et al. (2004), the submerging method was selected to accelerate MICP-treated samples deterioration caused by acid rain in this study. Eyssautier-Chuine et al. (2016) applied H_2SO_4 solution and HNO_3 solution to simulate acid rain on natural and artificial carbonate stones to assess the effect of acid rain on carbonate stones. They found that the H_2SO_4 solution was generally more aggressive than the HNO_3 solution when they were applied on both natural and artificial carbonate stones. Higher Ca losses were caused by the H_2SO_4 solution compared with the HNO_3 solution. To get relatively regular-shaped samples for mechanical testing, the relatively mild HNO_3 solution was selected to simulate acid rain in this study. The acid rain was prepared by diluting nitric acid (HNO_3) with distilled water. A corrosion-resisting rectangular tank, shown in Figure 5.33, was used to contain the acid rain solution and MICP-treated samples. After 7 days' MICP treatment and 48 h oven-dried, the MICP-treated samples were divided into four scenarios, denoted by A1, A2, A3, and A4, as shown in Table 5.9. A1 was used as a controlled scenario, which was maintained in the pH of 7.0. A2 through A4 were maintained in acid solutions with pH values of 5.5, 4.5, and 3.5, respectively.

Since the pH value of the solutions was changing because of the protons (H^+) being consumed by reacting with $CaCO_3$, dilute acid was added to the solution periodically and mixed well with the solution to maintain a constant pH. After being submerged in the solutions, three MICP-treated samples as one batch were taken out for testing at different immersion times, as shown in Table 5.9. The immersion time of 0 days shown in Table 5.9 means these samples were tested without an acid rain solution, which acted as controlled samples. The MICP-treated samples were first oven-dried for 48 h, followed by weighting, unconfined compression strength testing, $CaCO_3$ content testing, and scanning electron microscopy (SEM) analysis to determine the mass loss, unconfined compression strength, and $CaCO_3$ crystal distribution around the sand particle, respectively.

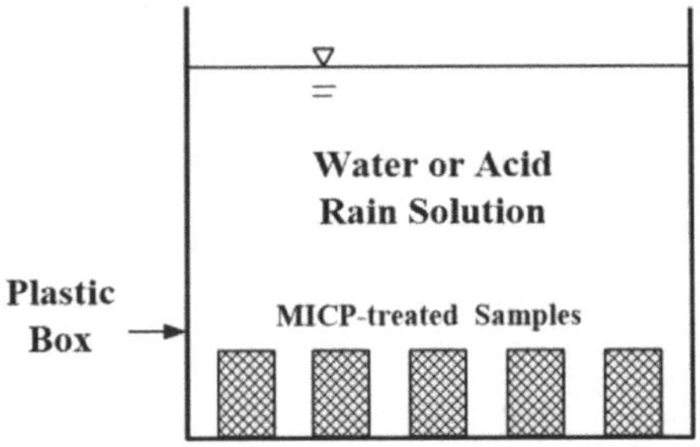

FIGURE 5.33 Sketch of tank for wet–dry cycle or acid rain simulation.

TABLE 5.9
Testing Details of Acid Rain Simulation

Designation	Exposing Condition	Solution Acidity	Sample Grouping	Immersion Time (Days)
A1	Pure water	pH 7.0	Three samples in a group, four groups	0, 5, 10, 15
A2	HNO_3	pH 5.5	Three samples in a group, four groups	0, 5, 10, 15
A3	HNO_3	pH 4.5	Three samples in a group, four groups	0, 5, 10, 15
A4	HNO_3	pH 3.5	Three samples in a group, four groups	0, 5, 10, 15

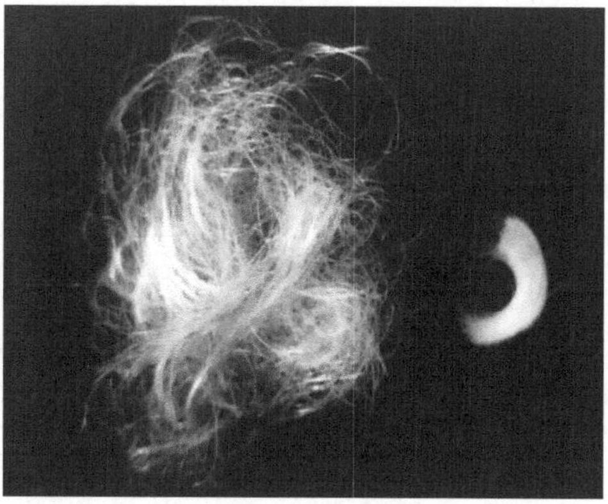

FIGURE 5.34 Photograph of synthetic fibers used in this study.

5.5.1.4 Fiber Reinforcement

Fibermesh 150e3 (from FIBERMESH), shown in Figure 5.34, was used in this study to improve the ability for MICP-treated sand to resist environmental deteriorations. It is a 100% uniform homopolymer polypropylene multifilament fiber with a specific gravity (Gs) of 0.91. It is chemically inert, with high acid salt resistance. The length and thickness of the fibers used in this study are 12 mm and 0.1 mm, respectively, with an aspect ratio of 120 between the length and the thickness of the fiber. Consoli et al. (2009b) used similar fibers of different lengths for the reinforcement of sand. It was concluded that fibers with an aspect ratio above 300 experienced strain-hardening behavior, which caused significant mechanical problems, such as fracture failures, during sharing of the soil and a significant decrease in the ductility behavior of the fiber. The aspect ratio of the fiber used in this study was lower than the upper limit determined in the previous studies, indicating that it should provide efficient reinforcing performance (Consoli et al. 2009b).

The optimum fiber content in the MICP-treated sand was found to be 0.2–0.3% (Li et al. 2015c). A 0.3% fiber content was applied in this study. The 140 ± 5 g Ottawa silica sand and fibers (0.3% by weight of dry sand) were uniformly mixed with 45 mL bacteria solution and then air-pluviated into the FCFM to reach a median dense condition (D_r in the range of 44–55%). In the preparation of all samples, the required bacteria solution was first mixed with the dry sand to prevent floating of the fibers in the soil matrix, and then the proposed content of fibers was mixed in small increments by hand to obtain a uniform mixture. It is important to ensure that all fibers are mixed thoroughly (Shao et al. 2014). Then the samples were put on the shelf and immersed in cementation media in the tank reactor. After 7 days of MICP treatment, the shelf was removed from the solution, each geotextile mold was cut, and the MICP-treated samples were removed.

5.5.2 Effect of Wet–Dry Cycles

The WD cycle process was accompanied by other physical changes: manual perturbation or increased temperature (Cosentino et al. 2006). The samples were carefully manipulated in such a way that no structural degradation and minimal temperature change occurred. The impact of WD cycles on MICP-treated samples was shown in the unconfined compression stress–strain curve (Figure 5.35a). The UCS of MICP-treated samples decreased with the WD cycles. The peak strength of un-weathered samples was about 1.2 MPa, which dropped to 0.25 MPa after one WD cycle. The drop was nearly 80%, but the failure strain did not change too much (around 1.0%). The mass loss rate had a leaping increase after the first WD cycle, as shown in Figure 5.35d. Water erosion can be accelerated by alternating between wetting and drying cycles (Xie et al. 2004). As water penetrates and inundates the material, the powdery bonds, unbonded minerals or loosely bonded calcite, acting as cementing agent, rapidly fall into suspension. The decrease of UCS was caused by this phenomenon in the first WD cycle period. A sketch representing such a mechanism is reported in Figure 5.35. When the soil sample was treated for MICP, some $CaCO_3$ precipitation was only attached to the surface of sand particles that may not help bond the sand particles, thus getting easily eroded during WD cycles. A lower percentage of total $CaCO_3$ crystals has been found precipitated in the gaps and acted as a bridge to bond sand particles (Cheng et al. 2012; Zhao et al. 2014a). However, when the second WD cycle was applied to the MICP-treated samples, the peak strength bounced back to 0.45 MPa, but the strain at the peak strength decreased from 1.0% to 0.55%. The bounced-back peak strength in the second

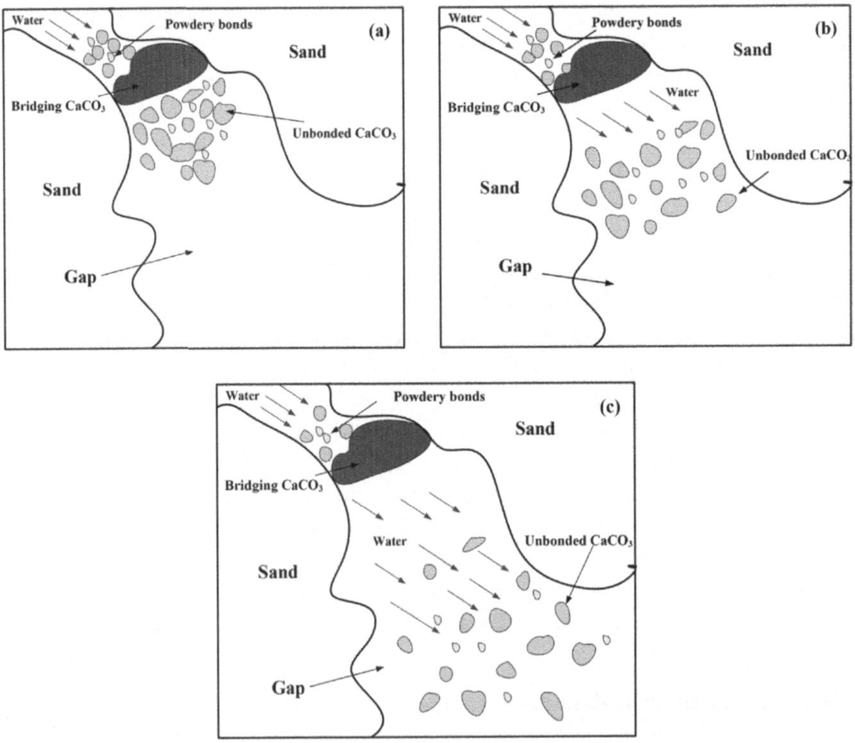

FIGURE 5.35 (a–c) Schematic representation of water erosion mechanism in WD cycles.

WD cycle period was caused by the hardening calcite formed in the MICP-treated sample. The decreasing strain after the second WD cycle indicated the strain-hardening phenomena. The MICP-treated samples became more brittle after the second WD cycle. With more cycles of WD (three, four, or five cycles of WD), the peak strength of MICP-treated samples decreased and tended to reach stability after the four cycles of WD with relative stable peak strength (0.2 MPa, with failure strain of 0.4%). The service life of this material is short, and its durability against WD cycles must be improved.

5.5.3 Effect of Freeze–Thaw Cycles

The unconfined compression stress–strain curves of the MICP-treated sand after 0, 5, 10, 15 FT cycles are shown in Figure 5.35b. After 5 FT cycles, only an 8.3% decrease of UCS occurred in the MICP-treated samples, which was like the finding of Cheng et al. (2013). Cheng et al. (2013) suggested that the high porosity or hydraulic conductivity of MICP-treated sand allows more rapid water mass transfer in the sand matrix, which can increase the FT resistance. The hydraulic conductivity of single-MICP-treated samples was around 5×10^{-3} cm/s (Wen et al. 2018b). This high hydraulic conductivity increased the FT resistance of MICP-treated samples, which was observed at an early stage (e.g., 5 FT cycles in this study). However, the UCS of MICP-treated samples decreased with more FT cycles. The reduction of UCS for MICP-treated samples was 42% and 58% after 10 and 15 FT cycles, respectively. Microcracks were formed and grown in more FT cycles. These microcracks were generated due to the tensile stresses around the sand particles when the pore water turned into ice during the freezing process. The microcracks could be the main reason to reduce the UCS of MICP-treated samples at a later stage of the FT cycles. Cheng et al. (2017) also had similar findings.

5.5.4 Effect of Acid Rain Conditions

Figure 5.36 shows the mass loss rate and UCS of MICP-treated samples as a function of immersion days in acid rain at four different pH conditions. As shown in Figures 5.36a and b, the mass loss rate of MICP-treated samples increased and the

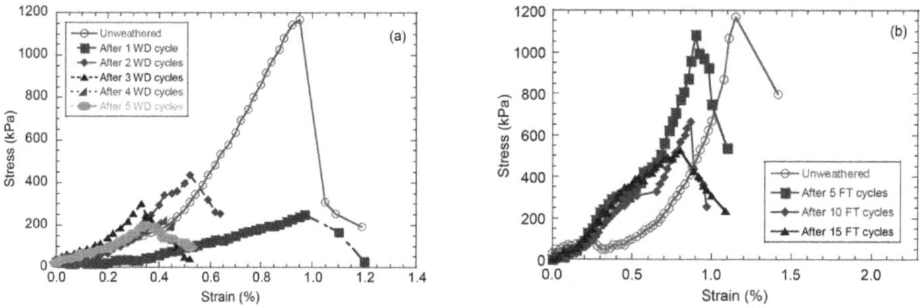

FIGURE 5.36 Unconfined compression stress–strain curve of MICP-treated samples (a) with effect of wet–dry cycles, and (b) with effect of freeze–thaw cycles.

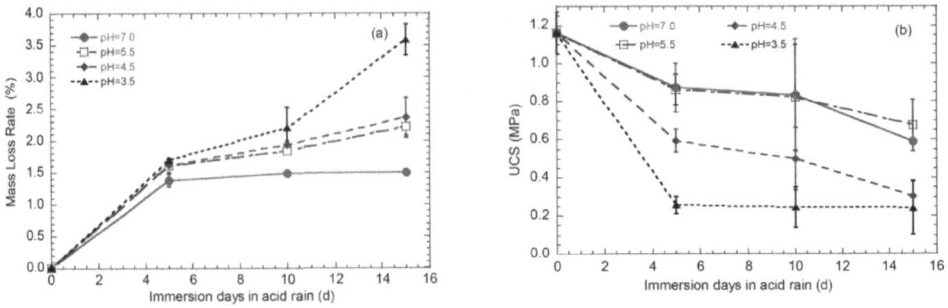

FIGURE 5.37 Effect of acid rain simulation on MICP-treated samples' (a) mass loss rate and (b) UCS as a function of immersion days in artificial acid rain.

UCS decreased with more immersion days in the acid rain solution. The chemical reaction between acid (nitric acid in this study) and calcium carbonate resulted in the dissolution of $CaCO_3$ that originally precipitated in the MICP-treated samples. Experimental data showed that the final pH value of the mixed solution was maintained around 7.2. It indicated that the protons (H^+) in the acid solution were consumed by reacting with $CaCO_3$. SEM image showed that almost no bridging $CaCO_3$ crystal left after 15 immersion days in an acid rain solution of pH 3.5, compared to these immersed in water of pH 7.0 (Figure 5.37). The strength of MICP-treated samples is related to the bridging of $CaCO_3$ crystal that bond sand particles together. Therefore, the dissolution of $CaCO_3$ led to the destruction of the connection between sand particles. More $CaCO_3$ crystals were eroded, and more sand particles were peeled off during longer immersion days in the acid rain solution. For severe acid conditions with a lower pH value, the MICP-treated samples lost more mass and strength. The UCS of MICP-treated samples decreased by about 83%, with 3.5% mass loss rate after 15 immersion days in acid rain solution of pH 3.5, as shown in Figure 5.37. Therefore, the durability of MICP-treated sand against acid rain conditions was relatively poor.

5.5.5 Enhancement of Durability for MICP-Treated Soil

Two methods, fiber reinforcement and multiple treatments, were applied in this study to improve the ability of MICP-treated sand to resist environmental deteriorations.

5.5.5.1 Fiber Reinforcement

In fiber reinforcement, fiber content was fixed as 0.3% (by weight of dry sand), which was the optimum fiber content from Li et al. (2015c). Figure 5.38a shows the SEM image of MICP-treated sample reinforced with 0.3% fiber. The fiber acted as a bridge and tension members between soil particles and calcium carbonate that improved the bonding between soil particles. When local cracks appeared on the MICP-treated sample with fiber reinforcement, fibers across the cracks took the tension within the soil due to the fiber–soil friction, which effectively impeded further development of cracks and thus improved the resistance of soil to the force applied. Moreover, fiber reinforcement induced more calcium carbonate (Figure 5.38b), and because of

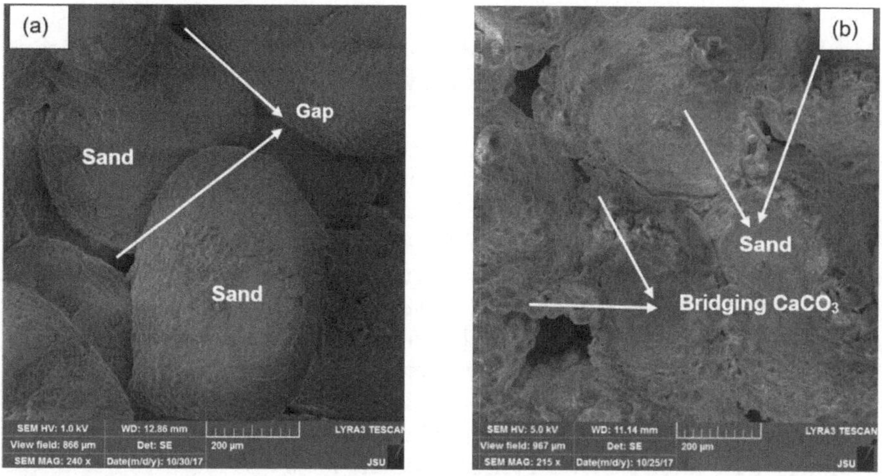

FIGURE 5.38 SEM images of MICP-treated sand after immersed in artificial acid rain simulation solution for 10 days: (a) pH = 3.5; (b) pH = 7.0.

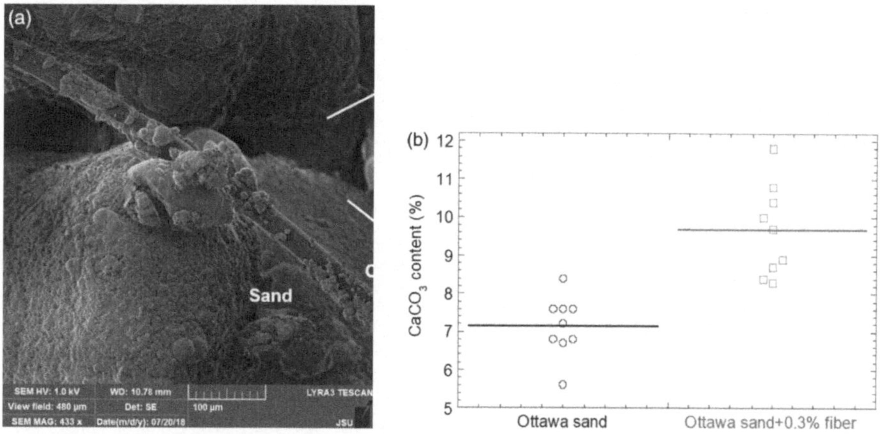

FIGURE 5.39 (a) SEM image of MICP-treated sand particles with fiber reinforcement. (b) $CaCO_3$ content of unreinforced and reinforced MICP-treated samples.

that, bacteria attach on fiber and thus form the mineral cores to attract more calcium carbonate crystals. The addition of fiber in the MICP process had the potential to increase the ductility of the MICP-treated soil and to prevent the loss of post-peak strength (Li et al. 2015c).

As shown in Figure 5.39a, the UCS of fiber-reinforced MICP-treated samples decreased to 50%, compared to an 83% decrease of unreinforced MICP-treated samples after five WD cycles. Figure 5.39b shows that fiber-reinforced MICP-treated samples had residual strengths of 0.1 MPa (at 15% strain) after five WD cycles.

The residual strength was more than half of their peak strength, while the residual strength was 0 kPa for unreinforced MICP-treated samples. The results indicated that fiber plays an important role in resisting WD cycles on the UCS of MICP-treated samples.

The effect of WD cycles on the failure strain of fiber-reinforced MICP-treated samples is shown in Figure 5.39c. The reinforced MICP-treated samples have much higher failure strain than that of unreinforced samples. The fiber-reinforced samples hold failure strain up to 1.6% after five WD cycles, compared to 0.4% of unreinforced samples. As 0.3% fiber was added, the failure strain of the fiber-reinforced MICP-treated samples increased 178%, 280%, 467%, 429% and 400% of that without fiber after one to five WD cycles, respectively. The results indicated that the ductility of MICP-treated soil can be significantly improved by adding a suitable number of fibers (0.3% in this study) to resist WD cycles.

In Figure 5.39d, the mass loss rate of fiber-reinforced samples remained stable at around 1.2% after 1~5 WD cycles, while the mass loss rate of unreinforced samples increased from 1.5% to 2.4% after 1~5 WD cycles. The fiber–soil friction and the addition bonding of $CaCO_3$ on fibers made soil particles difficult to change in positions, thus improving the bonding force between soil particles in WD cycles.

Figure 5.40 shows the effect of fiber reinforcement under FT cycles. Figure 5.40b shows that fiber-reinforced MICP-treated samples maintained a residual strength (0.2 MPa) of about 40% of their peak strength (0.5 MPa) after 15 FT cycles, while unreinforced MICP-treated sand did not have any residual strength. Fiber enhanced the failure strain against various FT cycles (212%, 150%, 178%)

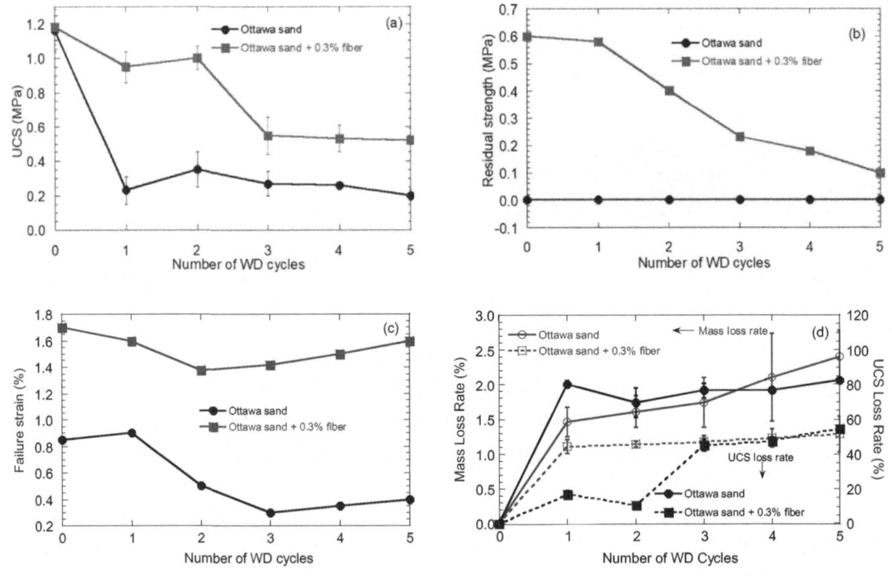

FIGURE 5.40 Effect of 0.3% fiber-reinforced samples under WD cycles: (a) peak strength of UCS; (b) residual strength; (c) failure strain; (d) mass loss rate and UCS loss rate.

compared to unreinforced samples. The results indicate that fiber improves the ductility of MICP-treated samples to resist FT cycles. But it can be seen from Figures 5.40a–d that fiber did not help resist UCS reduction during FT cycles. The fiber and 2.6% higher $CaCO_3$ content in the fiber-reinforced MICP-treated samples occupied sand pores, which would result in decreased permeability (as shown in Figure 5.38b). Therefore, fiber reinforcement has a negative effect on soil strength to resist FT cycles.

5.5.5.2 Multiple MICP Treatments

Multiple MICP treatments significantly improve the $CaCO_3$ content, especially in the first three treatments (Figure 5.41). These improvements enhanced the strength of MICP-treated samples. WD cycles, FT cycles, and acid rain conditions were applied on MICP-treated samples to examine the effect of multiple MICP treatments.

As shown in Figure 5.42a, the individual $CaCO_3$ crystals of single-MICP-treated samples had similar size and were separately distributed. However, for quadruple-MICP-treated samples, as shown in Figure 5.42b, the $CaCO_3$ shell coated the surface of the sand particles. The boundary between each MICP treatment was clearly observed. With more MICP treatments, $CaCO_3$ crystals filled in the gaps and acted as bridges among the sand particles that contribute to the strength of MICP-treated soil.

The UCS results of multiple-MICP-treated samples after one WD cycle are shown in Figure 5.43. The UCS of single-MICP-treated samples dropped from 1.16 MPa to 0.23 MPa (80% reduction) after one WD cycle. But for the double-MICP-treated

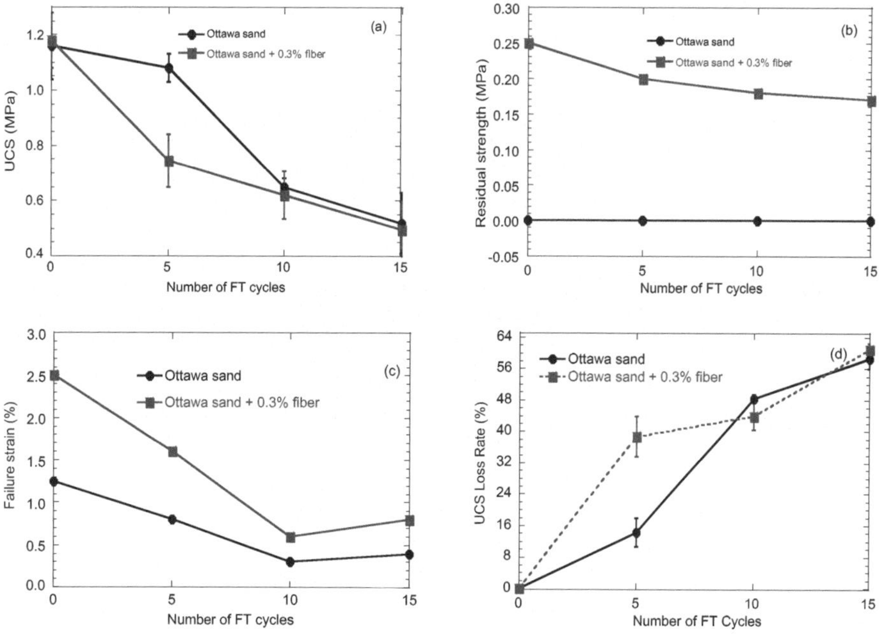

FIGURE 5.41 Effect of 0.3% fiber-reinforced samples under FT cycles: (a) peak strength of UCS; (b) residual strength; (c) failure strain; (d) UCS loss rate.

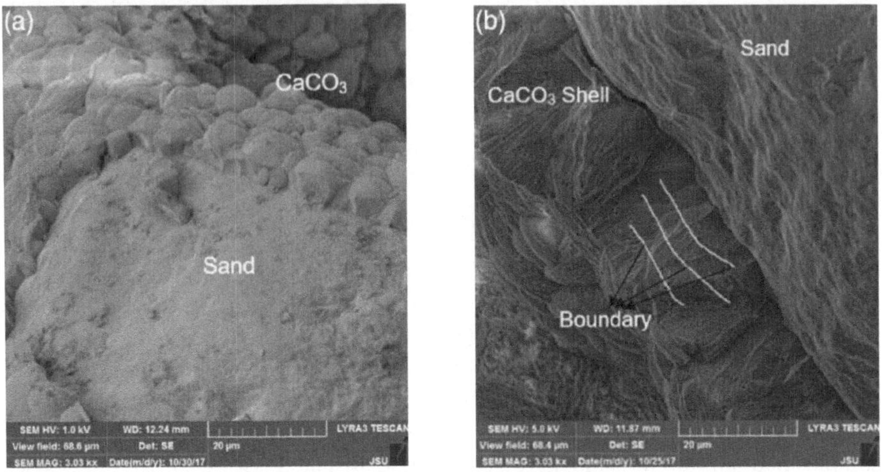

FIGURE 5.42 (a) $CaCO_3$ content of MICP-treated sand as a function of treatment times. (b) MICP-bonded sand particles after single MICP treatment. (c) MICP-bonded sand particles after quadruple MICP treatment.

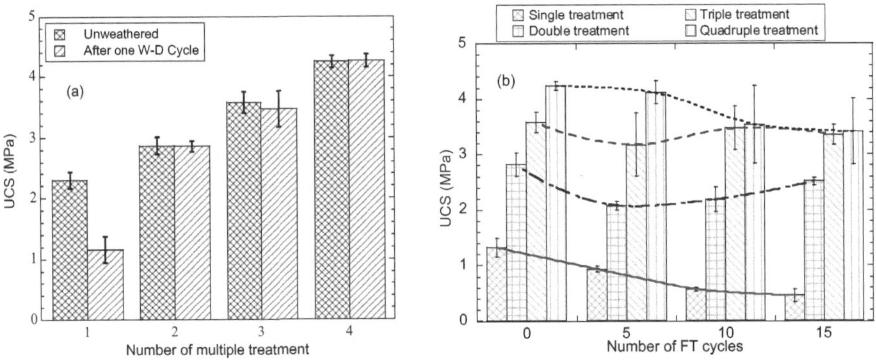

FIGURE 5.43 (a) UCS of multiple-MICP-treated samples before and after one WD cycle. (b) UCS as a function of FT cycles' number for multiple-MICP-treated samples.

samples, the UCS was almost maintained the same as 2.82 MPa after one WD cycle. The UCS of samples increased to 3.58 MPa and 4.24 MPa after triple and quadruple MICP treatments, and they all had reductions of nearly 0% after one WD cycle. The results indicated that the compact bridging $CaCO_3$ crystal induced by multiple treatments enhanced the durability of MICP-treated samples against WD cycles.

For FT cycles, the multiple-MICP-treated samples show a similar trend to WD cycles. As shown in Figure 5.43b, the UCS of multiple-MICP-treated samples was kept nearly constant after FT cycles. For double-, triple-, and quadruple-treated samples, the UCS was maintained around 2.5 MPa, 3.5 MPa, and 4.0 MPa during various FT cycles. The results showed that multiple treatments improved the durability of MICP-treated samples against FT cycles significantly. It can be explained by two reasons. First, the increasing $CaCO_3$ content could result in the reduction of water absorption to be less sensitive to the freezing of water for multiple-MICP-treated samples. The calcium carbonate precipitation effectively fills in the pores of samples during the MICP process; thus, the plugging of pores leads to the reduction of water absorption (Achal et al. 2013). For multiple-MICP-treated samples, more reduction of water absorption was caused by higher $CaCO_3$ content. Additionally, a thin layer of calcite formed on the surface of multiple-MICP-treated samples can act as natural calcite skin. The protective layer also induced the reduction of water absorption and protected the samples from degradation (Amidi and Wang 2015). Second, the increasing $CaCO_3$ could disperse the tensile force induced by FT cycles. Cheng et al. (2017) reported that more bridging $CaCO_3$ crystals can reduce the acting tensile stress and, hence, result in higher freeze–thaw durability.

However, for acid rain durability, the UCS and mass loss rate of multiple-MICP-treated samples kept deteriorating, along with lower pH value and longer immersion days (Figure 5.44). Even for quadruple-MICP-treated samples, only 51.9% UCS was left, with 2.3% mass loss rate after 15 immersion days in acid rain of pH = 3.5. SEM image (shown in Figure 5.45) indicated that the thickness of $CaCO_3$ shell decreased by around 50% after ten immersion days in acid rain solution of 3.5 pH, compared to samples

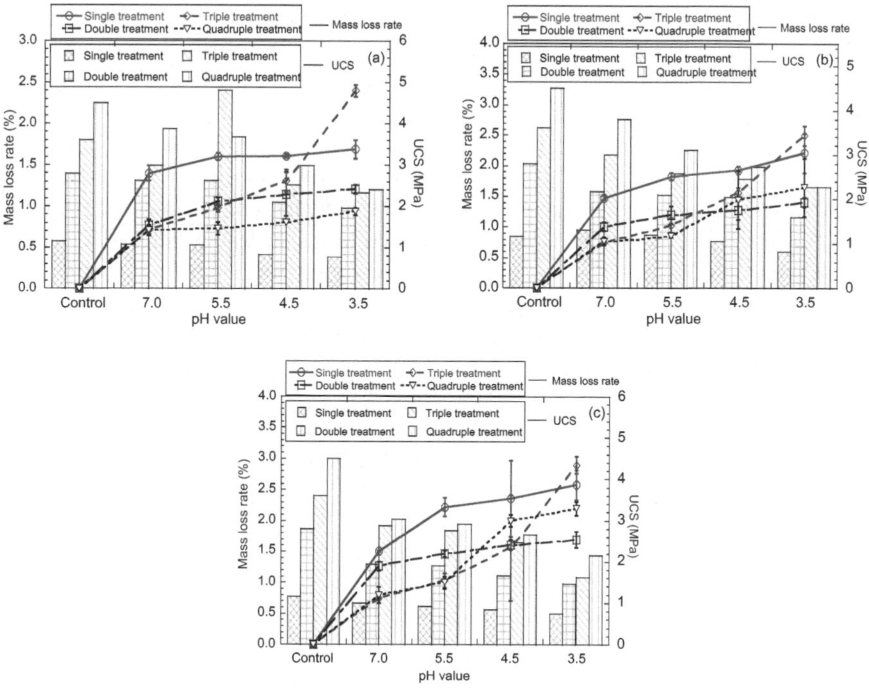

FIGURE 5.44 UCS and mass loss rate of multiple-MICP-treated samples immersed in acid rain solution as a function of pH value: (a) 5 immersion days; (b) 10 immersion days; (c) 15 immersion days.

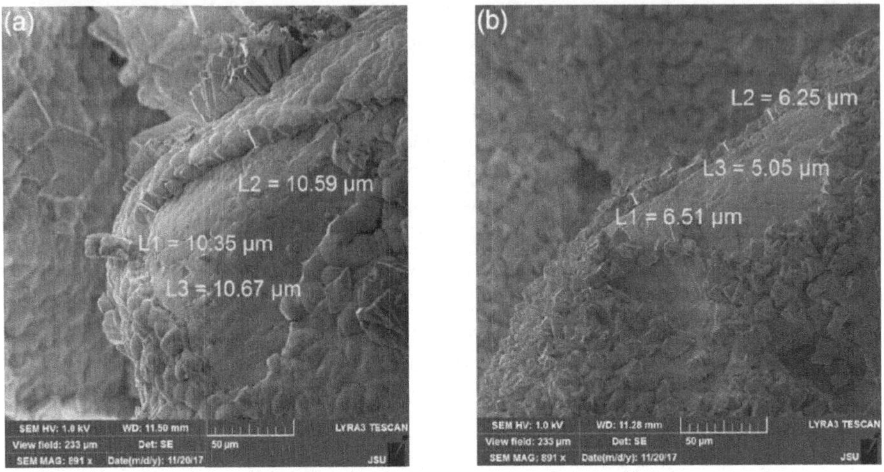

FIGURE 5.45 SEM image of quadruple-MICP-treated sand particles after ten immersion days in acid rain solution: (a) pH = 7.0; (b) pH = 3.5.

immersed in water of pH = 7.0. Since $CaCO_3$ between sand particles is soluble in acid, MICP-treated samples are weak in durability against acid rain erosion. The results indicated that even though multiple MICP treatment could significantly improve the $CaCO_3$ content and UCS, it was still limited to resist acid rain erosion.

Fiber reinforcement and multiple MICP treatment have different advantages to enhance the durability of MICP-treated sand against environmental deteriorations. Fiber reinforcement is more efficient to increase the ductility of the MICP-treated sand, which is more favorable to resist WD cycles. However, fiber reinforcement in MICP process is not easy to carry out in site implementation. Meanwhile, the multiple MICP treatment method is either beneficial to resist WD cycles or FT cycles. Nevertheless, multiple MICP treatment is more expensive compared to other soil improvement methods. As challenges, experiments about the implementation of fiber reinforcement in site and economic feasibility of multiple MICP treatment are worthwhile to carry out in the future.

6 Application of MICP in Environmental Remediation

6.1 BIOREMEDIATION OF COAL FLY ASH BY MICROBIALLY INDUCED CARBONATE PRECIPITATION

6.1.1 Bio-Modification of Coal Fly Ash through Urease-Producing Bacteria

6.1.1.1 Materials and Methods

6.1.1.1.1 Materials

Three types of CFA were collected from thermal power plant and used as received, with no treatment. The chemical compositions of CFA1, CFA2, and CFA3 were determined by energy-dispersive X-ray spectroscopy (EDX), and the results are shown in Table 6.1. According to ASTM C-618, CFA1 is classified as off-specification fly ash, CFA2 is classified as class F fly ash, and CFA3 is off-specification fly ash (International 2019). The pH of CFA1, CFA2, and CFA3 were 12.13, 10.3, and 11.1, respectively, according to the US Environmental Protection Agency (USEPA) method 9045D: Soil and Waste pH ((EPA) 1986). The total metal contents of CFA1, CFA2, and CFA3 were evaluated through the US EPA method 3050b ((EPA) 1996), and the environmental impact evaluations of CFA1, CFA2, and CFA3 were performed through US EPA method 1311, Toxicity Characteristic Leaching Procedure (TCLP) ((EPA) 1992b), and US EPA method 1312, Synthetic Precipitate Leaching Procedure (SPLP) ((EPA) 1992a). Ottawa silica sand which contains 99.7% SiO_2 was used to prepare cylinder samples and bio-brick. The median size (D50) of silica sand is 0.46 mm. CFA–sand mixtures were prepared with sand, 3% CFA, 6% CFA, 9% CFA, and 12% CFA based on the mass percent.

Sporosarcina pasteurii (ATCC 11859) was used as active bacteria for MICP, and its nutrition solution contains 20.0 g/L yeast extract and 10.0 g/L of $(NH_4)_2SO_4$. The pH of the nutrition solution was adjusted to 9.0 by Tris buffer. After 40 h culturing, the bacteria were harvested and adjusted to certain concentrations to conduct the bio-modification process. Optical density at 600 nm was measured by a spectrophotometer (ultraviolet (UV)-1700 UV-visible spectrophotometer, Shimadzu) to indicate the concentration of bacteria cells.

6.1.1.1.2 Bio-modification Process

To conduct the bio-modification process, an aliquot of 10.0 g CFA was mixed with 30.0 g/L urea solution. Then, bacteria liquor was added to the mixtures. The OD_{600} of

DOI: 10.1201/9781003519638-6

TABLE 6.1
Chemical Composition of Coal Fly Ashes Determined by EDX

Coal Fly Ash	SiO_2	Al_2O_3	CaO	CuO	SO_3	Fe_2O_3	Cr_2O_3	NiO	MgO	MnO	ZnO
CFA_1	43.3	19.4	19.2	N/A*	5.4	5.5	N/A	N/A	5.4	0.6	1.2
CFA_2	45.5	23.i	10.2	N/A	2.0	11.4	N/A	N/A	7.6	0.1	0.1
CFA_3	22.4	10.5	54.4	N/A	4.7	4.7	N/A	N/A	3.2	0.1	0.1

Note:
* N/A: not available.

bacteria was 0.8, which means the number of bacteria cells was about 6.34×10^7. The volume of bacteria liquor was 20 mL, and the total volume of reaction system was 100 mL. After mixing, the solutions were transferred to a 225 mL bottle for reaction and rotated at 30 ± 2 rpm. The reaction system was opened to exchange air for 1 h in each 24 h duration. Meanwhile, the pH and ORP of reaction solutions were recorded. Samples were harvested at 3 days, 5 days, 7 days, and 10 days for further analysis. Then, the mixture was oven-dried at 100 ± 5°C for 24 h and saved for leaching tests. The leaching behaviors of metals in samples before and after bio-modification were evaluated according to SPLP ((EPA) 1994).

6.1.1.1.3 Preparation of Cylinder Sample and Bio-Bricks

To prepare the cylinder samples, an aliquot of 150 g CFA–sand mixture was mixed with 40 mL bacteria liquor. The bacteria concentration was set as OD_{600} of 0.8. Then, the mixture with holder was immersed into the cementation media to perform the bio-cementation process. The cementation media consisted of 30.0 g/L urea, 73.5 g/L $CaCl_2{\cdot}2H_2O$, 2.12 g/L $NaHCO_3$, 3.0 g nutrition broth and 10.0 g/L NH_4Cl. The aeration system was involved in bio-cementation processes to maintain aerobic conditions for bacteria.

Bio-brick samples were prepared with 900 g CFA–sand mixtures and a volume of 250 mL bacteria liquor was added to the mixture. The bacteria concentration and cementation media recipe are the same as for the preparation of cylinder samples. The aeration system was also involved. After 7 days' reaction, bio-brick was ready to be harvested and oven-dried. The mechanical properties of bio-bricks were determined through ASTM method D6272 (D6272–17 2019).

6.1.1.1.4 Determination of Hydraulic Conductivity

The hydraulic conductivity was determined following ASTM (International 2003). Falling head, rising tailwater elevation method was performed by using a flexible wall permeameter.

6.1.1.1.5 $CaCO_3$ Content

The precipitated $CaCO_3$ content of bio-bricks caused by MICP treatment was determined by the acid washing method. The samples were weighed and washed with an

HCl solution (0.1 M) to dissolve precipitated carbonates. The weight of specimens was measured after rinsing, draining, and oven-drying. The difference in weight before and after acid washing was the weight of the carbonates precipitated during the MICP treatment.

6.1.1.1.6 Microstructure Analysis

The microstructure of the CFA was directly observed by scanning electron microscopy (SEM) (Lyra 3, TSECAN Inc.). The tested samples were oven-dried overnight at 105°C and mounted on the stubs with adhesive carbon conductive tabs before testing.

6.1.1.2 Sample Characterization

The SEM images of raw CFA were illustrated in Figure 6.1. CFA1 was dominated by amorphous-shaped particles with smooth surface, which is typical of aged fly ash after compaction. The particles of CFA2 were plate-like with needled materials. By comparison, CFA3 particles are mainly spherical in shape, and solid sphere, chemosphere, and irregular debris were all observed in CFA3 particles. Large void spaces were present in CFA1, CFA2, and CFA3, believed to provide a large surface area for leaching tests.

FIGURE 6.1 CFA photo images and SEM images: (a) CFA1, (b) CFA2, and (c) CFA3.

As shown in Table 6.2, the main components were Al, Ca, Mg, Mn, Si, and Zn, whereas some Ba, Cd, and Cr were detected. Compared with the US EPA limits for hazardous materials, TCLP results indicated that CFA1, CFA2, and CFA3 were not classified as hazardous wastes. However, high metal concentrations and leachability still pose negative risks to the environment. The leaching of metals from CFA modified by the biological process is still unclear. Bioleaching and bio-stabilization are two important processes which can occur simultaneously in the same sample. Therefore, the aim of this study was to discover the leaching behavior of metals before and after bio-modification and to figure out the possible influences for the reuse of CFA.

XRD test was utilized to illustrate the difference of CFA samples after bio-modification. As shown in Figure 6.2, compared with the raw CFA, carbonate compounds were generated after 7 days' bio-modification.

6.1.1.3 Variation of pH and ORP in a Bio-Modification System

The pH and ORP of the reaction solution were measured every 24 h, as shown in Figure 6.3. The pH of pure reaction solutions without CFA was 8.3. After mixing with CFA, the reaction systems became more alkaline. After 24 h, the pH for reaction solutions containing CFA1, CFA2 and CFA3 were 9.47, 9.40 and 9.46, respectively. As the reaction continued, the pH of the CFA1 reaction system increased gradually, while the pH of reaction solutions containing CFA2 and CFA3 were steady, with minor fluctuations. Ca and Mg in raw CFA particles were believed to influence the pH. Generally, it was observed that the pH increased as Ca and Mg were leached out. The pH also increased due to hydroxyl ions (OH-) produced by hydrolysis of urea. Till the end of the reaction, solutions were in alkalinity condition in the whole reaction process. The ORP of reaction solutions with respect to time is shown in Figure 6.3b. According to research by Patrick and Mahapatra, the aqueous condition can be classified into four general conditions at neutral pH as oxidized (>400 mV), moderately reduced (400 to 100 mV), reduced (100 to −100 mV), and highly reduced (−100 to −300 mV) (Patrick Jr and Mahapatra 1968). Metals in reaction systems could transform from one oxidation state to another to form compounds that might be more mobile or less. The reaction solutions in this study were apparently in highly reduced states, and the reduced conditions with groundwater systems are believed to promote the migration of metals (McLean et al. 1992).

6.1.1.4 pH and ORP Variation of Leachate

The pH and ORP variations of leachate solution are shown in Figure 6.4. The initial pH of SPLP extraction solution was 4.20 ± 0.05. After bio-modification, the pH of SPLP leachate for samples containing CFA1, CFA2, and CFA3 reached a steady state of approximately 8.0. The increase of pH for the leachate of SPLP revealed the high acid neutralization potential (ANP) of CFA samples after bio-modification. The system of bio-modified CFA samples transitioned from highly reduced status to reduced status.

6.1.1.5 Leaching Behavior of Metals

The bio-modification effects on the leaching of metals from CFA with and without bio-modification were evaluated by SPLP tests. By analyzing the leaching results, the

TABLE 6.2
Metal Contents and Leaching Behavior of Raw CFA

	CFA1			CFA2			CFA3				
Metal	**3050b (mg/kg)**	**TCLP (mg/L)**	**SPLP (mg/L)**	**3050b (mg/kg)**	**TCLP (mg/L)**	**SPLP (mg/L)**	**3050b (mg/kg)**	**TCLP (mg/L)**	**SPLP (mg/L)**	**EPA[1] (mg/L)**	**MDEQ[2] (mg/L)**
Ag	N/A[3]	N/A	N/A	N/A	N/A	N/A	N/A	N/A	N/A	5.0	0.5
Al	34,400	2.9	102	13,400	5.0	29.8	27,800	3.2	2.7	N/A	N/A
As	N/A	N/A	N/A	N/A	N/A	N/A	N/A	N/A	N/A	5.0	0.5
B	812	5.4	N/A	230	5.4	1.8	180	1.3	0.6	N/A	N/A
Ba	324	0.6	8.0	330	0.7	0.6	226	0.2	0.2	100	10
Ca	154,600	1,160	280	21,000	540	151.8	192,800	1,994	392	N/A	N/A
Cd	N/A	N/A	N/A	N/A	N/A	N/A	40	N/A	N/A	1.0	0.1
Cr	70	N/A	0.1	14.0	0.1	0.1	30	N/A	N/A	5.0	0.5
Hg	N/A	N/A	N/A	N/A	N/A	N/A	N/A	N/A	N/A	0.2	0.02
Mg	33,400	140	1.1	770	7.7	0.1	20,200	185.8	1.1	N/A	N/A
Mn	184.0	0.1	N/A	141.0	0.9	N/A	236	0.4	N/A	N/A	N/A
Ni	31.0	0.2	N/A	22.0	N/A	N/A	N/A	0.2	N/A	N/A	N/A
Pb	N/A	N/A	N/A	N/A	N/A	N/A	N/A	N/A	N/A	5.0	0.5
Se	N/A	N/A	N/A	N/A	N/A	N/A	N/A	N/A	N/A	1.0	0.1
Si	5,240	1,360	8.0	880	218	7	3,580	984	42.2	N/A	N/A
V	204	0.8	N/A	75	0.3	N/A	44.8	0.3	N/A	N/A	N/A
Zn	105.0	0.1	N/A	65.0	0.3	0.1	464	0.1	0.1	N/A	N/A

[1] US EPA regulation levels for classifying a waste as hazardous based on TCLP results.
[2] Mississippi Department of Environmental Quality (MDEQ) regulatory levels for beneficial use of solid wastes.
[3] Not available.

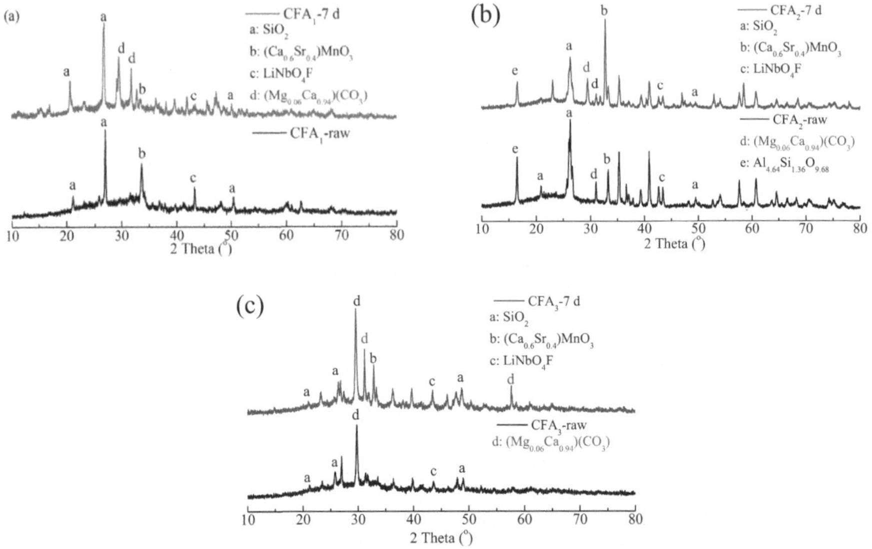

FIGURE 6.2 XRD patterns of (a) CFA1, (b) CFA2, and (c) CFA3.

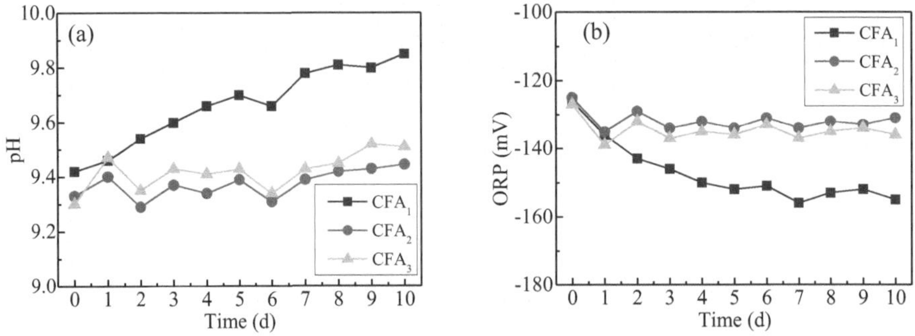

FIGURE 6.3 Variation of (a) pH and (b) ORP of reaction solution over time.

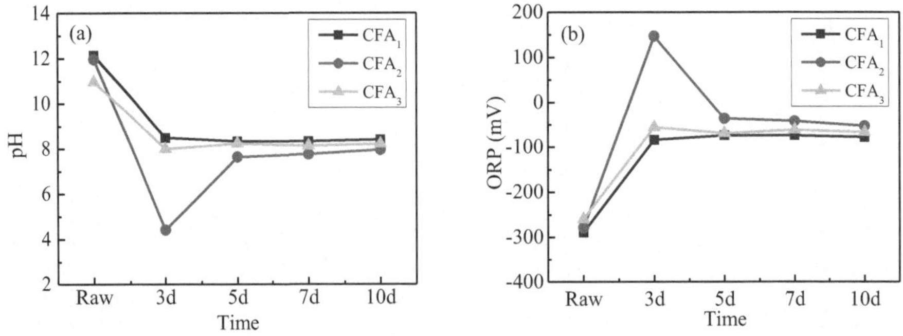

FIGURE 6.4 Variation of pH and ORP of SPLP leachates.

leaching behavior of metals from bio-modified CFA1, CFA2, and CFA3 was divided into two cases. For metals like Al, Ba, and Cr, leachate concentrations decreased after bio-modification. Leachate concentrations increased for metals such as Ca, Mg, Mn, and V. Therefore, it is concluded that Al, Ba, and Cr can be stabilized through the bio-modification process, while Ca, Mg, Mn, and V would be leached out.

As illustrated in Figure 6.5a, the leachate concentration of Al decreased after bio-modification. Specifically, the leaching concentration of Al decreased to near zero after reaction for 3 days. Afterward, the leachate concentration of Al remained stable, which indicated the formation of stable Al compounds. According to Astrup et al. (2006), the leaching of Al depends on the pH of the leachate and presents a U-shaped response with pH varying from acidic to alkaline conditions. They further stated that ettringite ($Ca_6Al_2(SO_4)3OH_{12}{\cdot}26H_2O$) controlled Al leaching at pH 9.5–10 (Astrup et al. 2006). As shown in Figure 6.3, the pH of reaction systems containing CFA1, CFA2, and CFA3 were 9–10, suggesting the presence of Al as ettringite. Similarly, the leaching of Ba from CFA decreased after bio-modification (Figure 6.5b). The concentration of Ba in raw CFA1 was 8.26 mg/L, while the leachate concentrations of Ba from bio-modified CFA1 were 0.10–0.65 mg/L, which means the removal efficiency of Ba was 92% to 99%. The solubility products of $BaCO_3$ and $Ba(OH)_2$ are 5.1×10^{-9} and 5×10^{-3}; therefore, the precipitation of $BaCO_3$ precedes the precipitation of $Ba(OH)_2$ when free CO_3^{2-} is available. The leaching behavior of Cr from CFA before and after bio-modification is illustrated in

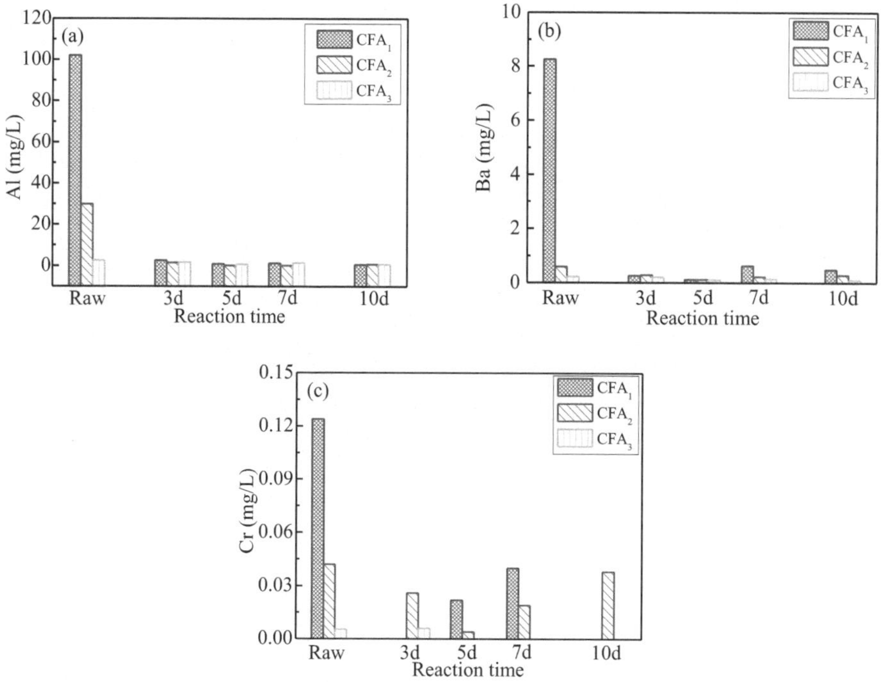

FIGURE 6.5 Leaching behavior of metals: (a) Al, (b) Ba, (c) Cr from the CFA mixtures.

Figure 6.5c. Compared to that in raw CFA, the decrease of Cr in CFA1 and CFA3 was significant, while the decrease of Cr from CFA2 was slight. The Cr ions in the leachate of CFA1 samples with 3-day and 10-day modification were not detectable. Cr in the leachate of CFA3 after 5-, 7-, and 10-day modification was also not detected. As shown in Figure 6.3, the reaction solutions of CFA during bio-modification were in alkaline and reduced states, which means Cr is present as Cr(III). The solubility product constant of $Cr(OH)_3$ is 6.3×10^{-31}, which is a low value. There is a conversion between $Cr_2(CO_3)_3$ and $Ca(OH)_2$, as illustrated in equation (6.1). Therefore, the precipitation of $Cr(OH)_3$ would be one reason for the decrease of Cr leaching concentration.

$$Cr_2(CO_3)_3 + Ca(OH)_2 \xrightarrow{\text{yields}} 2Cr(OH)_3 + CaCO_3 \tag{6.1}$$

Figure 6.6 shows the leaching behavior of Ca, Mg, Mn, and V. It is apparent that the leachate concentration of Mg, Mn, and V from CFA1, CFA2, and CFA3 increased after bio-modification. Meanwhile, the leachate concentration of Ca increased significantly after 3 days' treatment and continued to increase modestly after the initial 3 days. The precipitation of calcium carbonate induced by urease-producing bacteria is widely known to enhance the mechanical properties of materials (Daskalakis et al. 2015; Mahawish et al. 2018; Morales et al. 2019; Wang et al. 2019). However, calcium carbonate precipitation relies on the access of Ca2+ ions (DeJong et al. 2006;

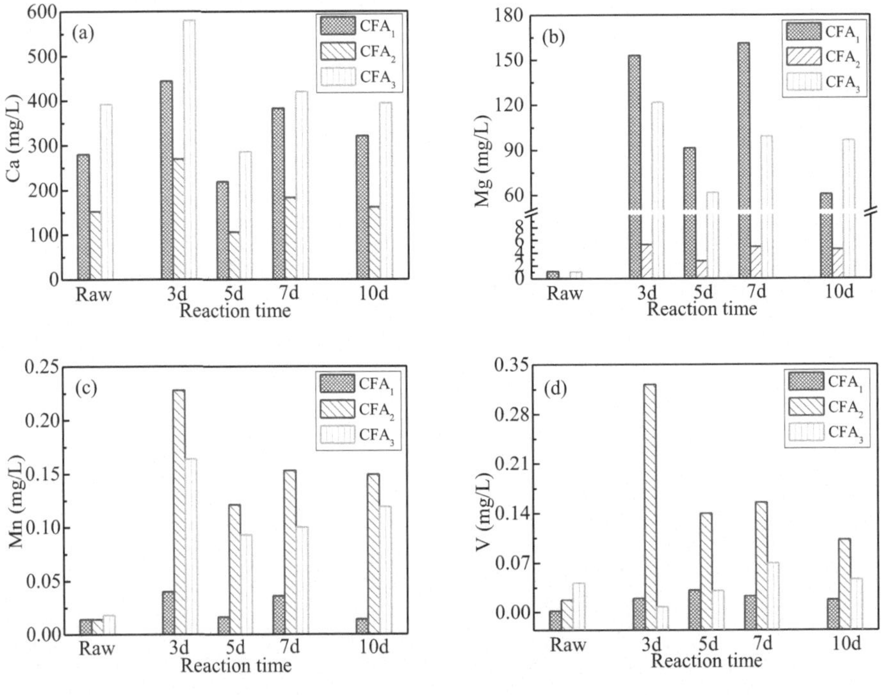

FIGURE 6.6 Leaching behavior of metals: (a) Ca, (b) Mg, (c) Mn, (d) V from the CFA mixtures.

Okwadha and Li 2010; Omoregie et al. 2019). In this study, Ca^{2+} ions were sealed in the CFA in aqueous solution, and weak-bonded Ca^{2+} ions were released to the solution and exposed to the carbonate environment produced by urease-producing bacteria. There is competition between metals for carbonate ions. The solubility product constant of carbonate products follows the order of $ZnCO_3$ (1.4×10^{-11}) < $MnCO_3$ (1.8×10^{-11}) < $FeCO_3$ (3.2×10^{-11}) < $CuCO_3$ (1.4×10^{-10}) < $CaCO_3$ (2.8×10^{-9}) < $BaCO_3$ (5.1×10^{-9}) < $CdCO_3$ (5.2×10^{-9}) < $MgCO_3$ (3.5×10^{-8}). Therefore, the precipitation of $CaCO_3$ contributed little to the decrease of leaching of Ca. The increase of leachate concentration of Ca after modification may be related to the bioleaching of bacteria. As shown in Figure 6.6b, the leachability of Mg from CFA was enhanced after bio-modification. For CFA1, the leaching concentration of Mg from bio-modified samples after 7-day reaction was 161.0 mg/L, which is 141.75 times that of the raw material. Mn in ash has been reported to be likely associated with ferromagnetic particles and/or the glass fraction, and it is insoluble under near neutral to alkaline conditions (Izquierdo et al. 2012). In this study, less than 0.02 mg/L Mn was detected in raw CFA. With the treatment of bio-modification, the mobility of Mn from CFA was enhanced, especially for CFA2 samples. The high leachability of Mn after bio-modification would benefit the recycle of Mn. V is a rare earth element which occurs in coal combustion products and a strategic material used in many industrials (Astrup et al. 2006; Engelsen et al. 2010). Several researches have tried to recover V from CFA (Astrup et al. 2006; Zhang et al. 2016). As shown in Figure 6.6d, the leaching behavior of V was influenced significantly by bio-modification. The mobility of V was enhanced after treatment, especially for CFA2. According to Cornelis et al. (2008), the leaching of V from fly ash is controlled by calcium vanadates, and V was readily leachable from a low-Ca fly ash (Cornelis et al. 2008). Compared to CFA1 and CFA3, CFA2 contains a lower content of Ca. Therefore, the high concentrations of V in the leachates of CFA2 are consistent to the observation of Cornelis et al. (Cornelis et al. 2008). The high mobility of V after bio-modification would enhance its recovery.

6.1.1.6 Leaching Mechanism of Bio-Modification of Metals

Based on the leaching behavior described earlier, the mechanism of metal bio-modification is proposed as follows. Firstly, bio-mineralization promotes the precipitation of metals as carbonate compounds via equations (6.2) and (6.4) and as hydroxide compounds via equation (6.3) and (6.5). Moreover, Cr, Zn, and Cd were confirmed to have the tendency to remain strongly adsorbed to calcite (Garcıa-Sánchez and Alvarez-Ayuso 2002). Therefore, the metal ions partitioned onto the generated precipitate, resulting in lower aqueous ion concentrations, as predicted by equation (6.6). The leachate concentrations of some metals after bio-modification by urease-producing bacteria increased, as observed in this study. This phenomenon related to the function of bioleaching. The increased leachability of metals from CFA indicates that MICP is a feasible method for valuable metal recovery. Thus, CFA recycling rates can be enhanced.

$$CO(NH_2)_2 + 2H_2O + Urease \xrightarrow{yields} CO_3^{2-} + 2NH^{4+} + Urease \tag{6.2}$$

$$CO_3^{2-} + H_2O \xrightarrow{yields} HCO^{3-} + OH^- \tag{6.3}$$

$$CO_3^{2-} + Mn^+ \xrightarrow{yields} M_2\left(CO_3\right)_n \tag{6.4}$$

$$Mn^+ + nOH^- \xrightarrow{yields} M\left(OH\right)_n \tag{6.5}$$

$$Mn^+ + mineral \xrightarrow{Adsorbs} \text{Mn}^+\text{mineral} \tag{6.6}$$

6.1.1.7 Physical and Mechanical Properties of Bio-Cemented CFA–Sand Samples

The cylinder samples containing different contents of CFA were tested to determine the hydraulic conductivity, and the bio-bricks were tested to determine flexural resistance. The photo images of bio-cemented samples are shown in Figure 6.7.

6.1.1.8 Mechanical Properties Development of Bio-Cemented CFA–Sand Samples

As shown in Figure 6.8a, bio-bricks containing sand only achieved the highest flexural strength of 902.1 kPa. The samples containing CFA1 exhibited a higher strength than samples containing CFA2 and CFA3. The highest flexural strength of samples containing CFA was 468.7 kPa by samples containing 12% CFA1. However, this value is only 52% of that of sand-only samples. Nevertheless, CFA will continue to find applications in lower-strength applications and as a lightweight aggregate additive. These applications will benefit from the decrease of leachable metals in bio-treated CFA.

6.1.1.9 Precipitated $CaCO_3$ Content

According to equations (6.2) to (6.5), the bio-cementation process depends mainly on the production of $CaCO_3$. Therefore, the magnitude of precipitated $CaCO_3$ is an important parameter for evaluating the efficacy of bio-cementation. The content of $CaCO_3$ precipitated during bio-cementation in this study is shown in Figure 6.8b. The $CaCO_3$ content of sand-only specimen was 6.37% of the total mass of specimen. For the specimens containing CFA, the $CaCO_3$ contents were in the range of 2.53% to 11.15%, which is consistent with the finding of Zhao et al. (2014a) and Li et al. (2015c). The $CaCO_3$ content in this study was higher than that reported by Lin et al. (2016) for an injection method. Bio-brick containing 9% CFA3 obtained the highest $CaCO_3$ content and achieved flexural strength of 144.8 kPa, which is 16% of that achieved by sand-only samples. The sand-only samples achieved the

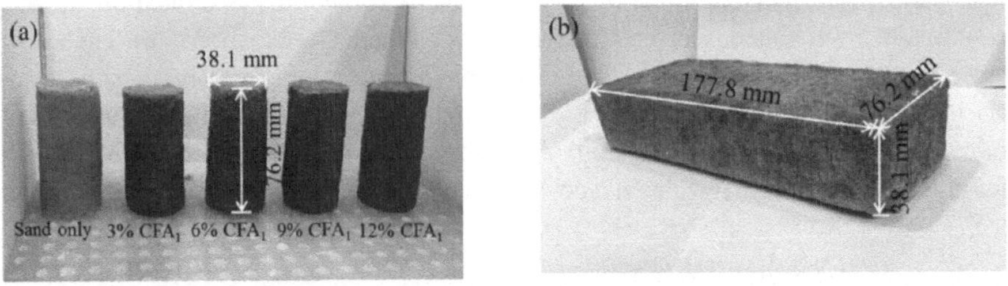

FIGURE 6.7 Photo images of bio-cemented samples: (a) cylinder samples, (b) bio-brick.

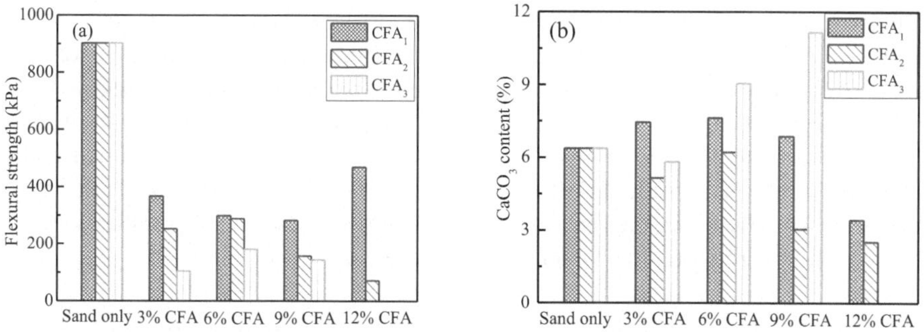

FIGURE 6.8 (a) Flexural strength of bio-brick samples containing CFA, and (b) precipitated $CaCO_3$ content.

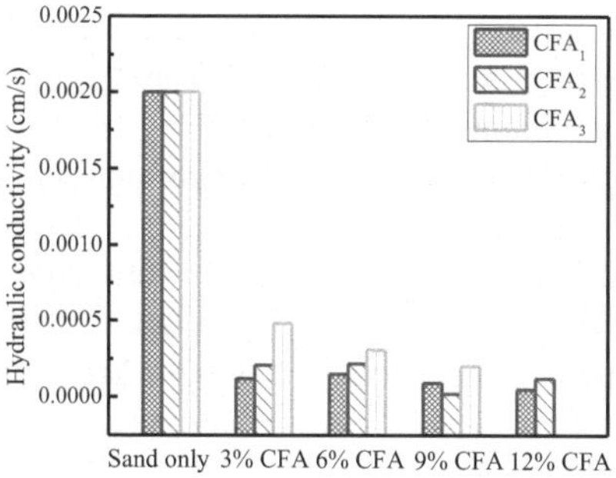

FIGURE 6.9 Hydraulic conductivity of sand and CFA mixtures.

maximum flexural strength in terms of obtaining 6.37% $CaCO_3$ content. Therefore, the enhancement of flexural strength of bio-cemented samples is not directly related to the precipitated content of $CaCO_3$. This contradicts the observations of Zhao et al. (2014a) and Yang et al. (2020), which concluded a positive synergistic relationship between $CaCO_3$ content and UCS. However, the material used in Zhao et al. (2014a) and Yang et al. (2020) was Ottawa silica sand (Zhao et al. 2014a). The inclusion of CFA would influence the distribution of pore size and precipitation of $CaCO_3$. As observed in this study, $CaCO_3$ was precipitated on the surface of samples. This partition of inefficient precipitated $CaCO_3$ did not lead to an increase of strength. Therefore, further research is needed into the effect of amount of effective precipitated $CaCO_3$.

6.1.1.10 Hydraulic Conductivity of Bio-Cemented Specimens

The hydraulic conductivities of specimens are shown in Figure 6.9. The bio-cemented sand-only specimen obtained the maximum hydraulic conductivity as 0.002 cm/s, while hydraulic conductivity decreased significantly with the inclusion of CFA. The hydraulic conductivity for specimens containing CFA ranged from 2.00×10^{-5} cm/s to 4.83×10^{-4} cm/s. Rowshanbakht et al. (2016) found that poorly graded sandy silica treated by bio-cementation achieved hydraulic conductivity ranging from 8.5 × 10^{-3} cm/s to 1.7 × 10^{-2} cm/s. Deng and Wang (2018) studied the hydraulic conductivity of fine and coarse sand and obtained the hydraulic conductivity ranging from 0.5 × 10^{-5} m/s to 100 × 10^{-5} m/s. Gomez et al. (2014) reported a significant drop of hydraulic conductivity when more fine content is contained in the specimens. The hydraulic conductivity ranging between 10^{2} and 10^{-4} cm/s represents a good drainage performance. Therefore, the specimens in this study possess good hydraulic conductivity even though more fine contents were contained.

6.1.2 Mechanical and Environmental Effects of MICP on Coal Fly Ash

6.1.2.1 Materials and Methods

6.1.2.1.1 Materials

The soil used in this study was Ottawa silica sand with median diameter of 0.46 mm. The sand has been widely used in MICP investigation (Lin et al. 2016; Wen et al. 2018a). Two types of coal fly ash (FA) obtained from local power plants were used, and their chemical compositions were analyzed by energy-dispersive X-ray spectroscopy (EDX-720, Shimadzu). The results are summarized in Table 6.1. According to ASTM C 618 (2017), FA1 is classified as class F fly ash, and FA2 is classified as off-specification fly ash. The pH of FA1 and FA2 were determined as 10.3 and 11.1 according to the US Environmental Protection Agency (EPA) method 9045D: Soil and Waste pH (1986). The size distribution of the two kinds of FA and sand is shown in Figure 6.10. As shown in the figure, FA2 contained many more fine particles than FA1. About 85% of FA1 particles were smaller than 0.1 mm, while about 100% of FA2 particles were smaller than 0.1 mm.

6.1.2.1.2 Bacteria Growing Culture

The bacteria used in this study was *Sporosarcina pasteurii* (ATCC 11859); more detail is shown in Section 3.1.1.

6.1.2.1.3 MICP-FA Enhanced Soil Improvement

Before the MICP process, 0%, 3%, 6%, and 9% of FA on weight basis were mixed into the sandy soil. After mixing thoroughly, an aliquot of 150 g FA–sand mixture was taken out and packed into a sample holder (38.1 mm diameter, 76.2 mm height).

TABLE 6.3
Chemical Composition of FA (%)

Composition	SiO_2	Al_2O_3	CaO	SO_3	Fe_2O_3	Cr_2O_3	NiO	MnO	ZnO	Fly Ash Classification (ASTM C 618)
FA_1	45.5	23.1	10.2	2.0	11.4	ND[a]	ND	0.1	0.04	F
FA_2	22.4	10.5	54.4	4.7	4.7	ND	ND	0.1	0.1	Off-specification

[a] Not detected.

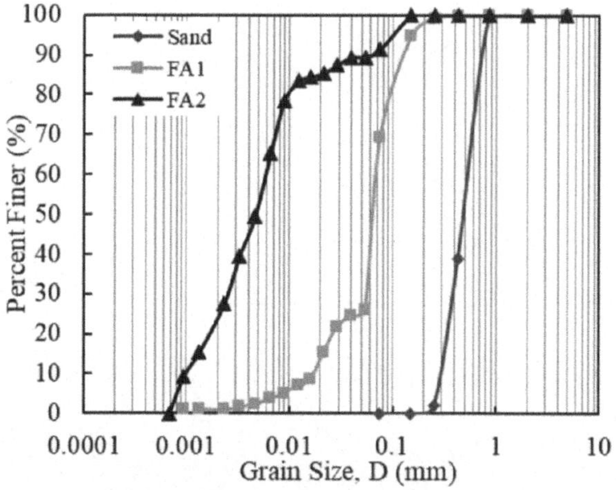

FIGURE 6.10 Grain size of sand, FA1 and FA2.

This sample holder was full-contact flexible mold made by geotextile, as described in Zhao et al. (2014a). Another set of samples prepared with sand only was used as control. For a single-MICP sample, a 40 mL bacteria suspension was mixed with FA–sand mixture. All the samples were placed on the shelf with opening holes. Then the entire shelf was slowly immersed into the batch reactor, which was filled with the cementation media. The cementation media was prepared using 0.5 M $CaCl_2$, 10 g/L NH_4Cl, 3 g/L nutrition broth (NB), 2.12 g/L $NaHCO_3$, and 30 g/L urea. Aeration condition was maintained for the reaction of MICP. After 7 days of continuous reaction, MICP-treated specimens were taken out for further test. Zhao et al. (2014a) show that after 7 days of reaction, the MICP-treated samples reached stability. The schematic diagram and image of MICP-treated samples are given in Figure 6.11.

6.1.2.1.4 Mechanical Properties of MICP-Treated Samples

Triaxial compression (TC) test was performed, followed by ASTM D7181–11 (2011). After 7 days of reaction in the MICP process, the samples were taken out from the holder and covered with a thin rubber membrane. The sample was then placed within a cylindrical chamber, followed by filling with water. The soil sample was subjected under confining stress through compressing fluid in the chamber. Deviator stress was applied to the sample by a vertical loading ram. In this study, the confining pressure was fixed as 100, 200, and 300 kPa, respectively, and constant axial strain rate was 0.02% strain/min. The test was halted after axial strain reached 15%. Based on the triaxial test, the internal friction angle and cohesion of MICP-treated samples can be determined.

6.1.2.1.5 Toxicity Leaching Test of MICP-Treated Soils

US Environmental Protection Agency (EPA) method 1311, Toxicity Characteristic Leaching Procedure (TCLP) (1992), was used to determine the concentrations of leaching metals in the leachate. TCLP is mainly used to determine whether a waste material is hazardous with respect to its leaching toxicity. It is also used to examine

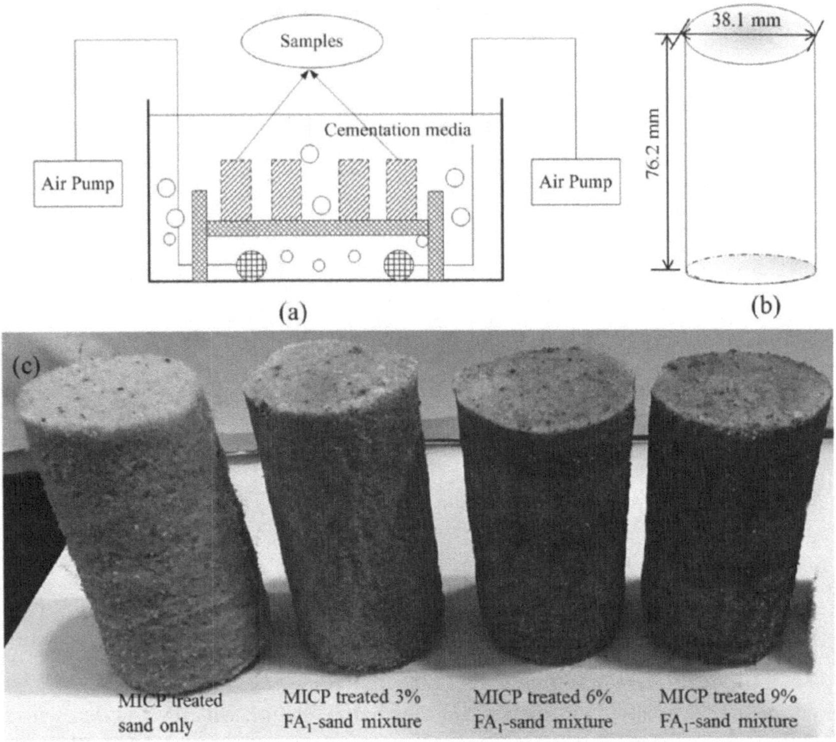

FIGURE 6.11 (a) Sketch of MICP cementation tank, (b) size of sample, and (c) pictures of MICP-treated samples.

the effectiveness of the improvement of hazardous waste (Li et al. 2009). TCLP extraction fluid #1 was made by adding 5.7 mL of glacial acetic acid and 64.3 mL of 1N NaOH solution to 1 L deionized water. An aliquot of 75 g MICP-treated soils was submerged in the extraction fluid with solid-to-liquid ratios of 1:20 in a 2 L HDPE bottle. The bottle was then agitated in a rotary extractor at 30 rpm for a period of 18 ± 2 h. The leachate was filtered through 0.75 μm glass-fiber filter, and then pH was adjusted to less than 2 for concentration determination.

To determine the fractions of metal content, a four-step modified sequential extraction test was applied in this study. A series of reagents was selected for their abilities to react with different major components of the matrix. A 40 mL 0.11 mol/L glacial acetic acid was used to extract the first fraction (F1). In the second step, a 40 mL 0.5 mol/L NH_2OH-HCl (adjust pH to 1.5 with HNO_3) was used to suspend the residual from the first step and extract the reducible fraction (F2). H_2O_2 (30% in water) and 40 mL 1 mol/L CH_3COONH_4 were applied to digest and extract the oxidizable fraction of metal (F3). Residual fraction (F4) was extracted with 10 mL aqua regia. In this test, 1 g solid was used, and the agitation speed was fixed as 30 rpm for 18 ± 2 h. After centrifuging at 3,000 rpm for 20 min, the leachate was filtered through 0.75 μm glass-fiber filter, and the pH adjusted to less than 2 immediately. More details can be seen in a former research by Su et al. (2019).

The leachate was analyzed using an inductively coupled plasma optical emission spectrometer (ICP-OES) (Shimadzu ICPE-9000) for the concentration (mg/L) of potential concerned trace metals.

6.1.2.2 Influence of FA on Deviator Stress of MICP-Treated Soil Mixture

The deviator stress curves of different samples are illustrated in Figure 6.12, where it can be seen that σd had a significant improvement with the addition of FA. When the confining pressure was 100 kPa (Figure 6.12a and b), the peak deviator stress of

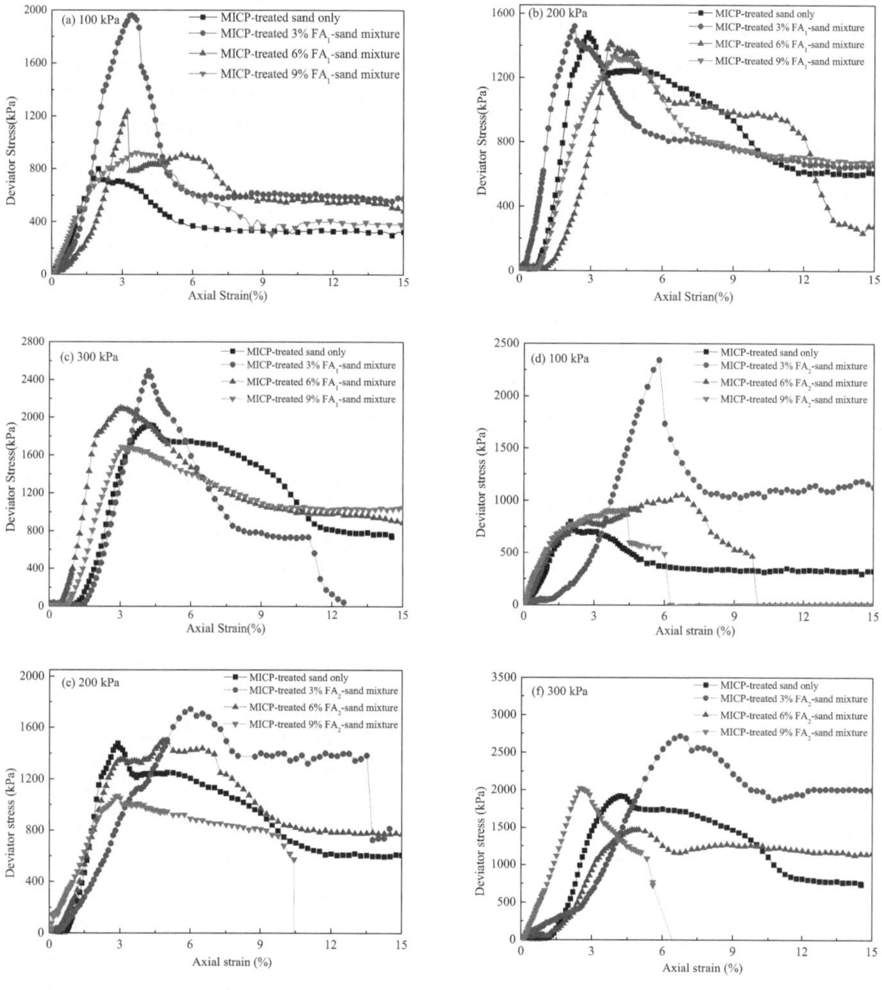

FIGURE 6.12 Effect of FA addition on deviator stress under different confining pressures of 100 kPa, 200 kPa, and 300 kPa: (a), (b), and (c) are for MICP-treated FA1–sand mixture; (d), (e), and (f) are for MICP-treated FA2–sand mixture.

MICP-treated sand-only sample was about 800 kPa. After mixing with 3% of FA1, peak deviator stress of the samples increased to 1,959 kPa.

The peak deviator stresses for MICP-treated 6% FA1–sand and MICP-treated 9% FA1–sand samples were 1230 and 900 kPa, respectively. Therefore, after being enhanced by fly ash, the overall peak deviator stress of the MICP-treated FA1–sand sample was higher than that of MICP-treated soil, although this increase trend grew slowly when the addition was higher than 3%. This phenomenon can also be found in Figures 6.12b and c. Figures 6.12b presented deviator stress curves in respect to 200 kPa confining pressure. The peak deviator stress of MICP-treated 3% FA1–sand sample was 1519 kPa, which increased slightly compared to 1477 kPa of MICP-treated sand-only sample. It is worth noting that when the additions of FA1 were 6% and 9%, the σd of the samples were 1368 kPa and 1336 kPa, even lower than that of MICP-treated sand-only sample. At 300 kPa confining pressure, the peak deviator stress for each sample was higher than that under 100 kPa and 200 kPa. The peak deviator stress for MICP-treated 3% FA1–sand sample and that of the MICP-treated 6% FA1–sand sample were 2600 kPa and 2200 kPa, both of which were higher than the 1800 kPa of MICP-treated soil. However, when the FA1 amount increased to 9%, the peak deviator stress for the sample decreased to 1700 kPa. It is hypothesized that the addition of FA1 (class F fly ash) would introduce the pozzolanic products to MICP-treated soils as the curing process goes on. Therefore, a higher FA1 content may benefit the enhancement of peak deviator stress. However, it is reported that the effect of the pozzolanic reaction of fly ash is long-term, and it generally occurs in a favorable environment pH (about 12). As mentioned in Section 2.1, the pH of pure FA1 was about 10.3, and the reaction media pH for MICP was about 8.0. Therefore, it seems that the pozzolanic reactions could not take place during the MICP process in this study. It can be speculated that the synergic effect of MICP and FA1's components was based on the decrease of void space and the densification. While the increase of FA1 content to 9% inhibits the movement of bacteria in the soil matrix, the non-uniformly distributed bacteria impact the distribution of calcium carbonate precipitation in the soil mixture, and thus the strength of soil specimen. The maximum σd was obtained when the FA1 addition was 3%.

Figures 6.12d, e, and f presented deviator stress curves of MICP-treated FA2–sand samples under 100 kPa, 200 kPa, and 300 kPa confining pressure, respectively. As shown in the figures, the deviator stress curves of MICP-treated FA2–sand samples showed similar trends as that of MICP-treated FA1–sand samples. MICP-treated 3% FA2–sand sample showed better improvement in the peak deviator stress compared to MICP-treated soil, MICP-treated 6% FA2–sand, and MICP-treated 9% FA2–sand samples. At 100 kPa confining pressure, the peak deviator stress of MICP-treated sand-only sample was about 800 kPa, while the peak deviator stress of MICP-treated 3% FA2–sand sample was up to 2343 kPa. When confining pressures were 200 and 300 kPa, the peak deviator stress of MICP-treated 3% FA2–sand sample remained higher than that of MICP-treated sand-only sample. For the samples with 6% and 9% FA2 addition, the enhancements of peak deviator stress were not as strong as that of 3% FA2. Moreover, as shown in Figures 6.12e and f, the σ_d of MICP-treated 9% FA2–sand sample at 200 kPa confining pressure and that of MICP-treated 6% FA2–sand sample at 300 kPa were lower than that of

MICP-treated sand-only sample. In this case, the optimal addition of FA2 was 3% as well.

6.1.2.3 Influence of FA on Shear Strength of MICP-Treated Samples

The effect of FA addition on shear strength is shown in Figures 6.13a and b. MICP-treated FA1–sand had notable increase in shear strength compared to that of MICP-treated sand-only sample (shear strength of 400 kPa), except for MICP-treated 9% FA1–sand sample at 200 kPa and 300 kPa. MICP-treated 3% FA1–sand sample had the highest shear strength at all confining pressure. At 100 kPa confining pressure, the shear strength of MICP-treated 3% FA1–sand sample was 809.6 kPa. The shear strength increased to 930 kPa and 1,090 kPa when the confining pressures were 200 and 300 kPa, respectively. The improvement in the shear strength was achieved with the enhanced bonding conditions between particles due to the bonding of precipitated $CaCO_3$. However, the shear strength of MICP-treated 9% FA1–sand sample was lower than that of MICP-treated sand-only sample at 200 kPa and 300 kPa confining pressures. This could be attributed to the excessive reduction of void space by filling of fine particles, which left not enough space for bacteria to go through when adding 9% FA1. In MICP-treated samples containing 9% FA1, the void space was filled by FA1, leaving limited space for bacteria to move. Therefore, the work of MICP

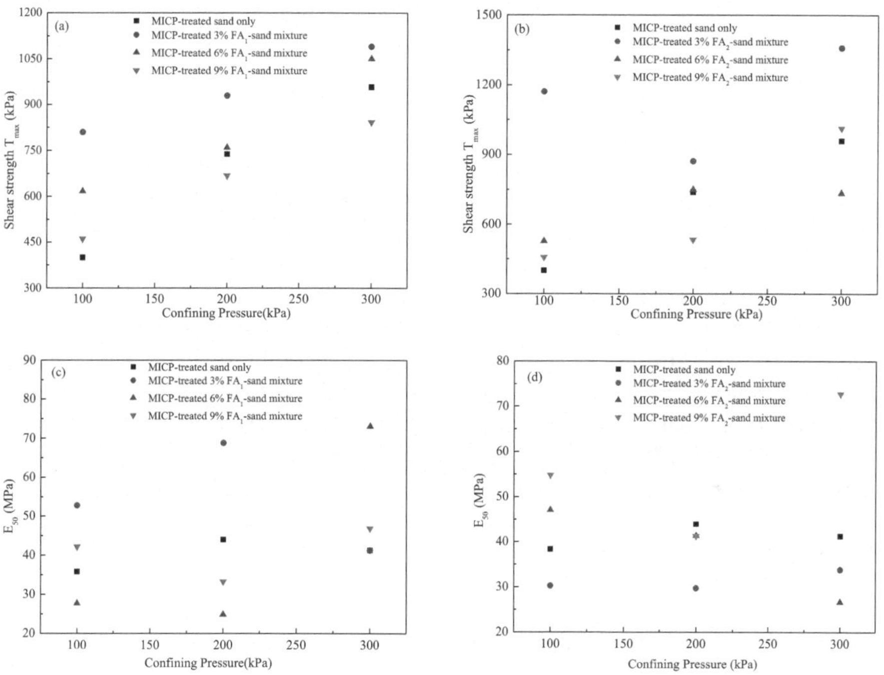

FIGURE 6.13 The effect of FA addition on shear strength and E50 under different confining pressures of 100 kPa, 200 kPa, and 300 kPa: (a) and (c) are for MICP-treated FA1–sand mixture; (b) and (d) are for MICP-treated FA2–sand mixture.

was not as excellent as that present in MICP-treated samples containing 3% FA1. The comparison of MICP-treated samples containing 3% FA1 and that containing 9% FA1 illustrated that densification by the inclusion of more FA1 did not lead to the increase of shear strength. Cheng et al. (2018b, 2019) found the loss of injected lubricant can rather occur while tunneling in poorly graded gravel than clayey gravel. The occasional gravel leads to the misleading reduction in the frictional coefficient. In this study, a similar observation occurred for the reduction of shear strength after adding more fine content. The decrease of void space for bacteria to go through is the reason to cause the decrease of shear strength of sample containing 9% FA1. As a result, the MICP process was restricted, and thus, shear strength was decreased. Therefore, the excessive addition of FA1 may limit the mechanical properties development of MICP-treated FA1–sand sample.

As shown in Figure 6.13b, with the addition of FA2, the shear strength presented a similar trend under varying confining pressure as that of FA1 addition, except that at 200 kPa confining pressure, the shear strength of MICP-treated FA2–sand samples increased greatly when the confining pressure varied from 100 kPa to 300 kPa. Meanwhile, the maximum shear strength of 1200 kPa was obtained by MICP-treated 3% FA2–sand sample at 100 kPa confining pressure. Compared to that of MICP-treated FA1–sand sample, the enhancement of shear strength by MICP-treated FA2–sand sample was superior. The higher content of CaO in FA2 could contribute to this enhancement. The fundamental reactions during lime ($Ca(OH)_2$) improvement can be attributed to four basic reactions: cation exchange, flocculation and agglomeration, lime carbonation, and finally, the pozzolanic reaction (James et al. 2008; Xu et al. 2019). The higher CaO content and pH (11.1) of FA2 made these reactions happen during the MICP process in the presence of moisture and, thus, bonded the adjacent soil particles together.

Secant modulus (E50) is one of the most important properties of modified soils which link the stress and the corresponding strain (Dutta et al. 2017; Luis et al. 2019). E50 as a function of confining pressures is presented in Figures 6.13c and d. All the values of E50 were calculated based on the ratio of the deviator stress at 50% of the ultimate load, with the axial strain at 50% of the ultimate load. E50 of both MICP-treated 3% FA1–sand and MICP-treated sand-only sample increased, and E50 of MICP-treated 6% FA1–sand sample and MICP-treated 9% FA1–sand sample decreased, when the confining pressure varied from 100 kPa to kPa. At 300 kPa, MICP-treated 6% FA1–sand sample and MICP-treated 9% FA1–sand sample obtained higher E50 than others. This phenomenon implies the effect of FA1 amount and confining pressure on E50 was coupled. The coupled effect also could be found in MICP-treated FA2–sand samples. It should be noted that the E50 of MICP-treated 3% FA2–sand sample was lower than that of 6% and 9% addition at 100 and 200 kPa.

6.1.2.4 Influence of FA on Friction Angle and Cohesion

Figure 6.14 presented the Mohr circle and failure envelope of MICP-treated 9% FA1–sand mixture sample under 100, 200, and 300 kPa confining pressures. The friction angle of MICP-treated 9% FA1–sand mixture was 41.01°, and the cohesion was 123 kPa. Table 6.4 showed the effects of FA1 and FA2 addition on the friction

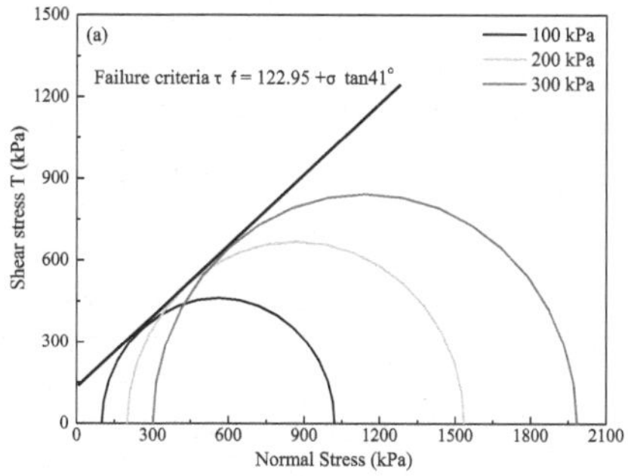

FIGURE 6.14 Mohr circle and failure envelope of MICP-treated 9% FA1–sand mixture.

TABLE 6.4
Cohesion and Friction Angle of MICP-Treated Sample with FA Inclusion

	0%	3% FA_1	6% FA_1	9% FA_1	3% FA2	6% FA2	9% FA2
C' (kPa)	46.79	343.03	173.66	122.95	635.63	168.49	69.49
<TO	47.44	35.72	43.16	41.01	28.94	39.23	47.35

angle and cohesion. With the inclusion of FA, friction angles of MICP-treated FA–sand mixture samples decreased compared with the MICP-treated sample without FA inclusion. When FA1 contents were 3%, 6%, and 9%, friction angle decreased to 35.72°, 43.16° and 41.01°, respectively, compared to 47.44° of MICP-treated sand-only sample. With 3%, 6%, and 9% FA2 inclusion, friction angle decreased to 28.94°, 39.23° and 47.35°, respectively.

The friction angle is mainly provided by extrusion and friction between particles. It means the addition of FA, which contained a large amount of fine content, smoothed the interface between soil particles and hence made it easy for soil particles to change position from one point to another point. These results have been manifested by the comparatively increased failure strain from Table 6.5.

The cohesions of MICP-treated samples with the addition of FA increased, as shown in Table 6.4 and Figure 6.14b. When 3% FA1 was added, cohesion increased by 730% compared to that of MICP-treated sand-only sample. The cohesions increased by 368% and 268%, with 6% and 9% FA1 contents. Meanwhile, the cohesions reached 635.6 kPa, 242.2 kPa, and 69.5 kPa, with 3%, 6%, and 9% FA2 inclusion, respectively, compared to 46.8 kPa of MICP-treated sand-only sample. During the MICP process, $CaCO_3$ may be deposited between soil particles, coat soil surfaces, or bridge soil particles, which improves the bonding force between soil particles. The increase in cohesion made it more difficult for soil particles to move under loading pressure. This could be the reason for the improvement of shear strength with the addition of FA. With the increase of fly ash addition, the enhancement of cohesion slowed down.

TABLE 6.5
Effect of FA Content on Failure Strain (εf) and Elastic Modulus (E) of MICP-Treated Samples under Triaxial Compressive Conditions

			FA$_1$						FA$_2$					
	0%		3%		6%		9%		3%		6%		9%	
Cell Pressure (kPa)	ε_f (%)	E (MPa)	ε_f (%)	E (MPa)	ε_f (%)	E (MPa)	ε_f (%)	E (MPa)	ε_f (%)	E (MPa)	ε_f (%)	E (MPa)	ε_f (%)	E (MPa)
100	2.0	39.97	3.4	57.64	3.2	38.59	3.6	25.60	5.8	20.35	6.8	7.89	3.6	12.54
200	2.9	50.93	2.3	66.05	3.8	37.31	4.0	33.4	6.0	14.50	4.8	15.58	2.9	18.40
300	4.1	46.77	4.2	59.39	3.0	70.03	3.5	48.84	6.8	20.11	4.7	15.58	2.5	40.43

6.1.2.5 Influence of FA Addition on Failure Strain and Elastic Modulus

The failure strain ($\mathcal{E}_f$) and the elastic modulus (E) of MICP-treated samples with different FA content under triaxial compressive conditions have been shown in Table 6.5. The failure strains of MICP-treated samples increased along with the addition of FA1 at 100 kPa and 200 kPa confining pressure. These results indicated that the ductility of MICP-treated samples was improved by the addition of FA, and the improvement was more significant with more FA content. The fine content of FA which may smooth the surface of particles would be the reason which contributed to the extension of the failure strain. However, this is different for the MICP-treated 6% FA1–sand mixture and the MICP-treated 9% FA1–sand mixture at 300 kPa confining pressure, as the failure strain stayed at 3% compared to 3.5% of MICP-treated sand-only sample. At high confining pressure, the brittleness of samples would be changed, making samples failure at low strain. For MICP-treated sand-only sample, the average elastic modulus was 45.9 MPa, with the maximum value of 51.9 MPa. With the inclusion of FA1, the average elastic modulus increased to 48.5 MPa, with maximum value of 70.0 MPa, which increased by 106% and 138%, respectively. Given the variation of elastic modulus is directly related to the variation of failure strength and failure strain, the increase of elastic modulus indicates that the increase of strength is more significant than the increase of strain. Moreover, the increase of elastic modulus demonstrated the strengthening effect of FA addition is mainly attributed to the increased elastic modulus, rather than the extended strain, which means the rigidity of samples was enhanced.

With the addition of FA2, the average elastic modulus of MICP-treated samples decreased to 18.4 MPa, with the maximum value of 40.4 MPa. Meanwhile, the failure strain of MICP-treated FA2–sand samples increased. MICP-treated 3% FA2–sand sample was broken when the strain reached 6.8% at 300 kPa confining pressure. The maximum strains of MICP-treated 6% FA2–sand mixture sample and MICP-treated 9% FA2–sand mixture sample against shear stress were 6.8% and 3.6% compared to the maximum strain of MICP-treated sand-only samples of 4.1%. The increase of failure strains indicated that the inclusion of FA2 prompted the enhancement of ductility of MICP-treated samples.

6.1.2.6 Leaching of Metals from MICP-Treated FA–Sand Mixture

As aforementioned, fly ash was criticized because it may release heavy metals during application. Therefore, the leaching concentrations of 11 elements were detected in accordance with US EPA method 1311 in this study, and the results are summarized in Table 6.6. The regulatory levels of EPA, Mississippi Department of Environmental Quality (MDEQ) (2005), and maximum contaminant level (MCL) are also listed in the table. It is apparent that the concentrations of 11 elements met the EPA, MDEQ, and MCL levels. Thus, it is safe to introduce FA1 and FA2 into the MICP process to enhance the mechanical properties of the soil. It is worth noting that the leaching concentration of several metals in the FA–soil mixture, such as Al, Ba, Ni, and Zn, increased after processing by MICP. If more FAs are contained in the samples, treatment for the stabilization/fixation of metals is needed.

In this study, a four-step European Community Bureau of Reference (BCR) leaching test was applied to recognize the environmental behavior of the metals. The results are presented in Figure 6.15. The potential leaching forms of metals are summarized

TABLE 6.6
Leaching Concentrations of Elements in Original Soil, FA1, FA2, and MICP-Treated Samples

Soil[1]	Al	As	B	Ba	Cd	Cr(III)	Cu	Fe	Mg	Ni	Zn
FA_1	ND	ND	0.01	0.01	ND	ND	ND	ND	7.37	ND	ND
S	0.003	ND	ND	ND	ND	0.009	0.009	0.009	0.17	0.01	0.016
M3	0.87	ND	ND	0.002	ND	0.008	0.008	0.15	1.36	0.005	0.012
M6	0.72	ND	ND	0.007	ND	ND	ND	0.2	1.88	0.006	0.006
M9	3.48	ND	0.04	0.03	ND	ND	ND	0.44	4.94	0.01	ND
FA2	0.18	ND	0.03	0.03	ND	ND	ND	ND	7.21	ND	ND
N3	0.68	ND	ND	ND	ND	ND	ND	0.11	1.1	ND	0.037
N6	0.53	ND	ND	0.005	ND	0.006	0.005	0.22	1.5	ND	0.007
N9	0.94	ND	0.04	0.01	ND	ND	ND	0.23	1.53	ND	ND
EPA[2]	NA	5.0	NA	100	1.0	5.0	NA	NA	NA	NA	NA
MDEQ[3]	NA	0.5	NA	10	0.1	0.5	NA	NA	NA	NA	NA
MCL[4]	NA	0.01	NA	2	0.005	0.1	1.3	NA	NA	NA	NA

[1] *FAi*: raw material; *FA2*: raw material; *S*: MICP-treated sand-only samples; *M3*: MICP-treated 3% FA2–sand mixture sample; *M6*: MICP-treated 6% FA2–sand mixture sample; *M9*: MICP-treated 9% FA sand mixture sample; *N3*: MICP-treated 3% FA2–sand mixture sample; *N6*: sample; *N9*: MICP-treated 9% FA2–sand mixture sample.

[2] US EPA Regulatory levels for classifying waste as hazardous based on TCLP results.

[3] Mississippi Department of Environmental Quality (MDEQ) regulatory levels for the beneficial use of solid wastes.

[4] MCL: maximum contaminant level (MCL); the highest level of a contaminant that is allowed in drinking water.

ND: not detected.

NA: not available.

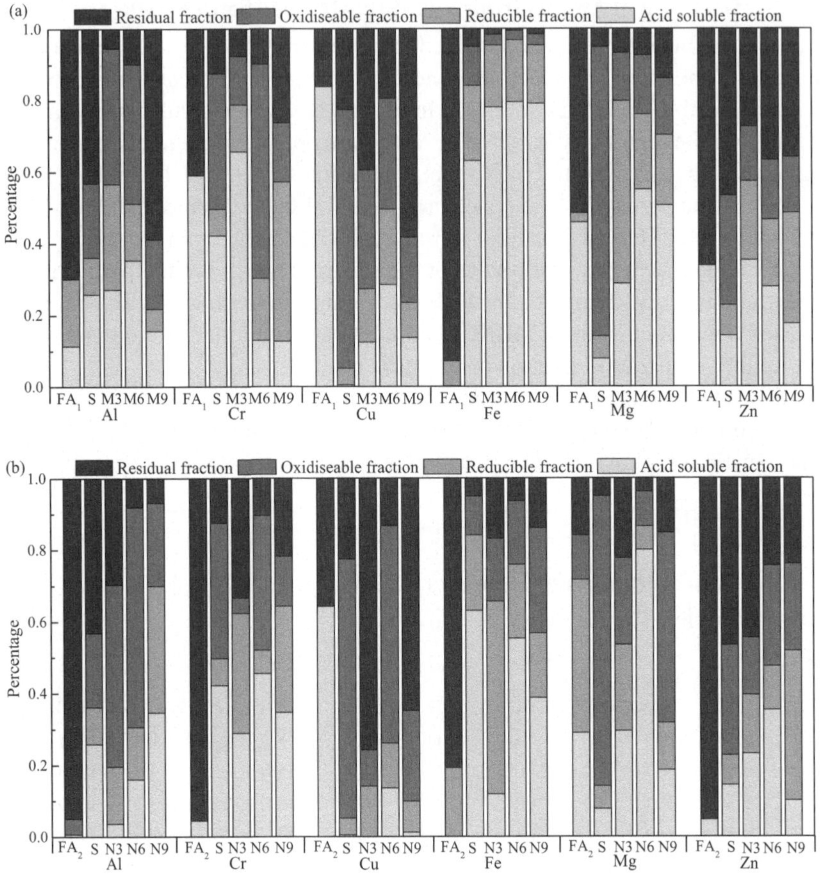

FIGURE 6.15 BCR sequential leaching results of (a) FA1-MICP-treated soil and (b) FA2-MICP-treated soil.

Note: *FA1*: raw material; *FA2*: raw material; *S*: MICP-treated sand-only samples; *M3*: MICP-treated 3% FA1–sand mixture sample; *M6*: MICP-treated 6% FA1–sand mixture sample; *M9*: MICP-treated 9% FA1–sand mixture sample; *N3*: MICP-treated 3% FA2–sand mixture sample; *N6*: MICP-treated 6% FA_2–sand mixture sample; *N9*: MICP-treated 9% FA_2–sand mixture sample.

in four fractions. F1 is designed to represent water-soluble, exchangeable, and carbonate-bound metals. Metals in F1 are more susceptible to pH changes. Reducible fraction F2 represents the amount of each element that would be released in reducing conditions. Metals existing in F2 are usually bonded to Fe-Mn oxide. F3 is related to the fraction which is sensitive to oxidative conditions, and metals are assumed to remain in the solid matrix for a long period but may be mobilized by the decomposition process. F4 is the most stable phase and, thus, is the most difficult to detach (Kartal et al. 2006). Therefore, the higher percentage of F4 metal obtained, the more stable the metal will be. As shown in Figure 6.15a, the concerned metals mainly existed in F4 in FA1, and therefore, the leaching concentrations of each metal in FA1 were extremely low. However, in the MICP-treated FA1–sand sample, metals in F4 decreased greatly. In the meantime, acid-soluble and oxidizable fraction of the metal increased significantly. For instance, about 95% of Fe existed in F4 in the raw FA1, while this number decreased to less than 5% after MICP process, and the acid fraction increased up to 80%. The same phenomenon could also be found in other metals. Therefore, this could be the main reason that after the MICP process, the leaching concentrations of some of the metals in TCLP leachate increased. As shown in Figure 6.15b, almost the same phenomenon was detected in the MICP-treated FA2–sand sample. Therefore, it is reasonable to conclude that the MICP process could induce heavy metals leaching out from MICP-treated FA–sand samples, though the concentrations of all the concerned metals met the relative regulatory levels in EPA, MDEQ, and MCL.

6.1.2.7 Microstructure Analysis

From a microscale view, the strength and stiffness increases of MICP-treated FA–sand samples were based on microstructure improvement. It is reported that there are three types of precipitated calcium carbonate in the MICP process: (a) calcite, (b) vaterite $CaCO_3$ crystals, and (c) amorphous $CaCO_3$. Factors such as urea hydrolysis rate and organic compounds may affect the crystallization process of $CaCO_3$, and thus the final production (Lin et al. 2016). It was observed by several researchers that the precipitated calcium carbonate nucleated around bacteria cells which attached to the sand particle surface. The produced $CaCO_3$ may work as contact cement (deposit at or near particle contact points), grain coating (coat the soil particles), or $CaCO_3$ bridge (fill the void space without contacting a soil particle or grow from the particle surface into the pore space) (Chew et al. 2004).

To investigate the microstructural properties of $CaCO_3$ produced in this work, a MICP-treated 6% FA1–sand mixture sample was chosen to carry out SEM tests. Details of SEM images have been presented in Figure 6.16 at the scale of 200 μm, 100 μm, and 10 μm, respectively. Figure 6.16c is an enlarged view of the selected part from Figure 6.16a, while Figure 6.16d is enlarged from the selected part of Figure 6.16b. The images of MICP-treated 6% FA1–sand mixture sample showed the clear presence of $CaCO_3$ crystals with different deposition. It is observed in Figure 6.16a that a large amount of precipitated $CaCO_3$ coated the particle surface and filled the void space. Small particles are buried and finally formed as aggregates. Thus, a high-cementation condition has been produced between particle matrixes. With the addition of FA1, a more stable soil structure was formed after MICP treatment. This is confirmed in the results that the strength of MICP-treated samples increased with the addition of FA1 compared to the MICP-treated sand-only samples. As shown in Figure 6.16d, irregular polyhedrons were formed in MICP-treated 6% FA1–sand sample, and the size of the polyhedrons was about 8 μm.

6.1.2.8 XRD Analysis

XRD analysis was conducted to further confirm the presence of $CaCO_3$ crystal. The expected peak angles of refraction, associated with silicon oxide (SiO_2) and calcium carbonate ($CaCO_3$), were marked as Si and Ca, respectively. As shown in Figure 6.17a,

FIGURE 6.16 SEM photos: (a) MICP-treated 6% FA1–sand mixture at 200 μm, (b) FA1 sample at 100 μm, (c) MICP-treated 6% FA1–sand mixture at 100 μm, and (d) MICP-treated 6% FA1–sand mixture at 10 μm.

for pure soil, the diffraction peaks at 2Θ = 20.9 °, 26.7°, 40.3°, and 50.1°, corresponding to crystal planes of (100), (101), (111), and (112) of quartz (PDF 00-046-1045). As shown in Figure 6.17b, quartz was also identified in the FA1–soil mixture. The XRD patterns of MICP-treated soil, shown in Figure 6.17c, with calcite at 2Θ of 36°, 39.4°, 43.1°, 47.5°, and 48.5°, corresponding to crystal planes of (110), (113), (202), (024), and (018) (PDF 00-005-0586), was formed after the MICP process. In the meantime, the intensity of peaks for $CaCO_3$ varied between samples. When the addition of FA1 was up to 9%, the peaks for $CaCO_3$ decreased greatly. This phenomenon, along with $CaCO_3$ content, would be very helpful in interpreting observation made in the mechanical properties of MICP-treated samples with or without FA addition.

6.1.3 Environmental and Mechanical Impact of MICP-Treated Coal Fly Ash–Soil Mixture

6.1.3.1 Materials and Methods

6.1.3.1.1 Materials

The Ottawa silica sand used in this study is one of the most widely used sandy soils, with D50 of 0.46 mm. The composition of Ottawa silica sand is shown

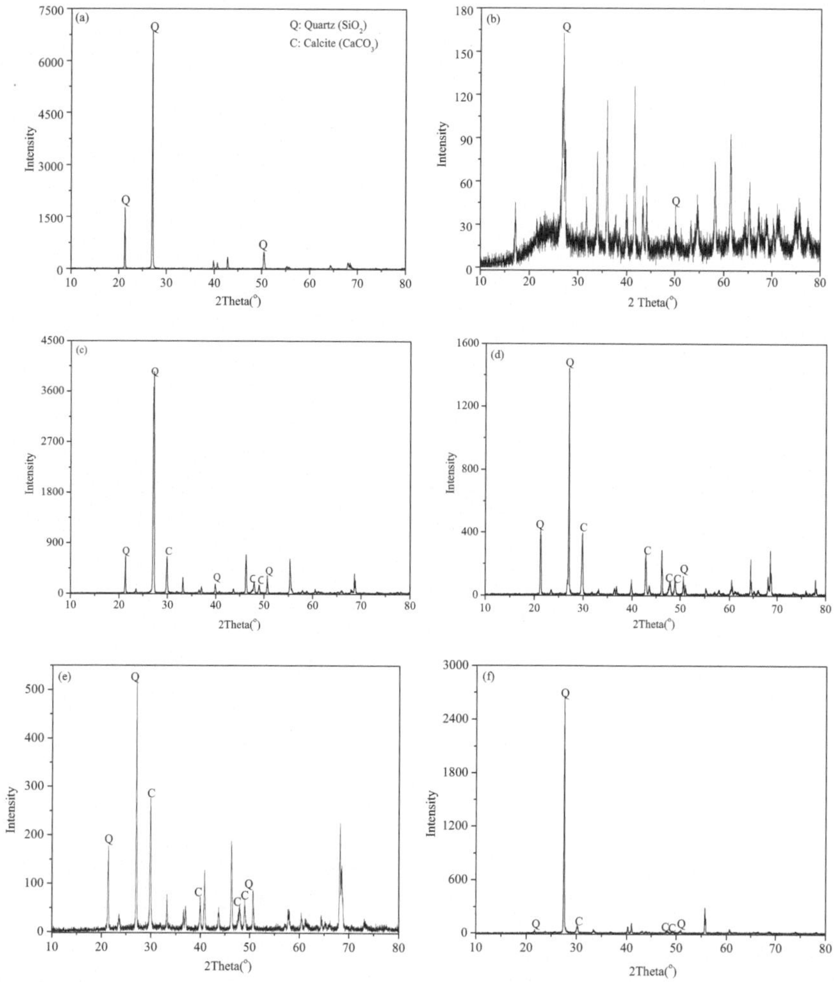

FIGURE 6.17 X-ray diffraction pattern: (a) sand, (b) FA1, (c) MICP-treated sand only, (d) MICP-treated 3% FA1–sand mixture, (e) MICP-treated 6% FA1–sand mixture, (f) MICP-treated 9% FA1–sand mixture.

in Table 6.7. Two types of CFA, named CFA1 and CFA2, obtained from thermal power plants were used. Their chemical compositions were analyzed through energy-dispersive X-ray spectroscopy (EDX-720, Shimadzu), and the results are given in Table 6.7. According to ASTM standard C 618, Standard Specification for Coal Fly Ash and Raw or Calcined Natural Pozzolan for Use in Concrete, CFA1 is classified as off-specification fly ash, and CFA2 meets class F criteria. The size distributions of the CFA and Ottawa sand are shown in Figure 6.18. It can be clearly seen that CFA1 and CFA2 are both fine-grade materials. The pH of the CFA was determined by the US Environmental Protection Agency (USEPA)

TABLE 6.7
Chemical Composition (%) of Coal Fly Ash

Material	SiO_2	Al_2O_3	CaO	SO_3	Fe_2O_3	Cr_2O_3	NiO	MnO	ZnO	Classification (ASTM C 618)
Sand	99.7	<0.1	ND	ND	<0.1	ND[a]	ND	ND	ND	NA[b]
CFA1	22.4	10.5	54.4	4.7	4.7	ND	ND	0.1	0.1	Off-specification
CFA2	45.5	23.1	10.2	2.0	11.4	ND	ND	0.1	0.04	Class F

[a] Not detected.
[b] Not available.

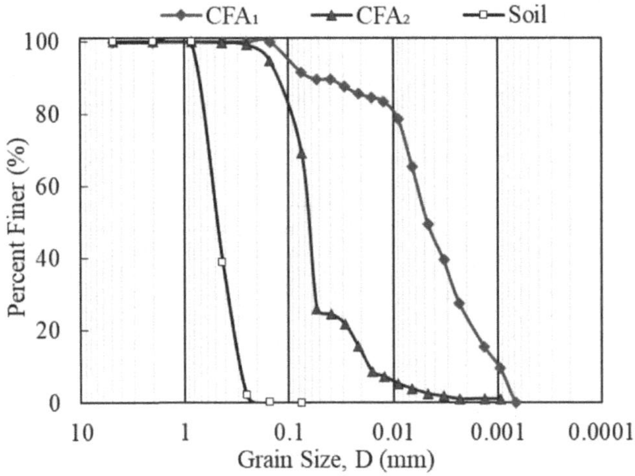

FIGURE 6.18 Grain size distribution curves of soil and two kinds of coal fly ash.

method 9045D: Soil and Waste pH. The pH of CFA1 was 11.1, and the pH of CFA2 was 10.3.

6.1.3.1.2 MICP Process

The bacteria used in this study was *Sporosarcina pasteurii* (ATCC 11859). More details about the MICP process are shown in Section 3.1.

Two kinds of coal fly ash (CFA1 and CFA2) were used in this study to prepare samples. The CFA–soil mixture was prepared by adding 0%, 3%, 6%, 9%, and 12% of CFA based on soil weight to achieve a total amount of 150 g. The 0% CFA addition means the sand-only sample is the control. The 3%, 6%, 9%, and 12% CFA additions were mixed with sand to prepare the CFA–sand mixture. Each sample was triplicated. The sample holder was full-contact flexible mode made by geotextile that has been developed by Zhao et al. (2014a). The sample prepared with sand and without CFA was used as control.

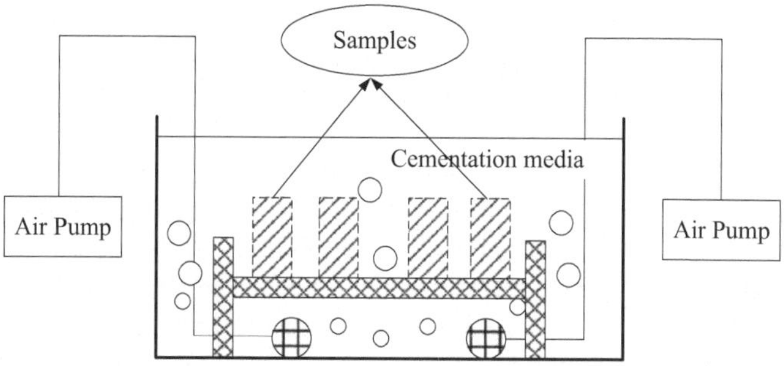

FIGURE 6.19 Diagram of MICP cementation tank.

TABLE 6.8
Cementation Media Recipe

Ingredient	NH_4Cl	Nutrition Broth	$NaHCO_3$	Urea	$CaCl_2·2H_2O$
Concentration (g/L)	10.0	3.0	2.12	30.0	73.5

A 40 mL bacteria suspension was mixed with the samples and then compacted into the mold. The samples were placed on the shelf, carefully immersed into the batch reactor. This batch reactor was filled with cementation media used to induce bio-cementation. The sketch of the batch reactor is shown in Figure 6.19. The treatment method followed Zhao et al. (2014a), which shows a uniform $CaCO_3$ precipitated inside the samples. Cementation media were prepared before the test, and the recipe is summarized in Table 6.8. The selection of the chemical concentration was based on the efficiency of the MICP process. The optimization of chemical concentration was done by Zhao et al. (2014a). The MICP process was conducted under aeration conditions. Following Zhao et al. (2014a), MICP-treated specimens were taken out after 7 days of continuous reaction. Zhao et al. (2014a) recommended that 7 days is the optimum span to reach the maximum UCS. Mechanical properties and leaching toxicity were then measured. As control, a set of samples with the same prescribed other than bacteria was prepared to indicate the work of MICP.

6.1.3.1.3 Unconfined Compression Test and Splitting Tensile Test

Columnar samples with 38 mm diameter and 76 mm height were prepared for UCS tests according to ASTM method D2166 (ASTM 2012). The test was performed under strain-controlled conditions at a uniform loading rate of 1.5%/min. The STS tests were conducted in accordance with ASTM D3967 (ASTM 2016b). The sample diameter was 63.5 mm, with a diameter-to-height ratio of 2.5:1. The compressive strength and STS were calculated using equations (6.7) and (6.8), as follows:

$$Q_c = P / A \tag{6.7}$$

$$\delta_t = 2p / \pi DL \tag{6.8}$$

where Qc is the compressive strength (kPa), P is applied load (N), A is cross-sectional area (mm^2), δ_t is splitting tensile strength (kPa), D is the diameter of the sample (mm), and L is the thickness of the sample (mm).

6.1.3.1.4 Leaching Toxicity of the MICP-Treated CFA–Soil Mixture

To examine the environmental impacts of CFA, water leaching test was conducted according to ASTM D3987 (2012). In this method, an aliquot of 5 g dried sample sieved through US No. 4 sieve (4.75 mm) was mixed with 100 mL deionized water in a 125 mL HDPE bottle. The mixture was rotated continuously on an end-over-end rotator at 30 rpm for 18 ± 2 h. After agitation, the specimens were filtered through 0.2 µm glass-fiber filter. The filtrates were acidified to a pH less than 2.0 using concentrated nitric acid and stored in 50 mL centrifuge tubes at 4°C. Triplicate leaching test was conducted on all samples. All the filtrates from batch leaching tests were analyzed using an inductively coupled plasma optical emission spectrometer (ICP-OES) (Shimadzu ICPE-9000). Seven possible heavy metals—Al, Cd, Cr, Cu, Mg, Ni, and Zn—were analyzed.

6.1.3.1.5 XRD, FTIR, and SEM-EDS Analyses

X-ray diffraction of the CFA–soil mixture before and after the MICP process was recorded using an X-ray diffractometer (Regiku Miniflex 600), and the functional groups of the samples were performed through FTIR (UTAR 2, PerkinElmer). For XRD tests, Cu-Kα radiation was used, and the 2Θ ranged from 10° to 80°, with a step width of 1°. FTIR analysis was conducted in the 1600–400 cm^{-1} (2.5–10 µm wavelength) region, with 4 cm^{-1} resolution. SEM-EDX (Lyra 3, TESCAN) analysis was performed to examine the morphology and chemistry composition of MICP products. The MICP-treated specimens were oven-dried at 105°C before tests, then, imaged by secondary electron detection.

6.1.3.2 Mechanical Properties of MICP-Treated CFA–Soil Mixture

After 7 days of continuous reaction in the cementation tank, the MICP-treated samples were harvested for further tests. The images of the MICP-treated samples are shown in Figure 6.20. It is apparent that the loose particles were cemented together by

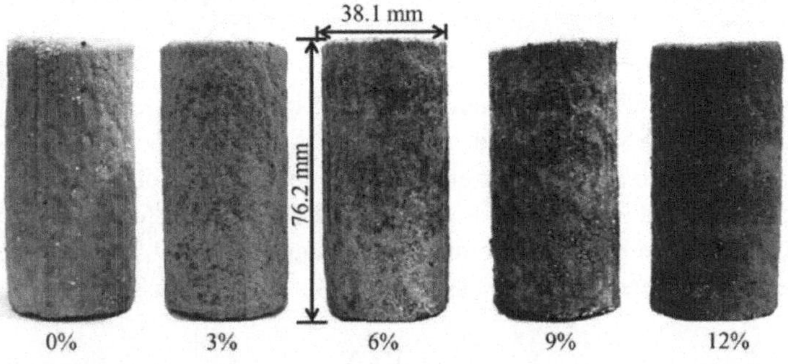

FIGURE 6.20 MICP-treated CFA1–soil mixture.

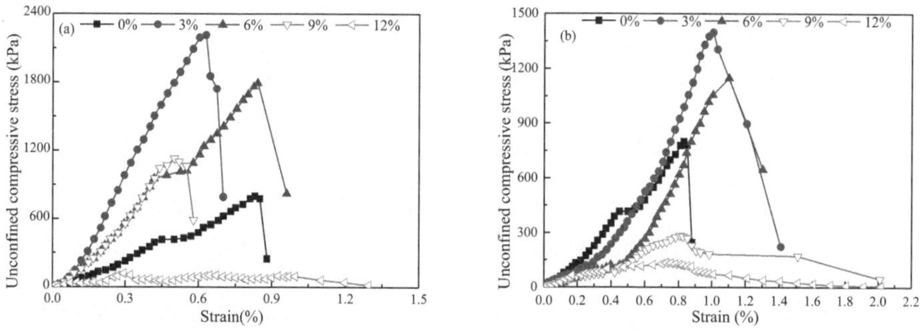

FIGURE 6.21 Unconfined compressive stress test of MICP-treated samples: (a) CFA1–soil mixture and (b) CFA2–soil mixture.

$CaCO_3$ produced in the MICP process, thus leading to the enhancement of mechanical properties. As shown in Figure 6.21a, the UCS of the MICP-treated soil-only sample was about 798 kPa. With the addition of 3% CFA1, the UCS of MICP-treated CFA1–soil mixture increased to 2200 kPa. However, when the content of CFA1 increased from 3% to 9%, the UCS of MICP-treated CFA1–soil mixture gradually decreased. With 6% and 9% of CFA1, the UCS of the MICP-treated samples were 1,800 and 1100 kPa, respectively. However, the UCS of MICP-treated samples with 6% and 9% CFA1 were still much higher than that of MICP-treated soil-only sample (798 kPa). Therefore, it is reasonable that a certain amount of CFA1 enhanced the cementation work and thus increased the UCS. The main reason could be fine content of CFA1 introduced into the MICP process, which filled the void space and thus promoted densification. When CFA1 amount was increased to 12%, the UCS of the MICP-treated CFA1–soil mixture decreased dramatically to about 100 kPa. Several researchers have investigated the effect of void samples on the MICP process by considering a different grading size and void ratio; they concluded that a suitable space of sample is vital for bacteria to survive and thus affect the quality of the MICP process. This is because (1) space affects the void volume and soil's saturated permeability, through which the bacteria access nutrient living in soil pores, and/or (2) lack of space for bacteria growth (Cheng et al. 2013; Stocks Fischer et al. 1999; Terzis et al. 2016). Therefore, too much CFA1 may prevent the mixture from providing enough void ratio and space for the bacteria to go through and, thus, result in a weak MICP process (van Paassen et al. 2010a). In this case, to enhance the UCS of a MICP-processed product, the amount of CFA1 should be no more than 9%, while 3% of CFA1 is recommended.

The effect of CFA2 on the UCS of MICP-treated CFA2–soil mixture is presented in Figure 6.21b. When the content of CFA2 was 3%, the UCS of the sample was 1398 kPa. The UCS decreased to 1147 kPa when the content of CFA2 increased to 6%. When CFA2 content increased to 9%, the UCS of the sample decreased to 285 kPa, which is much lower than that for MICP-treated soil-only samples (798 kPa). These results indicate that the participation of CFA2 can promote the solidification of CFA2–soil mixture when the content of the CFA2 is 3% and 6%, while too much CFA2 may limit the solidification and MICP process.

In the meantime, the control was investigated without bacteria. After 7 days of curing, this set of samples were not solidified, thus there were no data for the mechanical properties. In general, MICP-treated CFA1–soil mixture showed higher UCS than that of MICP-treated CFA2–soil mixture with the same percentage of CFA. This could be caused by CFA1 containing more CaO than CFA2, which can provide sufficient Ca^{2+} source and improve the MICP process. In addition, since CFA1 is classified as off-specification fly ash, it is expected that the pozzolanic reaction would happen in the cementation tank; therefore, the mechanic properties of the sample will be developed.

Tensile strength is one of the most important fundamental properties of construction material (Ibrahim et al. 2015). In this study, the splitting tensile test was conducted according to ASTM D3967 (ASTM 2016b).

MICP-treated CFA–soil mixtures with a diameter of 63.5 mm and thickness of 25.4 mm were used to evaluate the effect of CFA on STS development. Images of MICP-treated CFA–soil mixtures during STS test are presented in Figure 6.22. It revealed that the cracks were developed along the central axis of the samples, and the samples split into two pieces along the central axis at failure. For the sample containing 12% CFA1, it was observed that failure was manifested via central cracking because of splitting and multiple cracks developed because of possible weak bonds.

The splitting tensile stresses of samples are shown in Figure 6.23. The STS of MICP-treated soil-only sample was about 211 kPa at the displacement of 0.55 mm. The STS of MICP-treated soil can be considered mainly contributed by the presence

FIGURE 6.22 Images of splitting tensile stress test of MICP-treated CFA1–soil samples.

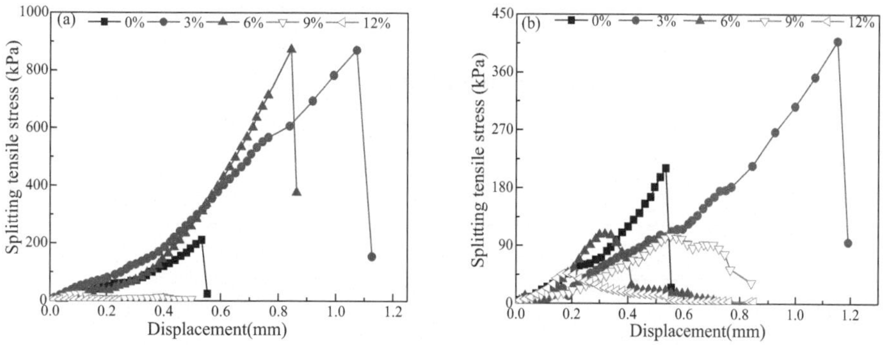

FIGURE 6.23 Splitting tensile stress of MICP-treated samples: (a) CFA1–soil mixture; (b) CFA2–soil mixture.

of bio-cement that bonds the grains. This phenomenon was also confirmed by other researchers (Cardoso et al. 2018; van Paassen et al. 2010a). By introducing 3% and 6% of CFA1, the STS of the samples reached 870 kPa and 873 kPa at 0.8 mm and 1.05 mm displacement, respectively, almost four times that of the MICP-treated soil-only sample. When the CFA1 amount reached 9% and 12%, the STS decreased sharply to 24 kPa and 0 kPa, respectively. Considering the mechanism of MICP process, enough void space for bacteria to work is the primary condition. As mentioned, too much CFA1 would confine the activity of bacteria and, thus, has a negative effect on the mechanical properties of MICP-treated CFA1–soil mixture.

The splitting tensile stress of MICP-treated CFA2-containing samples are given in Figure 6.23b. A sample that contained 3% obtained STS of 408 kPa, which was higher than that of MICP-treated soil-only sample (211 kPa). When the CFA2 content was higher than 9%, the splitting tensile stress of the samples decreased significantly, which is like the trend of CFA1-containing samples. When the CFA2 content was 12%, the STS of the sample was only 47 kPa.

6.1.3.3 Ratio of Compressive Strength to Tensile Strength

In order to study the influence of two kinds of coal fly ash and MICP on mechanical property improvement, the brittleness of the samples was examined. Brittleness has been suggested by several researchers as an important parameter for evaluating crack resistance (Hucka and Das 1974). Materials with a higher brittleness behave with low values of elongation and fracture failure. Moreover, brittle materials also have a high ratio of compressive to tensile strength and easy formation of cracks in indentations. The definition of *brittleness* based on the ratio of UCS and STS (UCS/STS) has been widely used in engineering analysis (Kahraman 2002; Li et al. 2016a). In this study, UCS/STS was calculated, and the results are summarized in Table 6.9.

From Table 6.9, the brittleness of the samples is 2.5 for MICP-treated sample containing 3% of CFA1 and 2.1 for a sample containing 6% of CFA2, both of which were lower than that of the MICP-treated soil-only sample with a brittleness of 3.8. However, other samples showed much higher brittleness, indicating the samples could be easily broken.

TABLE 6.9
Brittleness of MICP-Treated CFA–Soil Mixture

Sample	Soil	CFA1–3%	CFA1–6%	CFA1–9%	CFA1–12%	CFA2–3%	CFA2–6%	CFA2–9%	CFA2–12%
Brittleness	3.8	2.5	2.1	46.8	45.6	3.4	10.6	2.8	2.9

TABLE 6.10
Leaching Concentrations of Heavy Metals in MICP-Treated CFA–Soil Mixture

	CFA1–Soil Mixture (mg/L)						CFA2–Soil Mixture (mg/L)					MCLs[b]
Metal	0%[a]	3%	6%	9%	12%	100%	3%	6%	9%	12%	100%	(mg/L)
Al	ND[c]	ND	0.308	ND	0.209	2.17	ND	ND	ND	ND	29.2	NA[d]
Cd	ND	0.004	ND	ND	0.003	ND	0.026	ND	ND	ND	0.012	0.005
Cr	ND	ND	ND	ND	ND	ND	0.006	ND	ND	ND	0.006	0.1
Cu	0.415	0.282	0.419	0.484	0.317	0.475	0.372	0.322	0.244	0.266	0.385	1.3
Mg	1.23	2.38	2.54	6.38	0.686	0.835	0.146	0.072	0.168	0.256	0.065	NA
Ni	0.076	ND	0.09	ND	ND	ND	0.15	0.06	0.058	0.062	ND	0.1
Zn	0.237	0.225	0.087	ND	ND	0.035	0.026	ND	0.016	0.026	0.01	2.0
pH	8.28	8.46	8.53	8.64	8.55	11.1	8.31	8.35	8.43	8.52	10.3	6.5–8.5[e]

[a] Content of coal fly ash in the CFA–soil mixture on dry weight basis.
[b] *MCL*: maximum contaminant level; the highest level a contaminant is allowed in drinking water.
[c] Not detected.
[d] Not available.
[e] Secondary drinking water regulation. Non-enforceable federal guidelines regarding cosmetic effects of drinking water.

6.1.3.4 Leaching Concentrations of Heavy Metal from MICP-Treated CFA–Soil Mixture

According to ASTM D 3987 (2012), the final pH of the leachate is intended to reflect the interaction of the leaching fluid with the buffering capacity of the materials. As presented in Table 6.10, the pH of the original CFA1 and CFA2 were higher than the pH 6.5–8.5 regulated by the secondary drinking water regulation, while the pH of most samples treated by the MICP process falls into the range of 6.5–8.5 or slightly higher, thus posing less environmental effect.

The leaching concentrations of concerned heavy metals in the samples were analyzed based on ASTM D3987 (2012), and the results are summarized in Table 6.10. For the soil, Cu, Mg, Ni, and Zn were detected in the leachate at low levels (0.415, 1.23, 0.076, and 0.237 mg/L, respectively). The leaching concentrations of metals from CFA1 were all below the MCL limit, and Cd, Cr, and Ni were not detected in the leachate. For CFA2, a significant amount of Al was detected in the leachate, with a concentration of 29.2 mg/L. It is worth noting that Cd was 0.012 mg/L in the leached, exceeding the MCL limit.

Compared with the raw materials (soil, CFA1, and CFA2), the leaching concentration of Ni and Zn in the MICP-treated CFA1–soil samples decreased significantly. Cd detected in the samples that contained 3% and 12% of CFA1 was 0.004 and 0.003 mg/L, respectively, both of which were below the MCL limit. The leaching concentrations of Cd and Ni in the samples containing 3% CFA2 were 0.026 and 0.15 mg/L, respectively, both of which exceeded the MCL limit of 0.005 mg/L for Cd and 0.1 mg/L for Ni. When the CFA2 amount ranged from 6% to 12%, there was no Cd detected in the leachate. Meanwhile, the leaching concentrations of Ni decreased greatly and were much lower than the limit. As a result, there is no negative environmental effect derived from the MICP-treated CFA–soil mixture. Among the 11 types of MICP-processed mixture (0%, 3% CFA1, 6% CFA1, 9% CFA1, 12% CFA1, 100% CFA1, 3% CFA2, 6% CFA2, 9% CFA2, 12% CFA2, and 100% CFA2), 3% CFA1 is superior to other mixtures due to its better mechanical properties and relatively lower leaching concentrations.

6.1.3.5 XRD and FTIR Analysis

Based on the mechanical properties and water leaching results, XRD and FTIR analyses were conducted on the soil, CFA1, MICP-treated soil, and MICP-treated

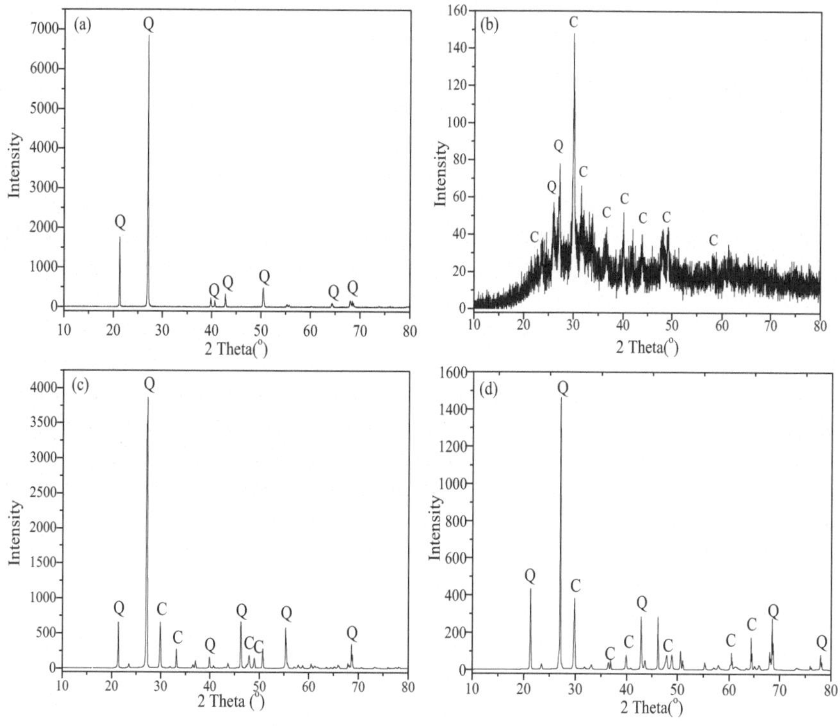

FIGURE 6.24 XRD patterns of (a) soil, (b) CFA_1, (c) MICP-treated soil, and (d) MICP-treated 3% CFA_1–soil mixture.

Note: *Q* = quartz; *C* = calcium carbonate.

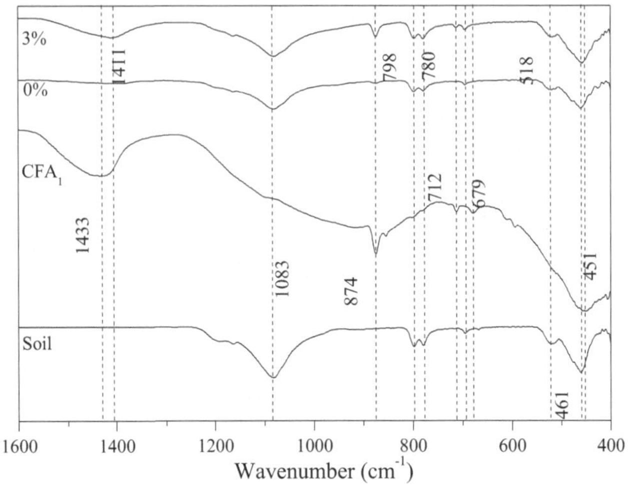

FIGURE 6.25 FTIR spectrum of the compound in different samples after MICP treatment (0% soil; 3% MICP-treated 3% CFA1–soil mixture)

3% CFA1–soil mixture to disclose the variation of the elemental composition of the samples before and after MICP treatment in this study. The results are illustrated in Figure 6.24 and Figure 6.25, respectively.

As shown in Figure 6.24, the original soil mainly consisted of quartz (labeled as *Q*), while calcite (named as *C*) and quartz were the two dominant compounds in CFA1. After the MICP process, calcite appeared in the treated soil. This is reasonable because the bacteria can induce CO_3^{2-} production, which then reacted with Ca^{2+} to form $CaCO_3$. Compared to CFA1, the 2θ of the main peak for MICP-treated 3% CFA1–soil mixture shifted from 30° to about 28°. This is evidence of structurally disordered compounds, with a high crystallization level (Bahoria et al. 2018; Hazra et al. 2016). Compared with the standard database, they are confirmed to quartz and calcite.

For the FTIR characterization, shown in Figure 6.25, the bands centered on 1,433, 874, and 712 cm^{-1} were originated by the presence of calcite (Bahoria et al. 2018; Hazra et al. 2016), and their intensities decreased with the decrease of CFA1 content. Bands at 1083 cm^{-1}, double bands at 798–780 cm^{-1}, and bands at 462 cm^{-1} were all assigned to quartz. These signals did not show significant changes; therefore, the crystalline phase did not affect the reaction conditions (Cristelo et al. 2018). Due to the overlapping of calcite produced during the MICP process and originated from CFA1, it is difficult to distinguish the source of signals in 3% of CFA1 addition specimens. The precipitation of calcite through the MICP process can be determined through the SEM graph.

6.1.3.6 Microscale Properties

According to the mechanism of MICP, the production of $CaCO_3$ is the main key to mechanical properties' development. $CaCO_3$ may be deposited at or near particles'

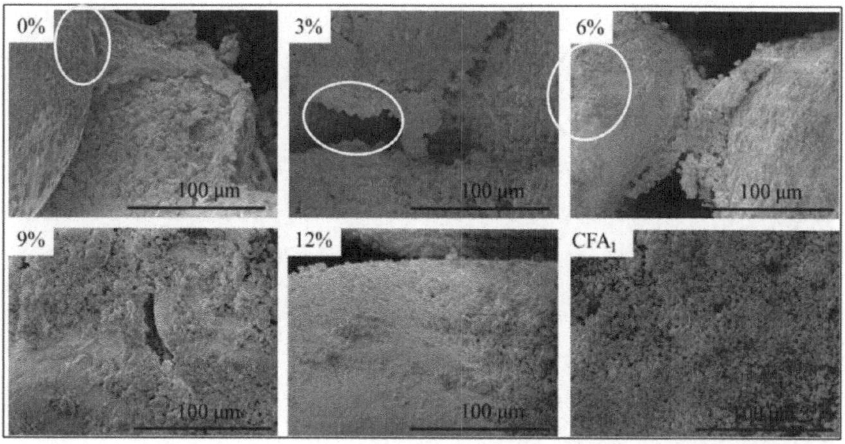

FIGURE 6.26 SEM images of MICP-treated samples under 800× magnification.

contact points, coating the particles. The $CaCO_3$ may also fill the void space without contacting a particle or grow from the particle surface into the pore space and eventually create a $CaCO_3$ bridge between different particles. Depending on the condition of the reaction that occurred, such as urea hydrolysis rate, calcite and vaterite $CaCO_3$ crystals or amorphous $CaCO_3$ are formed (Stuckrath et al. 2014; van Paassen et al. 2010a).

As shown in Figure 6.26, irregular $CaCO_3$ crystals and amorphous $CaCO_3$ were formed in MICP-treated soil-only sample. It was observed that formed $CaCO_3$ were either attached to the soil particles or deposited in pores between the particles. With the addition of CFA1, the fine particles of CFA1 took the advantage to fill the void. The formed $CaCO_3$ was encapsulated by CFA1, as shown in MICP-treated 6% CFA1–soil mixture. This combination between $CaCO_3$ and CFA1 made the void ratio much lower than the original MICP-treated soil, thus increasing their mechanical properties. However, scattered CFA1 samples located in the soil surface and filling the pores with less $CaCO_3$ crystals were detected in the samples with 9% and 12% of CFA1 additions.

More details about the microscale properties of MICP-treated samples have been illustrated in Figure 6.27. The $CaCO_3$ crystals can clearly be seen in the MICP-treated soil and MICP-treated 3% CFA1–soil mixture. While the flocculent structures were the dominant morphology of CFA1, the EDX pattern for a part of MICP-treated 3% CFA1–soil mixture indicated that $CaCO_3$ and SiO_2 were the dominant contents in the sample.

6.2 BIOREMEDIATION OF STEEL SLAG BY MICROBIALLY INDUCED CARBONATE PRECIPITATION

6.2.1 Environmental Investigation of MICP-Modified Steel Slag

6.2.1.1 Materials

The steel slag used in this study was provided by a local company in Jackson, Mississippi. The particle size distribution of SS has been determined through sieve analysis, and its median size (D_{50}) is about 0.65 mm. Particles which were between 0.425

FIGURE 6.27 SEM images of samples under 8,000× magnification

FIGURE 6.28 Materials used in this study: (a) raw SS; (b) MICP-treated SS–sand samples.

TABLE 6.11
Chemical Components of SS and Sand (wt%)

Components	MgO	Al_2O_3	SiO_2	SO_3	CaO	TiO_2	MnO_2	Fe_2O_3	NiO
SS	15.92	4.77	23.47	0.97	40.90	0.39	1.64	11.76	0.18
Sand	ND*	<0.1	99.7	ND	ND	ND	ND	<0.1	ND

* Not detected.

and 2.0 mm were utilized in this study (Figure 6.28). The chemical components have been detected by energy-dispersive X-ray analysis (EDX). As shown in Table 6.11, SS consists of 40.90% CaO, indicating its cementation potential. The amounts of SiO_2, MgO, and Fe_2O_3 in SS are 23.47%, 15.92%, and 11.76%, respectively. The pH of SS is 10.12, according to EPA method 9045b: Soil and Waste pH (USEPA 1986). Silica sand was used to prepare SS–sand mixtures. The chemical components in Table 6.11 indicate that sand consists of 99.7% SiO_2. SS–sand mixtures were prepared by 0% (sand-only), 10%, 20%, 30%, 40%, 50%, 75%, and 100% SS on weight basis.

6.2.1.2 Experimental Methods

6.2.1.2.1 Acid Digestion Test

US EPA method 3050B was used to conduct acid digestion test (ADT) for the determination of metal contents in SS (USEPA 1996). The strong acid digestion process of this test could make almost all elements environmentally available. In this test, concentrated trace metal–grade nitric acid (HNO_3) and 30% hydrogen peroxide (H_2O_2) were repeatedly added to an aliquot of 1.0 g sample until the general appearance was unchanged. A volume of 10 mL concentrated hydrochloric acid (HCl) was also added. The mixture was covered with a watch glass, placed on a hot plate, and refluxed at 95 ± 5°C. The digest was filtered and diluted with DI water to 100 mL for metal analysis.

6.2.1.2.2 Toxicity Characteristic Leaching Procedure

US EPA method 1311, Toxicity Characteristic Leaching Procedure (TCLP) (USEPA 1992a), is mainly used to determine whether waste material is hazardous according to its leaching toxicity. It is also used to examine the effectiveness of stabilization of hazardous waste. Some states, including Mississippi, use the TCLP test results to determine whether an industrial solid waste can be beneficially used (MDEQ, 2005).

The TCLP test specifies two kinds of extraction fluids. Extraction fluid no. 1 is prepared with 5.7 mL glacial CH_3CH_2OOH and 64.3 mL of 1 mol/L NaOH with DI water to 1 L volume. The pH of this fluid is around 4.93 ± 0.05. Extraction fluid no. 2 contains 5.7 mL glacial CH_3CH_2OOH with DI water to a volume of 1 L, with a final pH of 2.88 ± 0.05.

To determine the appropriate extraction fluid for TCLP, an aliquot of 5 g solid waste is weighted to mix with 96.5 mL DI water in a 500 mL baker. Then, the mixture is stirred vigorously for 5 min. The pH of the slurry is measured and recorded. If the pH is <5.0, extraction fluid no. 1 is used. If pH is >5.0, 3.5 mL 1 mol/L HCl is added to the slurry, which is heated to 50°C and held at 50°C for 10 min. When the solution cools to 23 ± 2°C, pH is measured and recorded. If pH is less than 5.0, extraction fluid no. 1 is used. Otherwise, extraction fluid no. 2 is used. Through the pretest for the extraction fluid of TCLP, extraction fluid no. 2 was used in this study.

An aliquot of 75 g solid sample was submerged in the extraction fluid with liquid-to-solid ratio (L:S) of 20:1 (L/Kg) in a 2 L HDPE bottle. The bottle was then agitated in a rotary extractor at 30 rpm for a period of 18 ± 2 hr. The leachate was filtered through 0.75 μm glass-fiber filter, and then, the pH and ORP were measured and recorded immediately. Then, pH was adjusted to less than 2 for the preservation and analysis of the concentrations of the concerned constituents.

6.2.1.2.3 Synthetic Precipitation Leaching Procedure

US EPA Standard Method 1312 was introduced to conduct synthetic precipitation leaching procedure (SPLP) (USEPA 1994). It was designated to determine the leachability of contaminants in wastes and soils. The extraction fluid consists of 60/40 weight percent mixture of sulfuric and nitric acid to DI water until pH reaches 4.20 ± 0.05. This extraction fluid is designated for wastes and soils from a site located east of the Mississippi River. An aliquot of 10 g solid samples was mixed with 200 mL extraction fluid and then rotated at 30 ± 2 r/min for 18 ± 2 hr. After agitation, the sample was filtered through 0.6–0.8 μm glass-fiber filter. The pH and ORP of leachates were measured immediately. After that, pH was adjusted to below 2 for metal determination.

6.2.1.2.4 Batch Water Leaching Test

Water leaching tests (WLT) were performed according to ASTM D 3987, Standard Test Method for Shake Extraction of Solid Waste with Water (ASTM D 3987–06, 2006). An aliquot of 10 g samples was mixed with DI water at an L/S ratio of 10:1. After 18 ± 0.25 h agitation period, the leachate was filtered. pH and the ORP of the leachate were measured. Then the pH was adjusted with 1 N HCl solution to pH below 2 for the preservation and determination of metals. The comparisons of TCLP, SPLP, and WLT about extraction fluid, solid-to-liquid ratio, and extraction time are listed in Table 6.11.

6.2.1.2.5 Metal Analysis

The metal concentrations in the leachate were determined by inductively coupled plasma atomic emission spectroscopy (ICP-AES) (ICPE-9000, Shimadzu, Japan). Commercially produced multi-element standard solutions were used to provide calibration curves. The minimum detection limits (μg/L) for metals are Al, 2.5; As, 50; B, 10; Ca, 0.5; Cd, 5; Cr, 5; Cu, 1.5; Mg, 0.5; Mn, 1; Ni, 5; V, 5; Pb, 50; Sn, 50; and Zn, 1, respectively.

6.2.1.2.6 Leachability Factor

As only certain parts of the total constituent of metals are leachable and available for organisms that have a toxic or deleterious effect, the leachability factor is employed to illustrate the mobility and availability of metals before and after the MICP process. The leachability factor is calculated by the total constituent concentration divided by the leachable constituent concentration, as shown in equation (6.9):

$$\text{Leachability Factor} = \frac{\text{Total Constituent Concentration}}{\text{Leachable Constituent Concentration}} \tag{6.9}$$

6.2.1.2.7 Pearson Correlation Coefficient

Pearson correlation coefficient can show the strength and direction of the association between two variables. Leaching concentrations from TCLP, SPLP, and WLT for the concentration of Ba, Cu, Fe, Mg, and Mn; SS content; and pH of effluent were set as the variables. Pearson correlation coefficients were calculated with SPSS 9.0 software. A two-sided p-value was used to test the hypothesis. If p-value is less than 0.05 and more than 0.01, the correlation between two variables is statistically significant. If p-value is less than 0.01, the correlation between two variables is statistically highly significant.

6.2.1.3 XRD Analysis of Raw Steel Slag

XRD results and mineral information of raw SS are shown in Figure 6.29. According to XRD analysis, it can be seen that SS was enriched in metals, and the main crystalline compounds were silica oxide (SiO_2), magnesium calcite synthetic ($Mg_{0.1}Ca_{0.9}CO_3$), copper selenate ($CuSeO_4$), magnesioferrite aluminian ($((MgA_{l0.74}Fe_{1.26})O_4)$), calcium gold germanium ($CaAu_{1.24}Ge_{0.76}$), and iron silicon oxide ($Fe_{5.36}Si_{0.64}O_8$). These results agreed with the EDX results (in Table 6.12), which showed SS consisted of 23.47% SiO_2, 40.90% CaO, 15.90% MgO, and 11.76% Fe_2O_3.

6.2.1.4 Contents and Leaching Characteristic of Metals

The leaching results of raw SS from ADT, TCLP, SPLP, and WLT tests are shown in Table 6.12. The related limits for metals from the Mississippi Department of

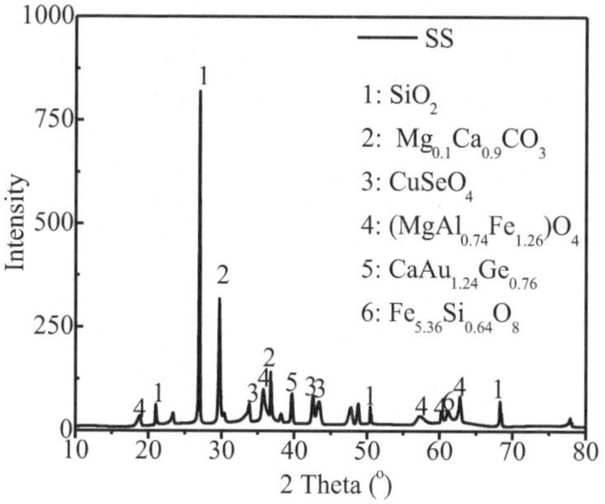

FIGURE 6.29 XRD patterns of raw SS.

TABLE 6.12
Comparison of Leaching Methods Used in This Study

Leaching Test	Method	Liquid-to-Solid Ratio (L/Kg)	Extraction Fluid
Acid digestion test (ADT)	US EPA Method 3050B	10:01	Concentrated nitric acid (HNO_3), 30% hydrogen peroxide (H_2O_2), and concentrated hydrochloric acid (HCl)
Toxicity characterization leaching procedure (TCLP)	US EPA Method 1312	20:1	Extraction fluid no. 2: 5.7 mL glacial CH_3CH_2OOH in 1 L DI water, pH = 2.88 ± 0.05
Synthetic precipitate leaching procedure (SPLP)	US EPA Method 1311	20:1	60/40 weight percent mixture of sulfuric and nitric acids in DI water, pH = 4.20 ± 0.05
Water leaching test (WLT)	ASTM D3987	10:1	DI water

TABLE 6.13
Contents and Leaching Concentrations of Raw SS

Elements	ADT (mg/kg)	TCLP (mg/L)	SPLP (mg/L)	WLT (mg/L)	EPA1 (mg/L)	MDEQ2 (mg/L)	MCL3 (mg/L)
Al	9772.5	ND[4]	0.289	0.269	NA[5]	NA	NA
As	ND	ND	ND	ND	5.0	0.5	0.01
Ba	323.5	1.41	0.543	0.553	100	10	2
Be	ND	ND	ND	ND	NA	NA	0.004
Ca	18,933.3	81.7	73.9	40	NA	NA	NA
Cd	0.48	ND	ND	ND	1.0	0.1	0.005
Cr	1,797.5	ND	ND	ND	5.0	0.5	0.1
Cu	201	0.112	0.832	0.124	NA	NA	1.3
Fe	14,6750	0.0142	0.0306	0.125	NA	NA	NA
Mg	5,0125	88.2	0.304	0.943	NA	NA	NA
Mn	21,275	0.761	ND	0.0345	NA	NA	NA
Ni	91.5	ND	ND	ND	NA	NA	NA
Pb	ND	ND	ND	ND	5.0	0.5	0
S	5,927.5	23.1	IA[6]	48.3	NA	NA	NA
Se	ND	ND	ND	ND	1.0	NA	0.05
Sn	1,907.5	4.76	ND	ND	NA	NA	NA
V	282.75	ND	ND	0.0582	NA	NA	NA
Zn	408	ND	ND	ND	NA	NA	NA

Note: *EPA* = US EPA regulatory levels for classifying waste as hazardous based on TCLP results; *MDEQ* = Mississippi Department of Environmental Quality (MDEQ) regulatory levels for the beneficial use of solid waste; *MCL* = maximum contaminant level (MCL), the highest level of a contaminant that is allowed in drinking water according to the National Primary Drinking Water Regulation; *ND* = not detected; *NA* = not available; *IA* = inappropriate.

Environmental Quality (MDEQ) and US EPA are also listed in Table 6.13. The metal contents (mg/kg) of raw SS followed the order of Fe (146,750) > Mg (50,125) > Mn (21,275) > Ca (18,933.3) > Al (9,772.5) > Sn (1,907.5) > Cr (1,797.5) > Zn (408) > Ba (323.5) > V (282.75) > Ni (91.5) > Cd (0.48).

The leaching results from TCLP, SPLP, and WLT revealed that As, Be, Cd, Cr, Ni, Pb, Se, and Zn were not detectable. The leaching concentration of Ba, Cu, and Fe met the regulation thresholds. Al, Ca, Fe, Mg, Mn, and Sn are not concerned with any regulations. Therefore, SS can be an eco-friendly material to be recycled.

6.2.1.5 pH of MICP-Treated Steel Slag–Sand Mixtures

The effluent pH of solid waste as an important safety criterion has been highlighted in many regulations. If the pH of a waste is either higher than 12.5 or lower than 2.0, it would be classified as hazardous waste due to the apparent corrosive characteristic based on the results of EPA 9045D (USEPA 1986). From Figure 6.30, the pH value

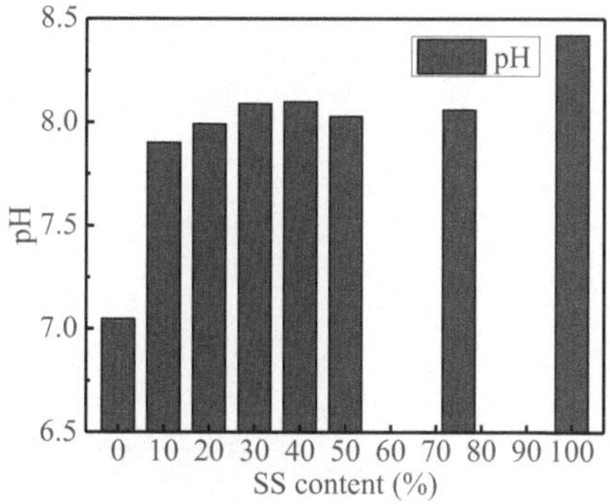

FIGURE 6.30 pH value of MICP-treated SS–sand samples

of raw steel slag was 10.12, which means there is no concern about corrosion. With the treatment of MICP process, the pH value of MICP-treated SS–sand samples was in the range of 6.5–8.5, which met the regulation of the US EPA National Secondary Drinking Water. As shown in Figure 6.30, the pH value of MICP-treated SS–sand samples increased with an increase of SS content, except samples that contain 50% and 75% SS. As stated by Mahedi et al. (2019), the pH of a material highly depends on the ratio of CaO and MgO contents and their dissolution rates in the aqueous solution. For MICP-treated SS–sand samples, SS is the unique origin of CaO and MgO. Therefore, it is reasonable that the pH of MICP-treated SS–sand samples was highly associated with the content of SS.

6.2.1.6 pH and Redox Potential (Eh) of Extract of TCLP, SPLP, and WLT Tests

The effluent pH could greatly influence the leaching behavior of metals, and thus, pH is an important factor to analyze the leaching trend of metals. Figure 6.31a indicates that the effluent pH of MICP-treated SS–sand samples increased significantly with the increase of SS content. When the SS content was higher than 30%, the effluent pH of MICP-treated SS–sand sample was kept steady, which means there was a dynamic equilibrium between H^+ provided by the extraction fluid and the alkalinity of MICP-treated SS–sand samples. The final pH was near 7.0 for TCLP effluent, 9.6 for SPLP effluent, and 9.7 for WLT effluent. Compared with the original pH of the extraction fluid, the huge increase of pH illustrates the strong acid neutralization characteristic and ability to resist acid corrosion for MICP-treated SS–sand samples.

Redox potential (Eh), also known as oxidation-reduction potential, is defined as the potential of a solution or environment to engage in either oxidation or reduction reactions (Zhang et al. 2016). The Eh of MICP-treated SS–sand mixtures is illustrated in Figure 6.31b. With treatment of the MICP process, the conditions of SPLP and WLT extract solutions were reductive, with the Eh value varying from −160 to −60 mV, while TCLP extract solution was near neutral, with the Eh value in the range of −20 to 100 mV.

The metal leaching behavior as a function of the pH of TCLP, SPLP, and WLT effluents is illustrated in Figures 6.32a, b, c. To clarify the effects of the pH of effluent

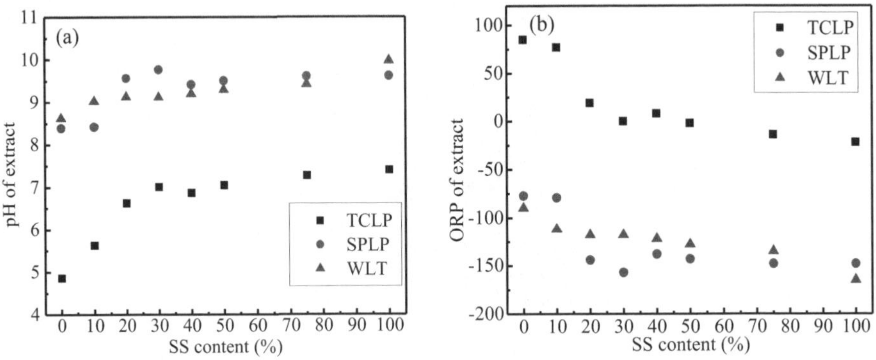

FIGURE 6.31 pH and Eh of TCLP, SPLP, and WLT effluents for MICP-treated samples: (a) pH, (b) Eh.

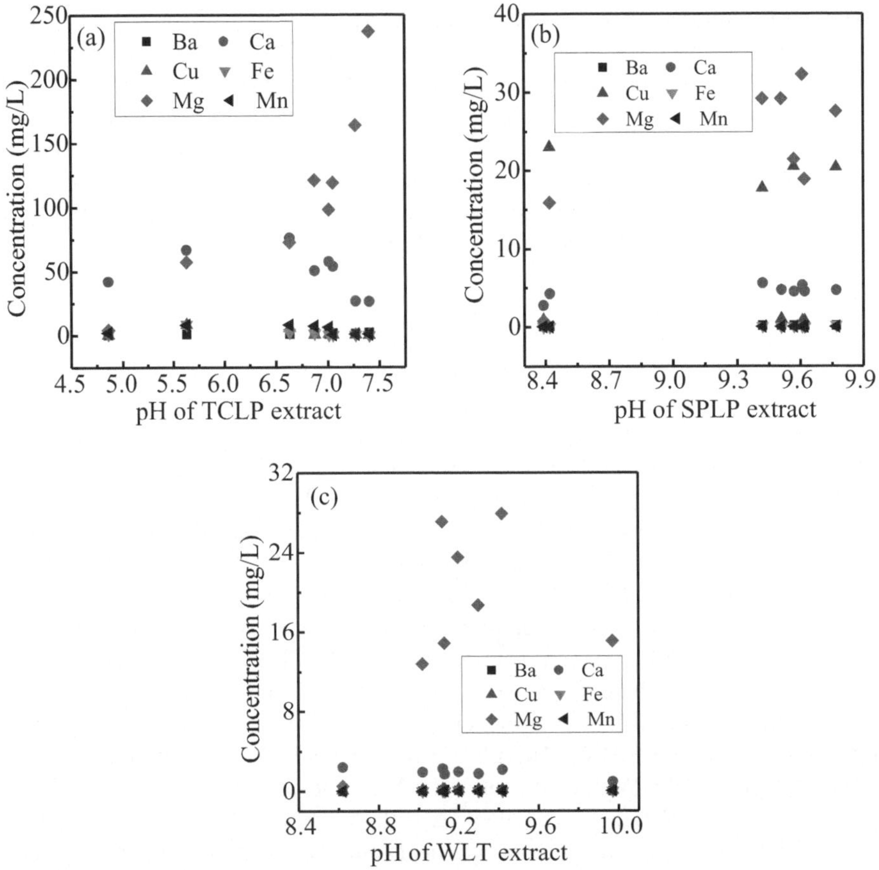

FIGURE 6.32 Leaching concentrations as a function of extract pH for MICP-treated samples: (a) TCLP extract, (b) SPLP extract, and (c) WLT extract.

on metal leaching behavior, correlations between leachate pH and metal concentrations in TCLP, SPLP, and WLT tests were calculated based on the Pearson correlation tests, as shown in Table 6.13. Komonweeraket et al. (2015b) found that the dissolution of CaO and MgO minerals from fly ashes is considered to contribute to the increase in the pH of the solution, while the dissolution of B_2O_3, $Fe_2(SO_4)_3$, and Al_2O_3 constituents in fly ashes decreases the alkalinity of the aqueous solution. In this study, the coefficients of leaching concentrations of Ba and Mg with the pH of TCLP and SPLP leachate were 0.804 and 0.893, respectively, which means the leaching of Ba and Mg from MICP-treated SS–sand samples is highly correlated with TCLP and SPLP leachate pH. The leaching concentration of Cu from MICP-treated SS–sand mixture in TCLP, SPLP, and WLT effluents was negatively related to the pH of the leachate solution. The leaching concentrations of Fe and Mn did not show any noticeable correlation to other factors. As Ca presented in this study was not merely from SS, it cannot be treated as the pH-controlling element.

6.2.1.7 Metal Leaching Behavior Evaluation

According to the US National Primary Drinking Water Regulation, Ba and Cu may cause concern with the drinking water system, and thus, both were measured in this study. There are also environmental concerns and human health hazards associated with the leaching of Fe and Mn. For instance, excess amount of Fe may cause tissue damage (Mahedi et al. 2019). The leaching concentrations of Mg and Ca were also measured for their abundant amounts in raw steel slag, and both may influence the precipitation of other metals.

The leaching concentrations of Ag, Al, As, Ba, Ca, Cd, Cr, Cu, Fe, Mg, and Mn in TCLP, SPLP, and WLT leachate are listed in Table 6.14, Table 6.15, Table 6.16. The selective regulations for TCLP and SPLP leachate are the US EPA regulation for classifying waste as hazardous based on TCLP results, and MDEQ regulation for the beneficial use of solid wastes based on TCLP results. The regulation compared with WLT results is the US EPA regulation for the highest level of contaminant allowed in drinking water. As shown in Table 6.14, Table 6.15, and Table 6.16, the leaching concentrations of metals met the regulation limits, which means MICP-treated SS–sand samples are environmentally friendly materials and can be beneficially used. The components detected in WLT effluents were below the regulation of MCL. Although

TABLE 6.14
Pearson Correlation Coefficients of Leaching Concentrations and Extract pH

Pearson Coefficient	Ba	Ca	Cu	Fe	Mg	Mn
pH of TCLP extracts	0.804*	−0.288	−0.378	−0.521	0.855**	−0.252
pH of SPLP extracts	0.893**	0.701	−0.054	0.338	0.796*	0.646
pH of WLT extracts	0.105	−0.825*	−0.181	−0.623	0.444	0.394

* Statistically significant (same with rest tables).
** Statistically highly significant (same with rest tables).

TABLE 6.15
Leaching Concentrations of MICP-Treated SS–Sand Samples in TCLP Leachates

	SS Content (%)									
Metal	**0**	**10**	**20**	**30**	**40**	**50**	**75**	**100**	**EPA[1]**	**MDEQ[2]**
Ag	ND[3]	ND	ND	ND	ND	ND	ND	ND	5.0	0.5
Al	ND	0.384	ND	ND	ND	ND	ND	ND	NA[4]	NA
As	ND	ND	ND	ND	ND	ND	ND	ND	5.0	0.5
Ba	0.452	0.697	0.796	0.480	0.440	0.023	0.039	0.048	100	10
Ca	55.2	66.8	75.9	57.5	50.5	3.8	6.48	6.23	NA	NA
Cd	ND	ND	0.003	0.009	0.001	ND	ND	ND	1.0	0.1
Cr	ND	0.020	0.002	0.001	0	ND	ND	ND	5.0	0.5
Cu	0.120	0.214	0.127	0.188	0.116	0.173	0.057	0.030	NA	NA
Fe	1.05	7.35	2.32	0.182	0.689	0.0045	0.0011	0.0006	NA	NA
Mg	4.25	36.3	45.6	66.9	55.1	4.41	12.0	15.8	NA	NA
Mn	2.05	8.06	7.83	6.00	6.83	0.314	0.676	0.680	NA	NA

Note: EPA = US EPA regulatory levels for classifying waste as hazardous based on TCLP results; *MDEQ* = Mississippi Department of Environmental Quality (MDEQ) regulatory levels for beneficial use of solid waste; *ND* = not detected; *NA* not available.

TABLE 6.16
Leaching Concentration of MICP-Treated SS–Sand samples in SPLP Leachates

	SS Content (%)									
Metal	**0**	**10**	**20**	**30**	**40**	**50**	**75**	**100**	**EPA[1]**	**MDEQ[2]**
Ag	ND[3]	ND	ND	ND	ND	ND	ND	ND	5.0	0.5
Al	ND	ND	ND	ND	ND	ND	ND	ND	NA[4]	NA
As	ND	ND	ND	ND	ND	ND	ND	ND	5.0	0.5
Ba	0.088	0.119	0.140	0.165	0.164	0.149	0.169	0.175	100	10
Ca	2.76	4.27	4.48	4.67	5.57	4.71	5.31	4.50	NA	NA
Cd	ND	0.032	0.044	0.062	0.068	0.069	0.080	ND	1.0	0.1
Cr	ND	ND	ND	ND	ND	ND	ND	ND	5.0	0.5
Cu	0.029	0.110	0.279	0.293	0.203	0.193	0.212	0.181	NA	NA
Fe	0.028	0.032	0.028	0.190	0.061	0.013	0.032	0.026	NA	NA
Mg	4.25	36.3	45.6	66.9	55.1	4.41	12.0	15.8	NA	NA
Mn	2.05	8.06	7.83	6.00	6.83	0.314	0.676	0.680	NA	NA

Note:

[1] EPA; US EPA regulatory levels for classifying waste as hazardous based on TCLP results.

[2] MDEQ; Mississippi Department of Environmental Quality (MDEQ) regulatory levels for the beneficial use of solid waste.

[3] ND: not detected.

[4] NA: not available

there is no government mandate to determine whether a solid waste was toxic or not based on WLT results, the positive results in this study, by comparing WLT results to MCL limit, still could illustrate that MICP-treated samples are environmentally friendly samples.

6.2.1.8 Major Elements

The concentrations of Fe leached from MICP-treated SS–sand samples are shown in Figure 6.33a. Komonweeraket et al. (2015b) revealed that Fe behaves as an

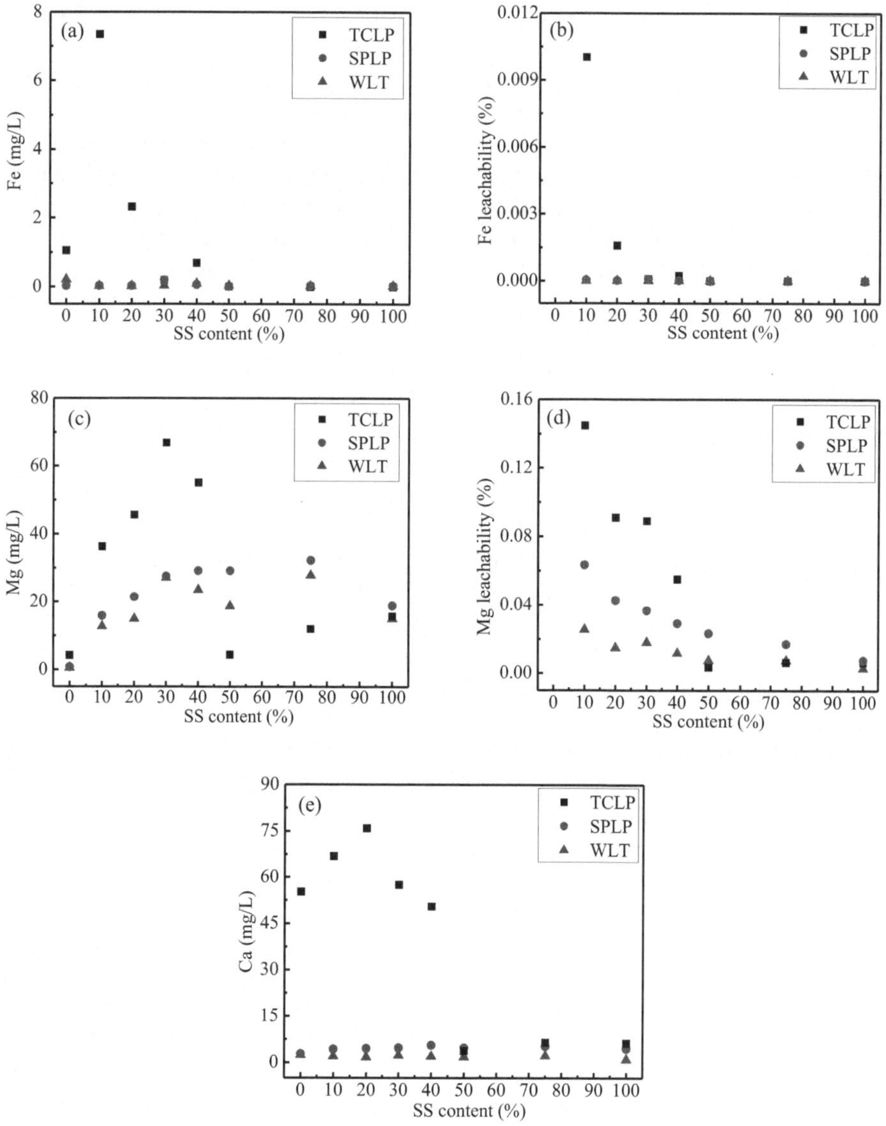

FIGURE 6.33 Leaching behavior of Fe, Mg, and Ca for MICP-treated SS–sand samples.

amphoteric metal, which can readily dissolve in leachate under both acidic and alkaline conditions. As shown in Figure 6.33a, the highest leaching concentration of Fe was 7.35 mg/L, obtained by TCLP effluent of MICP-treated sample containing 10% SS. As the content of SS increased, the leaching concentration of Fe decreased. Moreover, the leaching concentrations of Fe in TCLP, SPLP, and WLT effluents decreased to near zero when the contents of SS in MICP-treated samples were over 50%. The increased Fe content in samples and the decreased leaching concentration of Fe in effluents resulted in the low leachability of Fe. From Figure 6.33b, it is apparent that the highest leachability of Fe was 0.01% achieved by bio-modified sample containing 10% SS. Leachability for the rest of the samples was below 0.003%. This phenomenon illustrates negligible proportions of Fe were leachable. Fe in SS might present as amorphous oxyhydroxide, carbonate salt, magnetite, hematite, and glassy matrix. The spinel structures are highly stable and resistant to acid weathering, and therefore, they are not easily released to the environment (Moreno et al. 2005). Moreover, leaching of Fe is highly dependent on the pH variation of the extraction fluid. Kim et al. (2003) stated that less than 10 mg/kg of Fe are leachable in near neutral and alkaline solutions. As shown in Figure 6.31, the pH of samples modified by the MICP process is higher than 7.8. After reaction with the extraction fluid, the pH of the effluent for TCLP, SPLP, and WLT was in near neutral or slight alkalinity condition. These characteristics also promoted the stabilization of Fe in MICP-treated SS–sand samples.

As shown in Figure 6.33c, the leaching concentration of Mg is highly related to the extraction fluid. When the content of SS in MICP-treated samples was less than 50%, the concentration of Mg in the TCLP effluent was higher than that in SPLP and WLT effluents. However, as the SS content increased continuously, the trend reversed. The leaching concentrations of Mg in TCLP effluents were even lower than that in SPLP and WLT effluents. The highest leaching concentration of Mg was 66.9 mg/L in the TCLP effluent obtained by MICP-treated sample containing 30% SS. The concentration of Mg in SPLP effluents presented a high relation with the SS content in MICP-treated samples, except the sample containing 100% SS. However, the content of releasable Mg in WLT effluent presented a similar trend with that in TCLP effluents. The Mg concentration in WLT effluent increased first, and then decreased with the increase of SS content. Jones (1995) studied the leaching behavior of Mg from ash, and the results illustrated the release of Mg is pH-dependent. Mg appears to be slightly soluble at acidic pH and insoluble in water and alkalinity reagents. As shown in Figure 6.30, the pH of MICP-treated samples which contain more than 30% SS is around 8, which present weak alkalinity characteristic. Meanwhile, the effluent of TCLP, SPLP, and WLT reached neutral or alkalinity conditions. These conditions all contributed to the low leaching concentrations of Mg from samples containing more than 30% SS.

Dudas (1981) investigated the leaching of Mg in a column leaching test, and the results revealed that the leaching of Mg was enhanced as Ca was progressively washed out. Querol et al. (2001) also observed the leaching of Mg over time linked to the leaching of Ca. Compared with the leaching behavior of Ca in this study as shown in Figure 6.33e, the leaching of Mg over SS contents presented a lag trend to that of Ca leaching. The leaching concentration of Ca increased first, reached the maximum value at samples containing 20% SS, and then decreased when SS content increased continuously. While the leaching of Mg reached its peak point at samples containing 30% SS, and then decreased, it is confirmed in the conclusion of Dudas (1981) and Querol et al. (2001). The relative leachability of Mg in TCLP, SPLP, and WLT effluents for samples containing different SS decreased with the increase of SS. The leachability of Mg is below 0.16%, which means most of Mg was sealed in MICP-treated SS–sand samples.

Former research indicated that Ca in SS might be present in multiple mineral phases, primarily lime, anhydrite, calcite, and glassy matrix (Kim et al. 2003). Ca is the most readily released cation, regardless of the extraction fluid used. As shown in Figure 6.13, the concentration of Ca for raw SS in TCLP, SPLP, and WLT leachates were 81.7, 73.9, and 40.0 mg/L, respectively. Considering the high amount of Ca in raw SS, its leachability in TCLP, SPLP, and WLT test was about 8.63%, 7.81%, and 2.11%, respectively. Despite the different extraction fluids used in TCLP, SPLP, and WLT tests, a similar trend in the leaching behavior of Ca were observed. It sounds meaningless to calculate the Ca leachabilities of MICP-treated SS–sand samples considering the $CaCl_2$ as one of the reagents in reaction solution.

6.2.1.9 Trace Elements

The leaching behaviors of Ba from MICP-treated SS–sand samples are demonstrated in Figure 6.34. For raw SS, the leaching concentrations of Ba in TCLP leachate solution were higher than that in SPLP and WLT extract. Several researchers revealed that high CaO-, MgO-, and Fe-complexing oxides content could form calcium silicate lattices (Mombelli et al. 2016). And Ba could bind to this compound as β-Ca_2SiO_4 (Izquierdo and Querol 2012; Mombelli et al. 2015). Due to the high reactivity of calcium silicate with water, bounded Ba was released and thus enhanced its leachability.

After having been bio-modified by the MICP process, the leaching concentrations of Ba from TCLP, SPLP, and WLT tests were below 2.0 mg/L, which was the limit in drinking water regulated by the US EPA. As shown in Figure 6.33b, the leachability of Ba from MICP-treated SS–sand samples decreased with the increase of SS content. The high carbonate ions produced in the MICP process could capture the dissociative Ba ions and form $BaCO_3$. Leachability of Ba in WLT effluents for MICP-treated sample containing 20% SS increased, but it is acceptable, considering the pretty low values.

The limit regulated by national secondary drinking water for Cu is 1.3 mg/L, while the concentrations of Cu leached out from MICP-treated SS–sand samples were below this value. With the increase of SS content in MICP-treated samples, the leaching concentrations of Cu increased. The maximum leaching concentration of Cu was achieved with a MICP-treated sample containing 30% SS. When SS percent increased continually, the leaching concentration of Cu slightly decreased and fluctuated in the range of 0.05 to 0.15. The concentrations of leached Cu were not directly related to the ratio of SS. This trend was also observed by Chen et al. (2012), which studied the leaching behavior of Cu from recycled construction aggregates (RCA) based on the results of pH-dependent tests. Engelsen et al. (2010) found that the leaching of Cu was controlled by the dissolution/precipitation of tenorite (CuO) and malachite ($Cu_2(CO_3)(OH)_2$) in fly ash. Moreover, Cu has a very high affinity

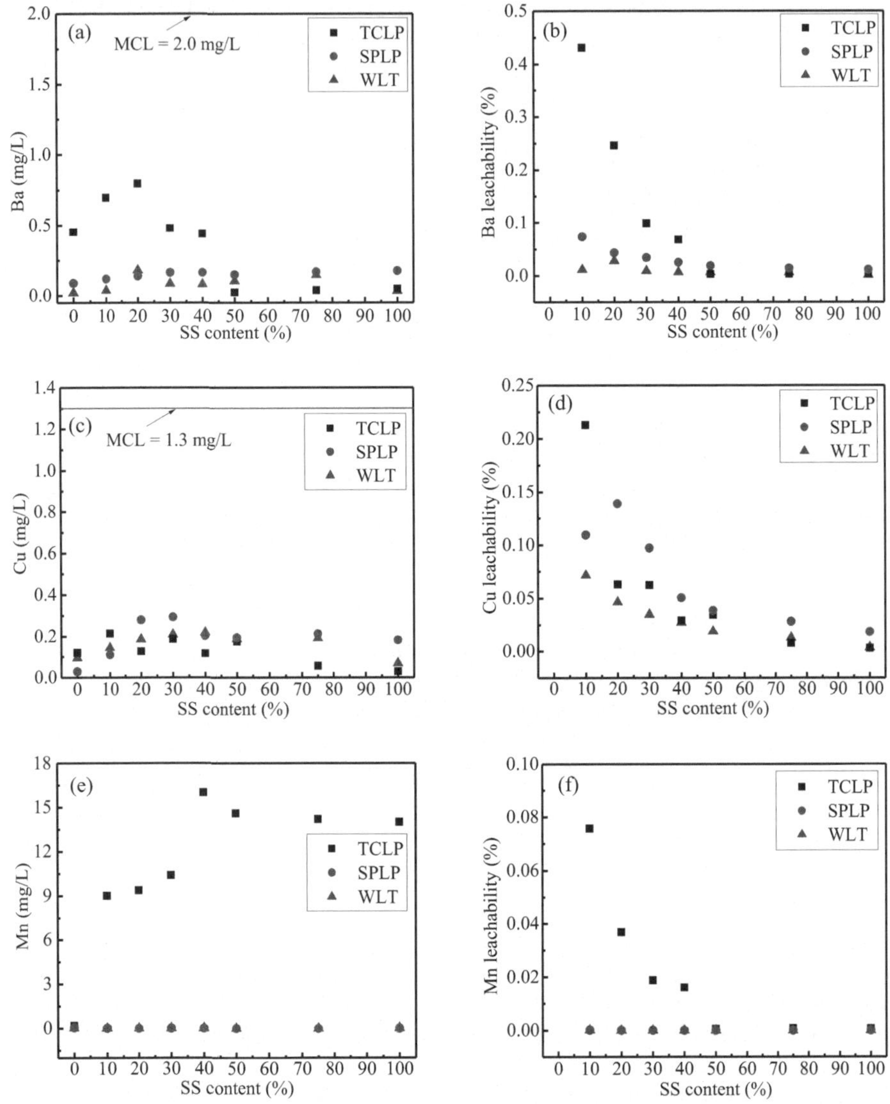

FIGURE 6.34 Leaching behavior of Ba, Cu, and Mn for MICP-treated SS–sand sample.

to organic ligands. Former research showed that existing dissolved organic carbon (DOC) could enhance the leachability of Cu under alkaline conditions (Engelsen et al. 2010; Mahedi et al. 2019). During the MICP process, organic ligands were produced through bacteria metabolism, and the high affinity between organic ligands and Cu forced the Cu to dissociate from the solid phase of MICP-treated samples. Therefore, the leachability of Cu increased. However, as the SS content increased continually, the limited organic ligands could not provide enough attaching points for Cu. The ratio of Cu released through the bioleaching mechanism decreased. Meanwhile, the produced CO_3^{2-} and the high alkalinity of the MICP reaction solution promoted the form of $CuCO_3$ and $Cu(OH)_2$, and thus, leaching concentrations and leachabilities decreased.

The leaching behaviors of Mn in TCLP, SPLP, and WLT leachates are shown in Figure 6.34e, f. It is apparent that the concentrations of Mn in SPLP and WLT effluents were very low, while the concentrations in TCLP effluents were high. State-of-the-art studies illustrated that Mn compounds are insoluble under near neutral to alkaline conditions (Kukier et al. 2003; Querol et al. 2001). The concentrations of Mn in TCLP effluents were not directly related to the content of the SS. The highest Mn concentration was achieved by samples containing 40% SS. Costa et al. (2016) studied the leaching behavior of Mn in SS and found the release of Mn was controlled by Mn carbonates, either in the amorphous or in the crystalline form. During the MICP process, carbonate was produced, and it might bond with the dissociated Mn in the solution and thus decreased the leachability of Mn.

6.2.1.10 Pearson Correlation Analysis

As shown in Table 6.17, Table 6.18, and Table 6.19, SS content was negatively related to the leaching concentrations of metals in TCLP leachate. The leaching concentration of Ba was positively related to the concentration of Ca and Mn. Meanwhile, the leaching concentration of Ca was highly and positively related to the leaching concentration of Mn. That coincides with the results illustrated in the last section.

TABLE 6.17
Leaching Concentration of MICP-Treated SS–Sand Samples in WLT Leachates

	SS content (%)								MCL[1]
Metal	0	10	20	30	40	50	75	100	
Ag	ND[2]	ND	ND	ND	ND	ND	ND	ND	NA[3]
Al	ND	ND	ND	ND	ND	ND	ND	ND	NA
As	ND	ND	ND	ND	ND	ND	ND	ND	0.01
Ba	0.019	0.036	0.180	0.085	0.081	0.104	0.147	0.034	2.0
Ca	2.42	1.93	1.72	2.26	1.96	1.78	2.19	0.961	NA
Cd	ND	ND	ND	ND	ND	ND	ND	ND	0.005
Cr	ND	ND	ND	ND	ND	ND	ND	ND	0.1
Cu	0.095	0.144	0.186	0.209	0.218	0.187	0.190	0.068	1.3
Fe	0.207	0.021	0.015	0.029	0.087	0.034	0.025	0.025	NA
Mg	0.519	12.8	14.9	27.1	23.5	18.7	27.9	15.1	NA
Mn	0.031	0.013	0.013	0.016	0.020	0.012	0.015	0.044	NA

Note:
[1] MCL, maximum contaminant level (MCL); the highest level of a contaminant that is allowed in drinking water according to the National Primary Drinking Water Regulation.
[2] ND: Not detected.
[3] NA: Not available

TABLE 6.18
Pearson Correlation Coefficients of Leaching Concentrations in TCLP Extracts

Pearson Correlation	SS Content	Ba	Ca	Cu	Fe	Mg	Mn
SS content	1	−0.798*	−0.811*	−0.724*	−0.573*	−0.276*	−0.609*
Ba	−0.798*	1	0.989**	0.539	0.659	0.617	0.912**
Ca	−0.811*	0.989**	1	0.551	0.589	0.680	0.916**
Cu	−0.724*	0.539	0.551	1	0.561	0.388	0.555
Fe	−0.573*	0.659	0.589	0.561	1	0.185	0.634
Mg	−0.276*	0.617	0.680	0.388	0.185	1	0.826*
Mn	−0.609*	0.912**	0.916**	0.555	0.634	0.826*	1

TABLE 6.19
Pearson Correlation Coefficients of Leaching Concentrations in SPLP Extracts

Pearson Correlation	SS Content	Ba	Ca	Cu	Fe	Mg	Mn
SS content	1	0.561	0.536	−0.576	−0.438	0.097	−0.054
Ba	0.561	1	0.123	0.247	−0.209	0.160	0.238
Ca	0.536	0.123	1	−0.810*	−0.506	0.748*	0.390
Cu	−0.576	0.247	−0.810*	1	0.424	−0.372	0.023
Fe	−0.438	−0.209	−0.506	0.424	1	−0.584	0.292
Mg	0.097	0.160	0.748*	−0.372	−0.584	1	0.252
Mn	−0.054	0.238	0.390	0.023	0.292	0.252	1

TABLE 6.20
Pearson Correlation Coefficients of Leaching Concentrations in WLT Extracts

Pearson Correlation	SS Content	Ba	Ca	Cu	Fe	Mg	Mn
SS content	1	0.375	−0.652	0.024	−0.235	0.489	−0.234
Ba	0.375	1	0.208	0.839**	0.227	0.972**	0.076
Ca	−0.652	0.208	1	0.212	0.280	0.031	0.096
Cu	0.024	0.839**	0.212	1	0.292	0.776*	0.249
Fe	−0.235	0.227	0.280	0.292	1	0.155	0.905**
Mg	0.489	0.972**	0.031	0.776*	0.155	1	0.018
Mn	−0.234	0.076	0.096	0.249	0.905**	0.018	1

SS content was positively related to the leaching of Ba in SPLP leachate, which means the released Ba in SPLP leachate increased with the increase of SS content. Considering that Ca was not merely from SS and the reaction solution was also an important source, this correlation was negligible. Meanwhile, the correlations of SS content and the concentrations of Cu and Fe were negative. And the correlation between SS content and the concentration of Mg was positive. But these correlations were not significant. It means the leaching concentrations of Cu, Fe, and Mg were not controlled by the total mass in samples. There are more complicated mechanisms behind the leaching process.

Fruchter et al. (1990) reported that Ba formed sparingly soluble compounds with carbonates and sulfates in fly ash. And its leachability was not pH-dependent but rather controlled by the ubiquitous Ca due to the competition for sulfates. As shown in Table 6.19, released Ba in SPLP leachate was positively related to the dissociated Ca ions in the leachate, even though it was a moderate correlation. Due to the sufficient CO_3^{2-} produced in the MICP process, the competition pressure for SO_4^{2-} between Ba and Ca weakened. Meanwhile, leaching concentrations of Mg were also strongly correlated to Ca concentrations. The correlation of Ca and Cu was negative and significant.

In WLT leachate, the correlation between SS content and metals was different from that in the SPLP effluent. There was no metal leaching significantly correlated to SS content. It means SS content has little effect on the leaching of metals under water erosion pressure. The correlation of Ba and Cu and the correlation of Ba and Mg were positive and significant. The Pearson correlation coefficient between Fe and Mn was 0.905, which is statically significant. It indicated that the leaching concentration of Mn might promote the leaching of Fe.

6.2.2 Mechanical Improvement and Sulfate Resistance of MICP-Treated Steel Slag

6.2.2.1 Materials and Methods

6.2.2.1.1 Materials

Ottawa silicon sand with median diameter (D_{50}) of 0.46 mm was used in this study. SS was provided by a local company (Jackson, Mississippi, USA). Before test, SS was stockpiled outdoors for more than 18 months to expose the material to moisture from natural precipitation. The chemical components had been measured by energy-dispersive X-ray analysis (EDX). As shown in Table 6.11, SS consisted of 40.90% CaO, indicating its cementation potential. The amounts of SiO_2, MgO, and Fe_2O_3 in SS were 23.47%, 15.92%, and 11.76%, respectively. The pH of SS was 10.12, according to EPA method 9045b: Soil and Waste pH (USEPA 1986). The purpose of such aging was to allow potentially destructive hydration and the associated expansion to take place prior to the use of the material in aggregate applications. The SS was sieved according to ASTM C136–01, Standard Test Method for Sieve Analysis of Fine and Coarse Aggregates (ASTM C136–01 2001). Group size below 0.85 mm was grouped as Part I SS, and group size between 0.85 mm and 2 mm was grouped as Part II SS, while Part III SS was size bigger than 2 mm. Part I and Part II SS were used in this study. The photo images of sand and SS are shown in Figure 6.35.

Part I (Size < 0.85 mm) | Part II (0.85 mm < Size < 2 mm) | Sand D_{50} = 0.46 mm

FIGURE 6.35 Materials used in this study: steel slag (Part I and II) and Ottawa sand.

6.2.2.1.2 Sample Preparation

SS–sand mixtures were prepared with Part I SS, Part II SS, and sand. Before tests, they were mixed thoroughly based on mass percentages and labeled as S0, S1, S2, S3, S4, S5, S6, and S7. The particle characteristics of mixtures were shown in Figure 6.36. D50 for SS–sand mixtures ranged from 0.48 mm to 0.58 mm, except that for S6 and S7 samples, which was 2.02 mm. According to the Unified Soil Classification System, S0, S1, S2, S3, S4, and S5 were uniformly graded, while S6 and S7 were well graded.

6.2.2.1.3 MICP Process

The bacteria used in this study was *Sporosarcina pasteurii* (ATCC 11859), which is rod-like, with a length of ~2 μm. The nutrition solution for the bacteria consisted of 130 mM Tris buffer (pH = 9.0), 10 g/L of $(NH_4)_2SO_4$ and 20 g/L yeast extract. After thoroughly mixed, the nutrition solution was autoclaved at 120°C for a period of 20 min. The bacteria liquor was incubated aerobically at 23 ± 2°C in a shaking bed at 200 rpm for 43 ± 2 h.

The harvest bacteria solution was maintained with a fresh growth medium to obtain the final optical density (OD 600 = 0.8) before use. The optical density of the suspension was measured by a spectrophotometer (UV-1700 Shimadzu) at 600 nm. A 200 μL of fivefold diluted bacteria solution was cultured in agar plate at 37°C for 2 days. The concentration of bacteria in this condition was about 7.5×10^7 colony-forming unit (CFU)/mL. Bacteria suspension was mixed with the samples. The wet samples were compacted into a fully permeable flexible mold that was developed by Zhao et al. (2014a). The samples with mold were placed on the shelf carefully and then immersed into the batch reactor. This batch reactor was filled with cementation media that was used to induce bio-cementation. Cementation media was prepared before the test. The recipe for the cementation media consists of 30.0 g/L urea, 73.5 g/L $CaCl_2{\cdot}2H_2O$, 3.0 g/L nutrition broth, and 10.0 g/L NH_4Cl. The MICP process was conducted under aeration conditions. Following the procedure of Zhao et al. (2014a), after 7 days' continuous reaction, MICP-treated specimens were taken out and flushed with tap water to remove the residual cementation media from the pore space. When the electronic conductivity (EC) of the solution was below 1000 μs, the samples were harvested for tests. As control, a set of samples was prepared with the same material and processed under the same conditions, except for bacteria.

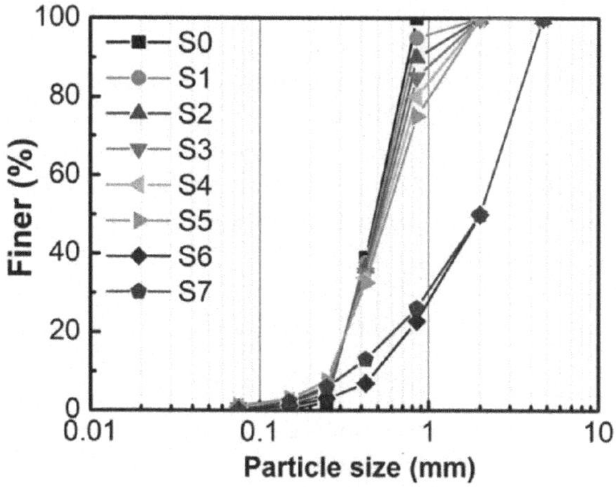

FIGURE 6.36 Particle size distribution of steel slag–sand mixtures.

6.2.2.1.4 *Water Absorption, $CaCO_3$ Content Determination, and Hydraulic Conductivity*

The water absorption of MICP-treated samples was determined following ASTM C 20–10, Standard Test Method for Apparent Porosity, Water Absorption, Apparent Gravity, and Bulk Density of Burned Refractory Brick and Shapes by Boiling Water (ASTM C20–10 2010). The mass of mixtures before MICP process was measured and labeled as M_0. The MICP-treated samples were oven-dried at 75 ± 2°C and weighed as M_1. Tap water was boiled and cooled to room temperature for the water absorption tests. Oven-dried MICP-treated samples were immersed into the water for 2 days till saturated. Then, the samples were taken out and wiped off the water absorbed onto the surface gently, and then weighed as M_2. Water absorption was calculated according to equation (6.10):

$$\text{Water adsorption: } \varepsilon = (M_2 - M_1)/M_1 \tag{6.10}$$

The precipitated $CaCO_3$ content caused by MICP treatment was determined by acid washing method according to Zhang et al. (2021b). The hydraulic conductivity was determined following ASTM D 5084–16a, Standard Test Methods for Measurement of Hydraulic Conductivity of Saturated Porous Materials Using a Flexible Wall Permeameter (ASTM D5084–03 2003). The falling head, rising tailwater elevation method was performed by using a flexible wall permeameter.

6.2.2.1.5 *Unconfined Compression Strength and Flexural Behavior*

Samples with a size of 38.1 mm in diameter and 76.2 mm in height were employed to perform an unconfined compressive strength (UCS) test according to the ASTM method D2166, Standard Test Methods for Measurement of Unconfined Compressive Strength of Cohesive Soil (ASTM D2166M-16 2016). The working condition of the equipment was fixed at loading rate as 1.5% strain/min. The flexural property of

SS–sand samples after MICP process was determined by four-point bending test following the standard test method ASTM D6272, Standard Test Method for Flexural Properties of Unreinforced and Reinforced Plastics and Electrical Insulating Materials by Four-Point Bending (ASTM D6272 - 17 2017). The samples were prepared by 900 g SS–sand mixture. The size of the bio-brick sample possessed a length of 177.8 mm, width of 76.2 mm, and depth of 38.1 mm. The vertical load was conducted under strain-control condition at a uniform loading rate of 1.5%/min. The stress was calculated according to equation (6.11):

$$\text{Stress: } S = PL/bd^2 \tag{6.11}$$

where S is stress in the outer surface throughout the load span (MPa), P is load at a given point on the load-deflection curve (N), L is supporting span (mm), b is width of beam (mm), and d is depth of beam (mm).

6.2.2.1.6 *Composition and Microstructure Analysis of Samples*

To determine the change of mineral compositions of the mixtures before and after the MICP process, X-ray diffraction (XRD) test was performed using the diffractometer (Regiku Miniflex 600). Cu-Kα radiation was used, and the 2Θ ranged from 10° to 80°, with a step width of 1°. The differences of functional groups before and after MICP process were determined using a Fourier transform infrared spectroscopy (FTIR) (UTAR two, PerkinElmer). FTIR analysis was conducted in the 2000–400 cm^{-1} wavenumber (2.5–10 μm wavelength) region with a resolution of 2 cm^{-1}. The microstructure of the samples was analyzed using a scanning electron microscope (SEM) (Lyra 3, TESCAN). The MICP-treated specimens were oven-dried at 105°C before tests.

6.2.2.1.7 *Erosion Resistance Test*

Th sulfate-bearing solution is assumed to lead to the failure of the cement-containing matrix by physical and chemical influence. To decouple the influence, powder samples were prepared to quantify the chemical influence of the sulfate solution, and the high concentration of Na_2SO_4 solution was used to accelerate the erosion in lab scale (Baldermann et al. 2018; Gat et al. 2016). In this study, the sulfate solution was prepared with analytical-grade, water-free Na_2SO_4 dissolved in DI water (18.2 MΩ at 25°C) to a concentration of 10 g/L. The reaction time was set as 30 min, 1 h, 2 h, 4 h, 8 h, 1 d, 2 d, 4 d, 7 d, and 10 d. Each aliquot of sample was prepared with 10.0 g grinded sample mixed with 100 mL sulfate solution (liquid-to-solid ratio = 10), and then rotated at 30 ± 2 rpm. After rotation, the extracted fluid was filtered through 0.4–0.6 μm glass-fiber filter. The pH and EC of the extracted fluid were measured, then pH was adjusted to less than 2 for metal determination. Tap water extraction tests were performed at the same liquid-to-solid ratio, the same working condition, and the same extraction time as control. Inductively coupled plasma atomic emission spectroscopy (ICP-AES) (ICPE-9000, Shimadzu, Japan) was used to detect the concentration of Ca^{2+} leached out from MICP-treated samples.

6.2.2.2 Unconfined Compression Strength Development

To investigate the stiffness development of MICP-treated samples containing SS, cylindrical samples with 2H:1D ratio (diameter of 38.1 mm and height of 76.2 mm) were prepared with SS–sand mixtures to perform UCS test. After 7 days' reaction, MICP-treated SS–sand samples were harvested. As illustrated in Figure 6.37a, loose SS and sand particles were cemented together. MS0, MS1, MS2, MS3, MS4, MS5, MS6, and MS7 represent MICP-treated S0 sample, MICP-treated S1 sample, MICP-treated S2 sample, MICP-treated S3 sample, MICP-treated S4 sample, MICP-treated S5 sample, MICP-treated S6 sample, and MICP-treated S7 sample, respectively.

The stress–strain curves of MICP-treated samples are shown in Figure 6.37b. It is observed that with the inclusion of SS, the UCS of MICP-treated samples increased compared to that of sand-only sample, except that of MS7. To illustrate the effects of SS addition on the development of stress, UCS as a function of samples was specified in Figure 6.37c. MICP-treated sand-only sample as MS0 achieved UCS of 793.10 kPa, while MICP-treated SS-only sample as MS7 achieved peak strength of 236.18 kPa, which is less than one-third of that for MS0. The maximum UCS was 2453.8 kPa, which is three times that achieved by MS0. It is obtained by MS5, which contains 25% Part I SS and 25% Part II SS. The UCS for MS1, MS2, MS3, MS4, and MS6 were also enhanced to different degrees. As shown in Figure 6.35, SS particles are in irregular shapes and have rough surface. Therefore, the frictions between

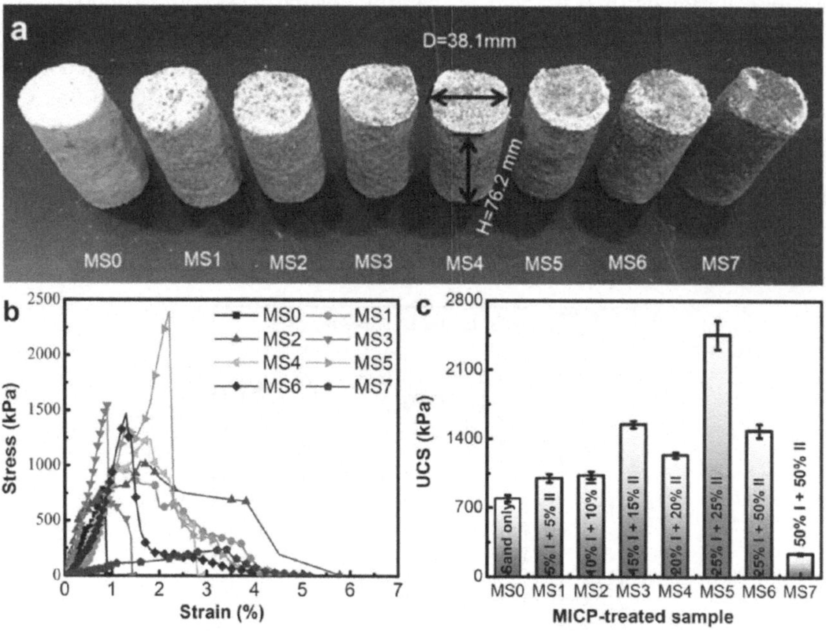

FIGURE 6.37 MICP-treated column samples: (a) photo image, (b) stress–strain curve, (c) UCS.

particles might be increased by the inclusion of SS content. As a result, it becomes more difficult for particles to move, leading to the increase of UCS. As discussed by previous research, the particle size or void space is an important factor which can affect the efficiency of the MICP process (Mahawish et al. 2017). Former research which introduced the coal fly ash to perform the MICP process illustrated that the inclusion of fine content narrowed the void space between particles and confined the activity of bacteria and thus weakened the work of the MICP process. Mahawish et al. (2017) observed that the void for coarse particles was too big to fill for the produced thin layer of $CaCO_3$, leading to the low efficiency of cementation. In this study, the results also confirmed the effects of particle size. As shown in Figure 6.35, the D50s for S0, S1, S2, S3, S4, and S5 were similar, while the median particle sizes for S6 and S7 were much bigger. Moreover, the median particle size for S7 was even bigger than that of S6. Even though S6 and S7 contained higher SS content, which might result in the increase of friction, the UCS of S6 and S7 decreased sharply compared to that of S5. The particle size of S6 and S7 could be attributed to this phenomenon. The thin layer of $CaCO_3$ produced in the MICP process could not fill the void space and thus resulted in a low strength to resist the loading.

6.2.2.3 $CaCO_3$ Content of Cylindrical Samples

As shown in Figure 6.38, the $CaCO_3$ content for MICP-treated samples ranged from 6.0% to 8.0%. Rowshanbakht et al. (2016) studied the calcite content of sandy silica soil column with injected bacteria suspension, and the calcite percentage varied from 4.5% to 8%. Mahawish et al. studied the effect of particle size distribution on the bio-cementation of coarse aggregates and revealed the average precipitated calcium carbonate varied from 5.7 to 6.9% (Mahawish et al. 2017). These results reported by previous research were in line with the content of precipitate $CaCO_3$ in this study. Meanwhile, the maximum $CaCO_3$ content was 7.7%, achieved by MS6, which is 20% higher than that of MS0. However, the $CaCO_3$ content of MS7 was difficult to obtain. Due to the weak cementation of particles, there was mass loss during the harvesting after MICP treatment.

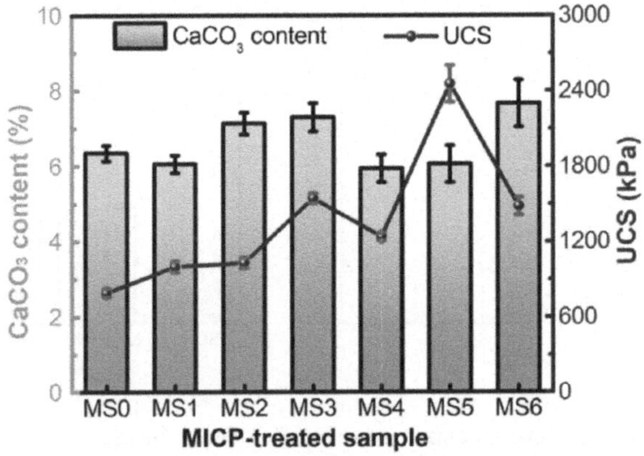

FIGURE 6.38 $CaCO_3$ content of MICP-treated SS–sand sample related to UCS.

The correlation between $CaCO_3$ content and UCS was also investigated, and the results are shown in Figure 6.38. The highest $CaCO_3$ content was achieved by MS6, while the maximum UCS was achieved by MS5. Moreover, there is no direct correlation between UCS and $CaCO_3$ content for MS0, MS1, MS2, MS3, and MS4. This is different from the results of some former research. Zhao et al. concluded that the higher the $CaCO_3$ content, the higher the UCS (Zhao et al. 2014a). Cheng et al. found that the UCS obtained by the produced bio-brick cube units was linearly related to $CaCO_3$ content (Cheng et al. 2020). Moreover, van Paassen et al. found their samples treated by MICP demonstrated an exponential correlation between UCS and $CaCO_3$ content (van Paassen et al. 2010a). The increase of UCS is mainly attributed to the produce of $CaCO_3$, which filled the voids between particles and increased the cohesion by cementing particles together. The immersion methodology for the MICP process employed in this study is a homogenous system which contributes to the uniform distribution of $CaCO_3$. However, the $CaCO_3$ crystal attached to the surface of the samples provided less-efficient cementation, while the $CaCO_3$ crystal deposited between particles may contribute greatly to the enhancement of mechanical properties. Therefore, the amount of effective $CaCO_3$ instead of the total amount of $CaCO_3$ content seems more responsible for the increase of UCS.

6.2.2.4 Flexural Behavior of Bio-Brick

Bio-bricks, through MICP, were prepared with SS–sand mixtures at the same contents as S0, S1, S2, S3, S4, S5, S6, S7. It has a length of 177.8 mm, width of 38.1 mm, and depth of 76.2 mm, as shown in Figure 6.39a. After 7 days' treatment by the MICP process, the samples made with S0, S2, S3, and S4 were successfully cemented together and named as BS0, BS2, BS3, and BS4, respectively, in the coming sections. A four-point bending test was employed to investigate the flexural resistant behavior. As shown in Figure 6.39b, samples were broken along the failure line, which is like

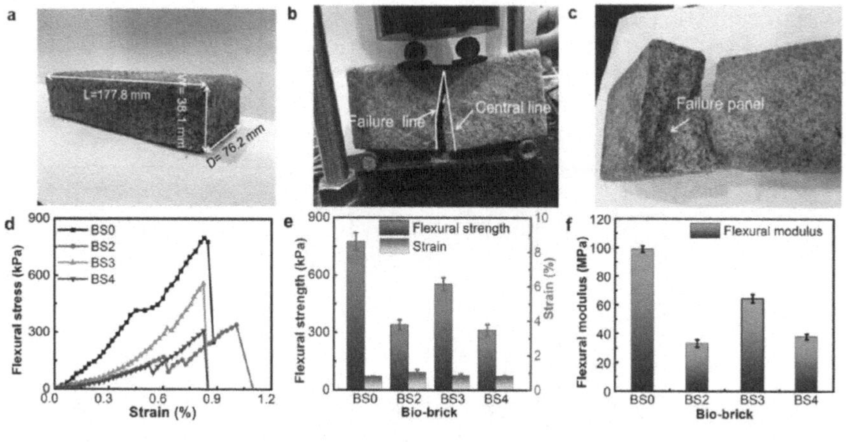

FIGURE 6.39 Flexural behavior of bio-brick: (a) dimension; (b) failure path under flexural stress; (c) cross-section of failure panel; (d) flexural stress as a function of strain; (e) flexural strength and flexural strain; (f) flexural modulus.

the central line of samples. The failure panel, as shown in Figure 6.39c, illustrated the even distribution of SS and sand particles.

As shown in Figure 6.39d, the failure of bio-brick was a kind of brittle failure, with no residual strength left after the sample was broken. The peak strength and strain are shown in Figure 6.39e. The maximum flexural strength was 775.83 kPa, obtained by BS0, which contains sand only. With the addition of SS, the flexural strength decreased. BS2 achieved a flexural strength of 339.43 kPa, and BS4 achieved a flexural strength of 312.20 kPa. BS3 achieved a higher flexural strength than BS2 and BS4 at 551.22 kPa, while it is only 70.0% of that of BS0. However, the failure strains achieved by bio-bricks containing SS were higher than those of BS0. As shown in Figure 6.39e, the flexural strain for BS2, BS3, and BS4 are 1.03%, 0.86%, and 0.82%, respectively, while the failure strain for BS0 is 0.78%. The decreased flexural strength and increased flexural strains achieved by bio-bricks made with SS–sand mixture revealed that the addition of SS could not enhance the flexural resistance characteristic of the bio-brick, but the ductility. This conclusion was further proved by the results of modulus, as shown in Figure 6.39f. With the addition of SS, the elastic modulus of the bio-brick decreased. The maximum modulus of the bio-bricks containing SS was achieved by BS3, which is 64.4 MPa and only 65.1% of that of BS0. This phenomenon may be related to the mechanism of MICP technology, which can produce $CaCO_3$ and connect the particles together. Therefore, cohesion between particles was enhanced, while the flexural resistance strength did not increase. To improve the flexural behavior, other assistant materials, that is, fiber, should be included (Wen et al. 2018a).

6.2.2.5 Water Absorption and Hydraulic Conductivity of MICP-Treated Samples

The water absorption performances of MICP-treated samples were shown in Figure 6.40a. Water absorption of MICP-treated samples ranged from 10.27% to 13.62%. The water absorption of MICP-treated sand-only sample was 10.27%. The water absorption of MICP-treated samples containing SS was higher than that of MS0. Though there is no ASTM standard for the material's water adsorption requirement, IS 1077 states that the 24 h cold water adsorption of masonry units shall remain

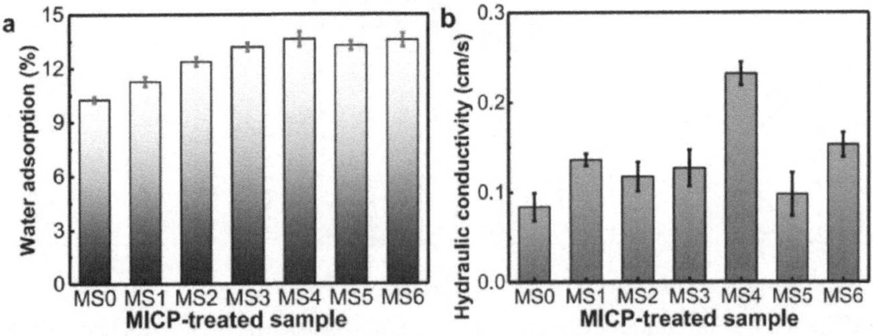

FIGURE 6.40 (a) Water adsorption and (b) hydraulic conductivity of MICP-treated samples.

under 20% of the considered weight (Cheng et al. 2020). Therefore, it is acceptable that the water adsorption property of MICP-treated SS–sand mixture samples satisfies this criterion. Former research illustrated that the water adsorbing process relies on the capillary water suction through the pores, and thus, water adsorption has a direct relation with the material's porosity (Mahawish et al. 2017). The honeycomb structure of SS might be attributed to the high water absorption. The hydraulic conductivity of MICP-treated samples showed high hydraulic conductivity.

6.2.2.6 Carbonate Product Identification

XRD analysis was used to illustrate the mineralogy of MICP-treated samples. As shown in Figure 6.41a, calcite and vaterite were observed in MICP-treated samples, while amorphous $CaCO_3$ was not detectable by XRD. This could be attributed to the weak sensitivity of XRD to the signal of amorphous $CaCO_3$ compared to that for quartz, calcite, and vaterite. Therefore, the signal of amorphous $CaCO_3$ was missing. The chemical components of samples were also observed by FTIR through the detection of functional groups. As shown in Figure 6.41b, calcite, vaterite, and amorphous $CaCO_3$ were detected. The humps centered at 714, 874, and 1392 cm^{-1} were reported to be calcite, while the hump centered at 776 cm^{-1} was attributed to vaterite, and the hump centered at 1,074 cm^{-1} was owned by amorphous $CaCO_3$ (Carlos et al. 2007; Dhami et al. 2016; Loste et al. 2003). It is apparent that calcite, vaterite, and amorphous $CaCO_3$ were also present in the raw SS. Therefore, it becomes difficult to determine the source of calcite, vaterite, and amorphous $CaCO_3$ by FTIR. However, the $CaCO_3$ detected by XRD has confirmed the work of the MICP process.

6.2.2.7 Microstructure Analysis

The microscale images of MICP-treated SS–sand mixtures are shown in Figure 6.42. In Figure 6.42a, the bigger particles which had a regular and smooth surface were silica sand, the irregular particle located in the left bottom corner was Part II SS, and the particles in the center were Part I SS. The fine size of Part I SS made it an ideal material to fill the void space of the mixtures and thus decreased the void spaces and facilitated a better cementation process. Figures 6.42b and c showed the 20-fold

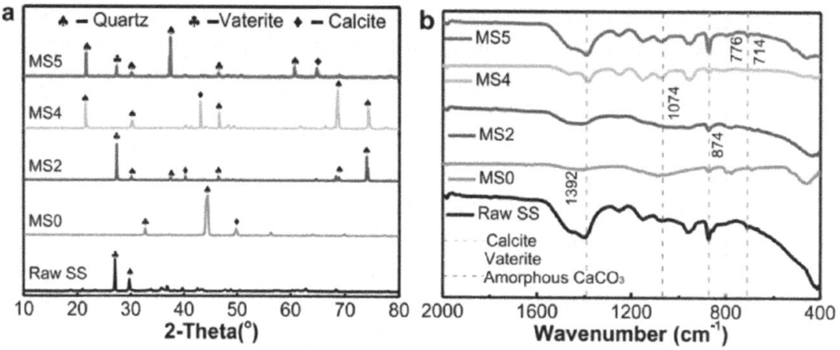

FIGURE 6.41 (a) XRD patterns and (b) FTIR spectral of MICP-treated samples.

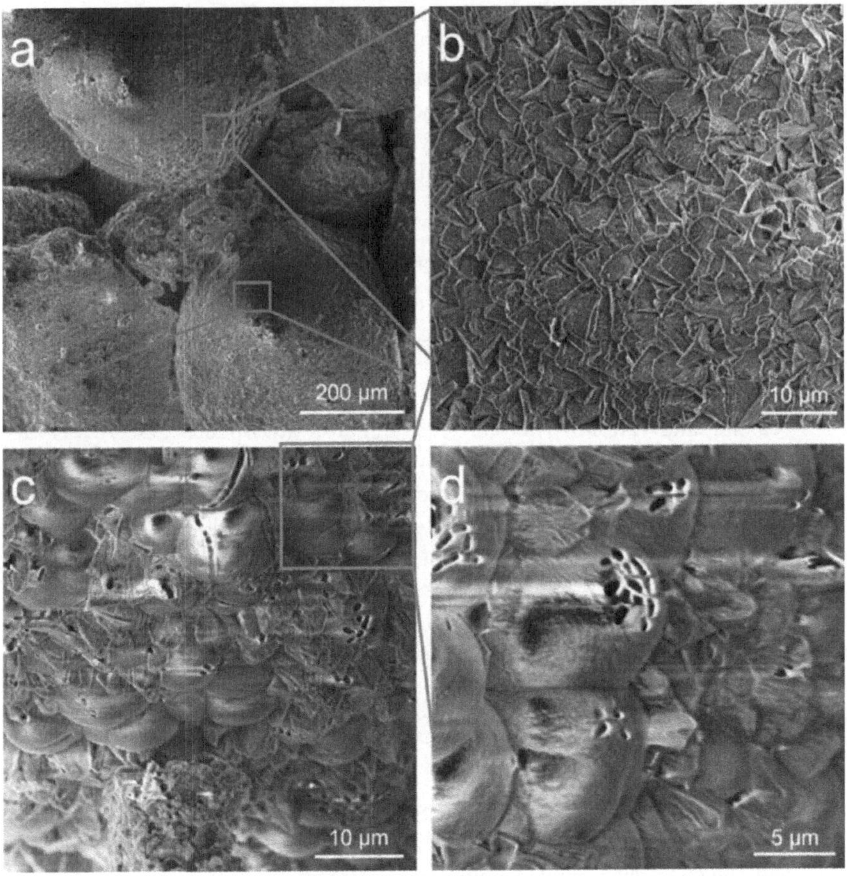

FIGURE 6.42 Microscale analysis of MS4: (a) SEM image at 200 μm; (b) part of *a* at 10 μm; (c) part of *a* at 10 μm; (d) part of *c* at 5 μm.

enlargement image of the square area in Figure 6.42a. It is obvious that a layer of calcium carbonate covered the surface of particles. Spherical crystals, rhombohedral, and amorphous crystals were uniformly distributed on the surface. Former studies reported that there are three types of $CaCO_3$ produced in the MICP process (De Muynck et al. 2020). *Calcite* is a rhombohedral crystal $CaCO_3$, vaterites are $CaCO_3$ crystals in spherical shape, while amorphous $CaCO_3$ was present in irregular shapes. As shown in Figures 6.42b and c, three types of $CaCO_3$ crystals were all observed in the MS4 sample. Wen et al. studied the growth process of crystals during the MICP process induced by bacteria. They found that the spherical crystal would grow into rhombohedral crystal after 72 h of reaction (Wen et al. 2019). Li et al. reported that amorphous $CaCO_3$ formed at first, then transformed to vaterite, and finally presented as calcite (Li et al. 2013). The polymorph of $CaCO_3$ crystal and the transformation between formations were also observed by Chen et al. (2009). Rod-shaped imprints of bacteria were clearly observed on the crystals, as shown in Figure 6.42d. It is reported that bacteria cells work as the nucleus in the formation of precipitates

(Balam et al. 2017; De Muynck et al. 2020). The negative charge of the bacterial surface attracts the positively charged calcium ions concentrated on its surface. When the carbonate ionic group was produced by the enzymolysis of urea encountered with calcium, deposition occurred around the bacteria. The bacteria holes remained after the bacteria died.

6.2.2.8 Erosion Resistance Property under Sulfate-Bearing Solution

The MS4, MS5, and MS6 were selected to perform the erosion resistance tests. The pH and EC variations of samples in tap water and Na_2SO_4 solution are illustrated in Figure 6.43. The pH of solutions was in the range of 8.0–11.5, which means the leachate of MICP-treated samples was in alkalinity conditions. At first, the pH of tap water and Na_2SO_4 solution increased gradually, then the pH stayed stable or slightly decreased, except that of MS6 in Na_2SO_4 solution. In the meantime, the EC which reflects the variation of ions in the solution for MS4, MS5, and MS6 are shown in Figures 6.43d, e, and f. It is apparent that after a period of fluctuation, EC reached a stable state, which means the release and deviation of ions reached equilibrium. Ca leaching concentrations over time are illustrated in Figure 6.44. It is apparent that the

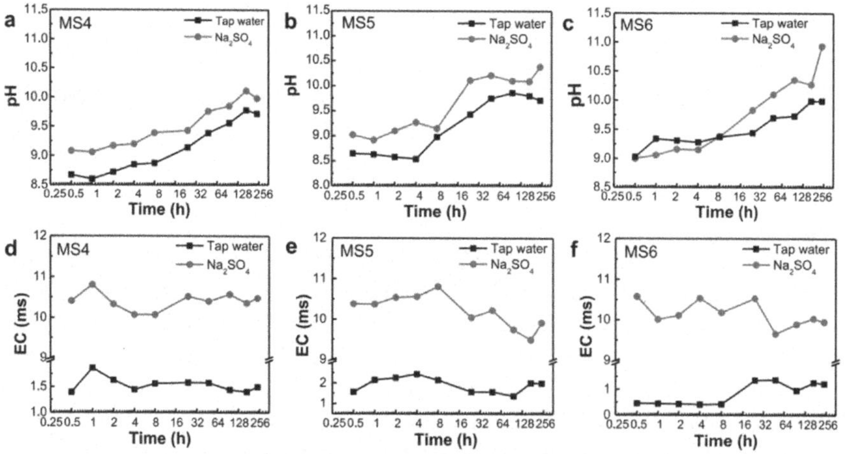

FIGURE 6.43 pH and EC over the test duration of sulfate solution and tap water solution: (a) pH of MS4, (b) pH of MS5, (c) pH of MS6, (d) EC of MS4, (e) EC of MS5, (f) EC of MS6.

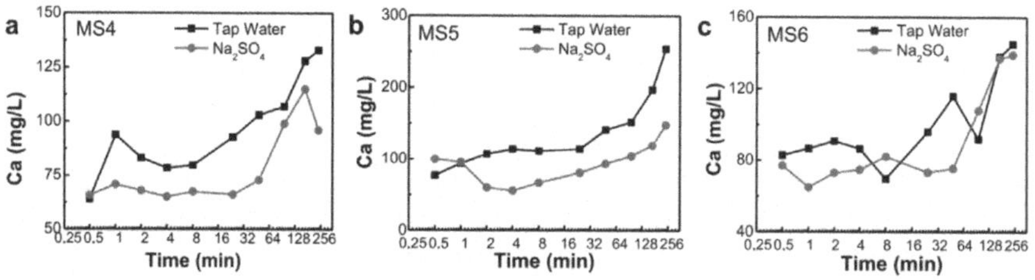

FIGURE 6.44 Ca leaching behavior under sulfate attack: (a) MS4, (b) MS5, (c) MS6.

leaching concentration of Ca increased with the time going on. However, the concentrations of Ca leaching out from the samples in the sulfate solution were lower than that in tap water overall. This observation illustrated that the erosion effect of the sulfate solution on the MICP-treated samples is unremarkable, even weaker than the effect of tap water. Steindl et al. determined the Ca leaching behavior under sulfate attack to evaluate the chemical resistance property of lime–concrete (Steindl et al. 2019). Their results showed the increase of Ca concentration in sulfate solution from 150 mg/L to 600 mg/L within 10 h reaction for samples CEM 100%, which is composed of 67% cement. After that, the concentration of Ca began to decrease due to the precipitation of ettringite, calcite, gypsum, and possible Ca-bearing compounds. In this study, the main Ca-bearing compounds in the MICP-treated samples were calcite, vaterite, and amorphous $CaCO_3$. Moreover, the solubility product for calcite is 3.36×10^{-9}, while the solubility product for gypsum ($CaSO_4{\cdot}2H_2O$) is 3.14×10^{-5}. Moreover, the pH of bio-modified SS products was in the range of 8.0–11.5, which would benefit the stability of $CaCO_3$ products. Therefore, there are a few chances for the transformation from calcite to gypsum to take place. Meanwhile, the low availability of Al ions also confined the product of ettringite. Therefore, the low content of Ca was leached out from the MICP-treated samples within 10 days' reaction under high concentration of Na_2SO_4 solution (10 g/L). Therefore, MICP-treated samples possess a high ability to resist the erosion from sulfate attack.

6.2.3 Bioremediation of Stainless Steel Pickling Sludge through MICP

6.2.3.1 Materials

6.2.3.1.1 Stainless Steel Pickling Sludge

SSPS was provided by a stainless steel factory. According to the energy-dispersive X-ray test, SSPS contains 36.1% CaO, 27.9% SO_3, 27.3% Fe_2O_3, 4.8% Cr_2O_3, 2.3% NiO, 1.3% MnO, and 0.4% ZnO. Acid digestion test was performed on raw SSPS according to the US Environmental Protection Agency (EPA) 3050B, Acid Digestion of Sediments, Sludges, and Soils (USEPA 1996). The results, as shown in Table 6.1, illustrated thagt the total content of Cr, Ni, and Ca reached 18,940, 22,350, and 178, 450 mg/kg, respectively. The content of Cr(VI) in the SSPS was found to be 328 mg/kg, which means most of the Cr presented in SSPS is Cr(III). TCLP test results showed the leaching concentration of Cr from raw SSPS was 5.73 mg/L, which is higher than the limit of the US EPA for the regulation of hazardous waste. Though there is no limit for Ni in the regulation for hazardous waste, the toxicity effects of Ni posed on the health of the living make it be a most concerning element.

6.2.3.2 Methods

6.2.3.2.1 Investigation of Bacteria Activity

Bacteria activity was evaluated through the variation of pH and electric conductivity (EC) of urea solutions over time. The urea solution was prepared with urea and deionized water. An aliquot of 10 mL bacteria solution with the concentration (OD_{600}) of 0, 0.6, 0.7, 0.8, 0.9, and 1.0 were added to 90 mL urea solutions. After that, the initial urea concentration for the reaction solution was 30 g/L. The pH and EC of the reaction solution be recorded over time.

6.2.3.2.2 *Toxic Effects of Cr(III) on Bacterial Activity*

Cr(III) solution was prepared with chromium nitrite ($Cr(NO_3)_3$). An aliquot of 26 mL culturing media with 3 mL bacteria solution was mixed with 1 mL chromium nitrite ($Cr(NO_3)_3$) liquor. The final concentrations of Cr(III) were 0, 10, 20, 30, 40, 50, and 60 mg/L. Then, the mixtures were placed on a horizontal oscillation culturing instrument. The rotation speed was 200 rpm. After culturing for about 43 ± 2 h, the bacteria solution was harvested. Then, the OD_{600} of Cr solutions was determined. Meanwhile, the viable cell numbers after cultured in Cr solution were also determined through agar plate culturing. An aliquot of 100 μL of bacteria was inoculated into the agar plate, which consists of 1.5% agar and the culturing solution at 30°C for 24 h. Then, the colonies in the agar plate were counted and expressed as CFU (colony forming unit). Viable cell numbers were calculated by the CFU multiplied by the dilution factor. The urease activity of cultured bacteria in Cr solution was determined through the variation of the EC of the urea solution. An aliquot of 1 mL of harvested bacteria liquor or bacteria lysis liquor (BLL) was poured into 9 mL 1.11 M urea. The EC of the reaction solution was recorded at 1 min and 8 min after mixing (Jiang et al. 2019). Then, the urease activity of Cr solution-cultured bacteria can be estimated through the empirical relationship, as shown in equation (6.12) (Jiang et al. 2019):

$$Urease\ activity = (EC8 - EC1)\ /7 \times 10 \times 11\ (mM\ urea \cdot min\text{-}\ 1) \qquad (6.12)$$

Where, EC1 and EC8 are the EC at 1 min and 8 min, respectively.

6.2.3.2.3 *Bio-Mineralization Process*

To conduct bacteria-based bio-mineralization, five aliquots of the bacteria liquor were prepared with OD_{600} as 0.6, 0.7, 0.8, 0.9, and 1.0, respectively. Bacteria and urea mixtures were prepared with 5 mL bacteria liquor and 45 mL urea solution, which made the final concentration of urea be 30 g/L before the bio-mineralization experiments. Then, the bacteria–urea reaction matrix was poured into the SSPS to make a homogenous solution, immediately. The solid-to-liquid ratio was set as 1:2 (g/mL). The BLL-based bio-mineralization process was performed with ultrasonic-treated bacteria solutions. BLL was harvested with mature bacteria cells treated by an ultrasonic extraction instrument at 25 HZ for 10 min. The concentrations of bacteria used

TABLE 6.21
Leaching Concentration of Cr, Ni, and Ca in Raw SSPS

Leaching Methods	Cr	Ni	Ca
Acid digestion test (mg/kg)	19,140	20,200	172,600
TCLP (mg/L)	5.73	18.7	1540
SPLP (mg/L)	11.1	0.048	547

to prepare BLL were also 0.6, 0.7, 0.8, 0.9, and 1.0 (OD_{600}). After that, the BLL was mixed with the urea solution and SSPS and performed the biomineralization process. The solid-to-liquid ratio was 1:2 (g/mL) as well. To evaluate the effects of Ca addition on the performance of biomineralization, 0.01 M, 0.02 M, 0.03 M, 0.04 M, and 0.05 M $CaCl_2$ were added to the bio-mineralization reaction system, respectively. After 7 days' reaction, the reaction was halted. Samples were oven-dried in 100 ± 5°C for over 24 h. Then, the samples were harvested and grounded for leaching tests.

6.2.3.2.3.1 Toxicity Characterization Leaching Procedure

The US EPA method 1311, Toxicity Characteristic Leaching Procedure (TCLP) (USEPA 1992a), is employed to evaluate the leaching behavior of metals and the bio-mineralization performances. After pH determination, an extraction fluid which contains 5.7 mL glacial CH_3CH_2OOH in 1 L solution was used. An aliquot of 75 g solid was mixed with the extraction fluid with solid-to-liquid ratios of 1:20 (g/mL) in a 2 L HDPE bottle. The bottle was then agitated in a rotary extractor at 28 ± 2 rpm for a period of 18 ± 2 h. The leachate was filtered through the 0.75 μm glass-fiber filter, and then the pH was adjusted to less than 2 for the preservation and analysis of concentrations of the concerned constituents.

6.2.3.2.4 pH-Dependent Leaching Procedure

US EPA Standard Method 1313, Liquid–Solid Partitioning as a Function of Extract pH Using A Parallel Batch Extraction Procedure, was employed to determine the leaching behavior of metals in biomineralized samples (USEPA 2012). Dilute solutions of nitric acid (HNO_3) and potassium hydroxide (KOH) were used to adjust the pH of the extraction solution. The solid-to-liquid ratio was 1:10 (g/mL). After the samples were mixed homogenously, the mixture was rotated at 28 ± 2 rpm for 24 ± 2 h. Then, the mixture was filtered through a 0.6–0.8 μm glass-fiber filter. The extract was harvested, and the pH and oxidation-reduction potential (ORP) were recorded. After that, the pH of the extract was adjusted to below 2 and saved for metal determination.

6.2.3.2.5 Metal Determination

The metal concentrations in preserved leachate samples were performed by an inductively coupled plasma optical emission spectrometer (ICP-OES) (ICPE-9000, Shimadzu). Commercially produced multielement standard solutions were used for calibration. The detection limits for metals are Ca (0.5 μg/L), Cr (5 μg/L), and Ni (5 μg/L), respectively.

6.2.3.2.6 Geochemical Modeling

The geochemical model Visual Minteq was used for the calculation of the equilibrium composition of natural aqueous systems and aqueous solutions in a laboratory setting. The equilibrium mass distribution was calculated under conditions that include a gas phase with constant partial pressure for dissolved species and multiple solid phases. The ions were allowed to precipitate when they reached their supersaturation. The ions concentration of Ca, Cr, and Ni were input to the model software with the pH of the extract solution.

6.2.3.3 Toxic Effect of Cr on the Activity of Bacteria

In the MICP process, urea is hydrolyzed into ammonium (NH^{4+}) and carbonate (CO_3^{2-}) ions. With the ionization of urea solution proceeding, the pH and EC will change accordingly. According to former study, there is a positive relationship between the urea hydrolyzation rate and EC (Krishnamurti and Kate 1951). Therefore, the variations of pH and EC are usually employed to evaluate the activity of bacteria. The pH of the reaction solution as a function of time with different concentrations of bacteria is illustrated in Figure 6.45a. Without bacteria, the pH of the urea solution increased slightly in the first 3 h and then decreased. The maximum pH was lower than 9. However, with the addition of bacteria solution, the pH of the reaction solution increased dramatically to about 9.30 in the first 3 h. Moreover, the pH remained above 9 in the rest 24 h reaction. EC as a function of time with different bacteria concentrations is shown in Figure 6.45b. It is seen that synchronous change occurred between pH and EC. The EC of the solution increased sharply at first 17 h in the presence of bacteria, followed by a slight increase. EC reached 3.4 with some flocculation after 24 h reaction for the solutions with bacteria. By comparison, the EC of the urea solution without bacteria experienced a slight increase over time (Figure 6.45b), indicating that only a small amount of urea was hydrolyzed in this reaction. In addition, comparing the performance of urea hydrolyzation by bacteria in different bacteria concentrations, it is apparent that the higher the concentration of bacteria, the faster the hydrolyzation rate of urea. Considering the balance of cost and consumption, the effect of Cr on bacteria activity was performed at initial OD_{600} of 0.8.

As shown in Figure 6.46a, OD_{600} varied from 1.35 to 1.36 with the cultivation of bacteria and Cr. However, the viable cell number was considerably influenced by the addition of Cr. With the Cr concentration increasing from 0 to 40 mg/L, the viable cell number decreased from 3.2×10^7 to 2.4×10^7, indicating the loss of workable bacteria cells. OD typically reflects the total number of cells rather than the worked one. Therefore, agar plate culturing (CFU number) was preferred in this study (Mugwar and Harbottle 2016) to evaluate the worked bacteria. As shown in Figure 6.46a, the viable cell numbers increased to 2.4×10^7 when the concentration of Cr increased

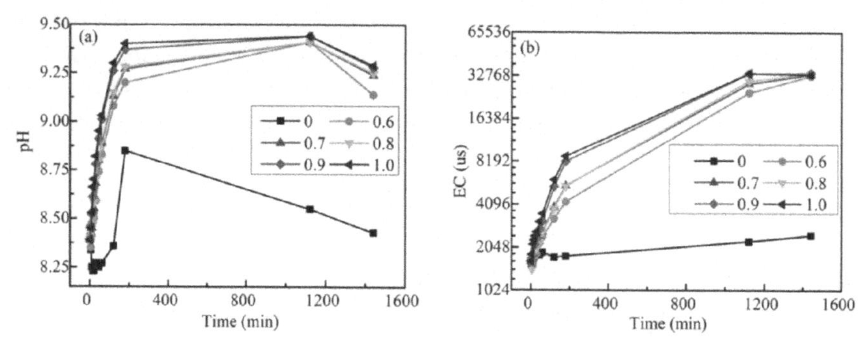

FIGURE 6.45 Effects of bacteria concentration (OD_{600} = 0, 0.6, 0.7, 0.8, 0.9, 1.0) on the variation of (a) pH and (b) EC of the reaction solution.

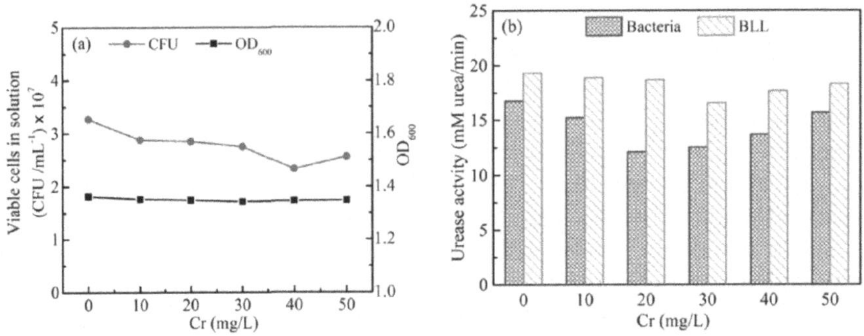

FIGURE 6.46 Toxicity effects of Cr on (a) viable bacteria cell and (b) urease activity.

from 40 mg/L to 50 mg/L, which means bacteria could coexist with 50 mg/L Cr. Even though the viable cell numbers decreased when the Cr solution was added, there were a considerable number of viable cells that survived. Therefore, considering the concentration of Cr in SSPS (19,140 mg/kg as total content and 5.7 mg/L in TCLP extraction solution), the bacteria used in this study would be workable for the bio-mineralization of SSPS.

The urease activity is the key factor to MICP efficiency. Therefore, it was utilized to evaluate the bacteria activity of bacteria solution and BLL solution in this study. As presented in Figure 6.46b, the urease activity varied in the range of 16.6–19.3 mM urea/min, with the addition of Cr for the BLL solution. However, the concentration of Cr had a significant influence on the urease activity of bacteria solution. The urease activity of the BLL solution was superior to that of the bacteria solution. When the concentration of Cr was 20 mg/L, the urease activity of the BLL solution was 18.7 mM urea/min, while it was 12.1 mM urea/min for the bacteria solution. It meant that the urea hydrolyzation was much faster with the BLL solution rather than that with bacteria solution. This result confirms the hypothesis that the breakdown of the bacteria cell wall would promote the access of urease to the urea in the solution and thus promote the hydrolysis of urea. However, this result did not mean the urease activity of BLL solution was better than that of the bacteria solution till the end of bio-mineralization. Since the results of urease activity were achieved in 8 min and the advantage of bacteria solution was to produce urease in the long span while the urease in BLL solution would be consumed in a short time, when the concentration of Cr increased from 20 to 50 mg/L, the urease activity of bacteria began to increase. The back reason for this phenomenon would be that the high pressure of Cr concentration stimulated the metabolism of bacteria and thus promote the production of urease.

6.2.3.4 Effect of BLL on the Performance of Bio-Mineralization

The concentration of Cr in control sample was 5.7 mg/L, which is higher than the regulation concentration (5.0 mg/L) of the US EPA for hazardous material. The leaching concentration of Ni in the control sample was 18.7 mg/L, and the leaching concentration of Ca was 1540 mg/L. As shown in Figure 6.47a, with treatment

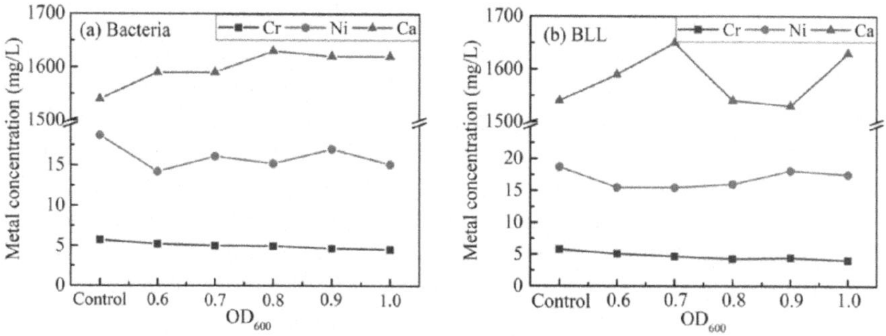

FIGURE 6.47 Effect of bacteria concentration (OD_{600}) on the leaching of metals: (a) bacteria, (b) BLL.

of the bacteria-based biomineralization process, the concentration of Cr decreased. Moreover, the higher the concentration of bacteria consumed, the lower the leaching concentration of Cr. When the concentration of bacteria was 1.0 (OD_{600}), the leaching concentration of Cr was 4.5 mg/L, which is 78.8% of that in the control sample. Impressively, when the concentration of bacteria was higher than 0.8 (OD_{600}), the leaching concentration of Cr can meet the regulation of US EPA for hazardous materials (USEPA, 1990). The concentrations of Ni in bacteria-based bio-mineralization products were in the range of 14.2–17.0 mg/L, which were 76.0–91.9% of that in the raw SSPS. However, Ca in bacteria-based bio-mineralization products is presented to be more likely to leach out than that in the raw SSPS. Considering the current laws and regulations of solid waste, the change in Ca is acceptable.

As shown in Figure 6.47b, with the treatment of BLL-based bio-mineralization, a high efficiency of Cr immobilization/stabilization was achieved. When SSPS was treated with BLL at the bacteria concentration of 0.6, the concentration of Cr decreased to 5.1 mg/L. Moreover, a higher concentration of BLL resulted in a better bio-mineralization performance of Cr. The leaching concentration of Cr was 4.1 mg/L, which is only 70.7% of that in raw SSPS, when the concentration of BLL increased to 1.0. Notably, when the concentration of BLL was 0.7, the leaching concentration of Cr was 4.7 mg/L, which meets the regulation of US EPA for hazardous material. Compared with the results of the bacteria-based bio-mineralization process that the bacteria concentration was at least 0.8 to meet the regulation of US EPA, fewer bacteria cells were consumed in BLL-based bio-mineralization process. Moreover, higher bio-mineralization efficiencies were achieved by the BLL-based bio-mineralization process when the same amounts of bacteria cells were used. This phenomenon revealed that the BLL-based bio-mineralization process could improve stabilization efficiency and cut down consumption as well. Previous studies have reported that the biosorption process from the biomass contributed to the metal removal in MICP treatment (Oliveira et al. 2014; Zhang et al. 2021a, 2021b). Many bio-macromolecules, for example, polysaccharides, proteins, glycoproteins, and phospholipids, were reported to be predominant functional groups for binding sites, and the binding sites of heavy metals and organic matters would also serve as nucleation sites for the initial precipitation based on supersaturation in microenvironments (Oliveira et al. 2014; White et al. 1995). Even though the formation of abundant carbonates contributed to a considerable portion of heavy metal removal in the MICP process, the metal–organic matter coordination associated with carboxyl, alkane, and amide groups served as the preliminary step for subsequent precipitation (Qian et al. 2017). Therefore, the improved accessibility of BLL achieved excellent biosorption efficiency over the bacterial cell and thus led to the high efficiency of bio-mineralization. Moreover, there were obvious stabilization effects observed for Ni in BLL-based bio-mineralization products, while the bio-mineralization of Ca was not observed.

6.2.3.5 Effect of Ca Addition on the Performance of Bio-Mineralization

As stated in former literature, Ca could promote the stabilization of metal ions by co-precipitation or adsorption of metals (Zhang et al. 2021a). Therefore, Ca ion was introduced to the bacteria-based and BLL-based bio-mineralization process at the bacteria concentration of 0.8 (OD_{600}). The effects of Ca addition on the leaching behavior of metals were evaluated through the TCLP test as well. As shown in Figure 6.48a, the bio-mineralization performances of Cr were improved greatly with the addition of Ca for the bacteria-based bio-mineralization process. When 0.01 M Ca was supplied, the leaching concentration of Cr was 3.2 mg/L, which means more than 44.6% Cr was stabilized compared to that in raw SSPS of 5.7 mg/L. Meanwhile, the value of 3.2 meets the requirement of US EPA hazardous waste regulation of 5 mg/L (USEPA 1990). Moreover, the leaching concentration of Cr for bacteria-based bio-mineralization products was 1.0 mg/L when the addition of Ca increased to 0.05 M. Compared with the leaching concentration of Cr for bacteria-based bio-mineralization products without Ca, the bio-mineralization performance was improved dramatically. The leaching concentration of 1.0 mg/L is only 20.0% of 5.0 mg/L, which is achieved by the bacteria-based bio-mineralization process without Ca addition at the same bacteria concentration. The improvement of the bio-mineralization effect for Ni by the addition of Ca was not as excellent as that for Cr. The leaching concentrations of Ni were in the range of 14.6–20.1 mg/L. The stabilization effect of the bacteria-based bio-mineralization process for Ca was not

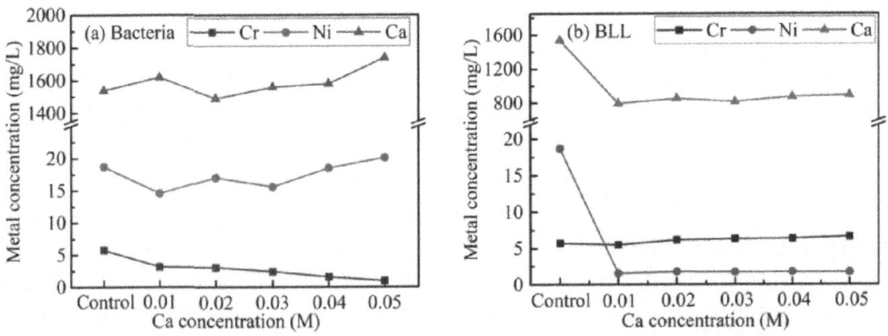

FIGURE 6.48 Effect of Ca addition on the leaching behavior of metals: (a) bacteria, (b) BLL.

satisfying as well. The leaching concentration of Ca fluctuated in the range of 1490–1740 mg/L for the samples treated with the addition of Ca, while the concentration of Ca in bacteria-based bio-mineralization products without Ca addition was in the range of 1540–1630 mg/L. Therefore, the promotional effect of Ca addition on the bacteria-based bio-mineralization process presented the likelihood of selectivity of target metals.

The effect of Ca addition on the BLL-based bio-mineralization process is illustrated in Figure 6.48b. The leaching concentration of Cr increased in BLL-based bio-mineralization products, which means the bio-mineralization efficiency was restricted other than improved with the addition of Ca. Hua et al. (2007) studied the incorporation of chromate into a calcium carbonate structure during coprecipitation, and the results revealed that the incorporation of chromate into calcium carbonate precipitates only takes place during the process of calcium carbonate formation, not after precipitates were formed. With the addition of Ca, the formation of calcite lattice was accelerated. However, the dissociation of Cr from the sludge matrix into the reaction system and incorporating it into the calcite lattice need more times. Therefore, the co-precipitation of Cr into the lattice of calcite was restricted. Meanwhile, the bioleaching effect from the organic acid matters and other bio-macromolecules of bacteria prevailed. Therefore, more Cr was leached out than that in the bacteria-based bio-mineralization process. However, the addition of Ca would promote the stabilization of Ni and Ca. The leaching concentration of Ni and Ca decreased dramatically. The leaching concentration of Ni in BLL-based bio-mineralization products was in the range of 1.6–1.8 mg/L, which is about 8.3–9.5% of that in raw SSPS. The leaching concentration of Ca was in the range of 795–893 mg/L, which is about 51.6%–58.0% of that in raw SSPS.

6.2.3.6 Leaching Behavior of Metals in pH-Dependent Tests

The acid neutralization capacity (ANC) was investigated through the pH-dependent test and is shown in Figure 6.49. The negative number indicates the addition of base in terms of meq/g weight dries solid. Bacteria-based bio-mineralization products with a different addition of Ca presented similar ANC. The natural pH without the addition of acid or base of bio-mineralized samples was in the range of 7.9–8.5. All the samples obtained wide neutralization capacities of approximately 2 meq/g acid or base consumed to change 1 pH unit. The large amounts of alkalinity products have contributed to the high buffering property of bio-mineralized samples. In the pH range of 5–8, the high buffering property of bio-mineralized samples was much more obvious. According to former research by Chen et al. (2009), calcium carbonate, gypsum, and gibbsite are most likely responsible for the buffering capacities in the pH range of 5–6. The variation of equilibrium pH with acid addition of 4 meq/g observed in Figure 6.49 illustrated the different mineral phases formed in the bio-mineralization process.

The oxidation-reduction potential (ORP) is used to determine if oxidizing or reducing conditions are prevalent in bio-mineralization products. Furthermore, ORP is one of the most important factors that can affect the leaching behavior of elements (Zhang et al. 2016). As shown in Figure 6.50, the ORP of bio-mineralized samples decreased linearly with the increase of pH. Moreover, no noticeable differences were

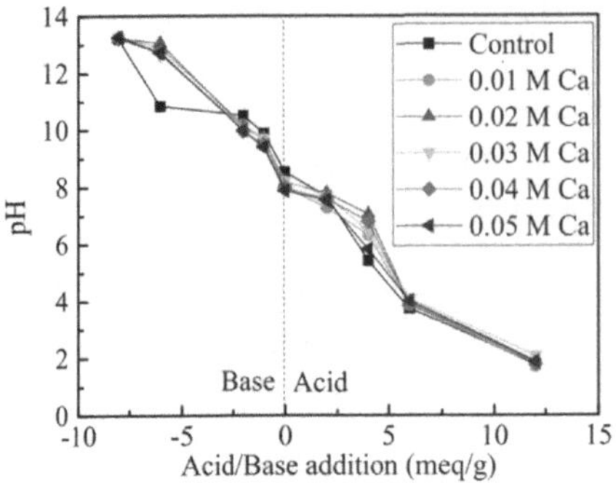

FIGURE 6.49 ANC curve of bacteria-modified samples with the addition of Ca.

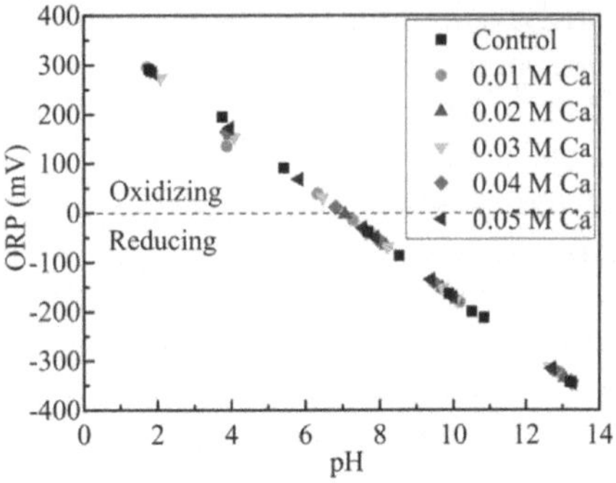

FIGURE 6.50 ORP as a function of pH for bacteria-modified samples with the addition of Ca.

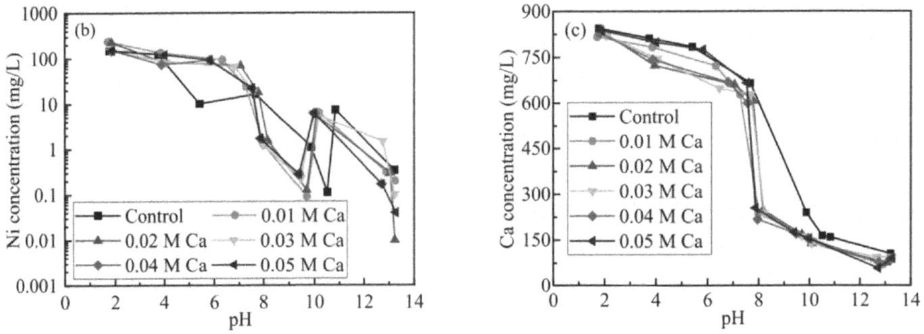

FIGURE 6.51 Leaching behavior of (a) Cr, (b) Ni, and (c) Ca as a function of pH.

observed for the variation of ORP versus pH for different bio-mineralization products. The metals in the bio-mineralization products existed in reducing states, when the pH was in the range of 7–9.

The leaching behavior of metals follows three distinct patterns, which include cationic, oxyanions, and amphoteric patterns in general (Zhang et al. 2016). As shown in Figure 6.51a, the leaching of Cr exhibited a typical amphoteric pattern, which had high concentrations in acidic and basic pH conditions and low concentrations in neutral pH conditions. Dissolution of Cr-carrying oxide/hydroxide minerals has a great influence on the solubility of Cr. Cr_2O_3(s) is expected to control the release of Cr in neutral pHs, while the dissolution of $Cr(OH)_3$(s) is responsible for the increase of Cr concentration in acidic and basic pHs. The leaching behavior of Cr in different bio-mineralized products presented consistency with the variation of Ca addition in acidic conditions. The difference of Cr leaching concentration in neutral and basic conditions illustrated the difference in bio-mineralization efficiency. A strong resistance to a basic solution was observed for the samples treated by the bacteria-based bio-mineralization process with 0.05 M Ca addition, as the leaching concentration of

Cr was the lowest in the pH range of 8–14. The leaching of Ni followed the cationic pattern, for which the leaching concentration increased with the decrease of pH. Ni-carrying carbonate and hydroxide minerals were assumed to control the dissolution of Ni in alkalinity conditions. The leaching of Ca is shown in Figure 6.51c. It followed typical cation leaching pattern with strong buffering properties in acidic and basic solutions. The adsorption of Ca, oxide mineral, hydroxide mineral, and carbonate minerals of Ca may contribute to the dissolution of Ca. The bio-mineralization process improved the base resistance of Ca in the pH range of 8–13.

6.2.3.7 Geochemical Modeling

Aqueous-phase equilibrium composition and saturation index of all leachates concerning solids or minerals were computed by allowing aqueous complexation reactions at fixed pH. Calculated log activities of the Cr, Ni, and Ca obtained from Visual Minteq and their corresponding leachate pH are graphed in Figure 6.52. These graphs were employed to determine if the element is solubility-controlled. If the calculated log activities of the elements are consistently close to the stability/solubility line of the mineral or solid, the element is mineral solubility–controlled. In contrast, elements that demonstrate over- or under-saturation for the stability/solubility line are not solubility-controlled (Komonweeraket et al. 2015a, 2015b). As shown in Figure 6.52a, the log activity of Cr in the pH range of 5–14 was consistently close to the

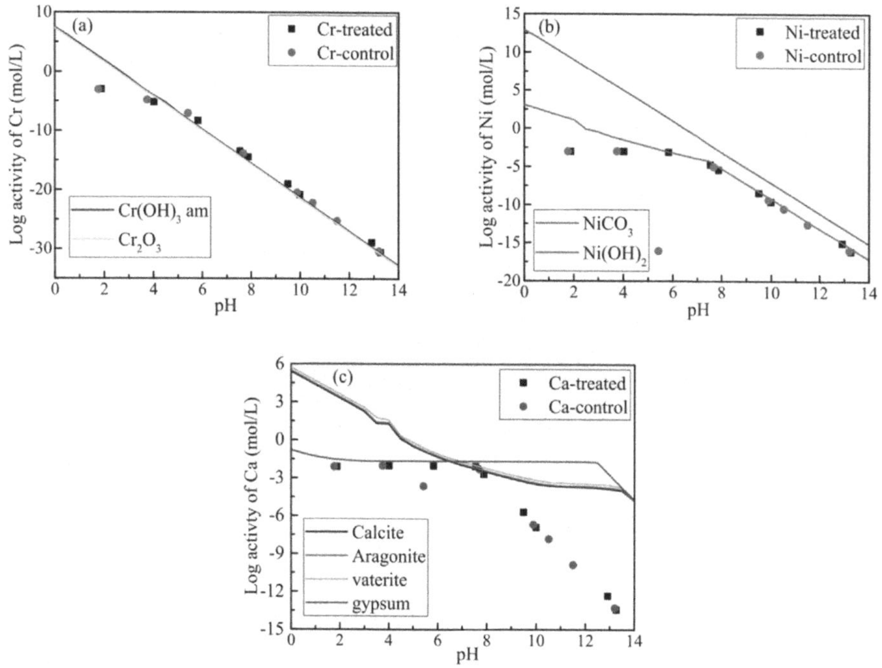

FIGURE 6.52 Log activity of (a) Cr^{3+}, (b) Ni^{2+}, and (c) Ca^{2+} in the effluent of pH-dependent experiments from control and treated sample.

solubility line of $Cr(OH)_3$ a.m. and Cr_2O_3, which means Cr in the bio-mineralized and raw samples were solubility-controlled, as the original pH of raw SSPS and treated SSPS was in the range of 6–9. However, the log activity of Cr in the pH range of 2–4 was below the solubility line of $Cr(OH)_3$ a.m. and Cr_2O_3. The high-solubility product of Cr-carrying minerals may be responsible for the dissociation or solubility of Cr. The geochemical model of Ni leaching in bio-mineralized and raw PS illustrated that $NiCO_3$ contributed to the solubility of Ni in the pH range of 6–14, while another Ni-carrying product was responsible for the solubility of Ni in the pH range of 0–5. As shown in Figure 6.52c, Ca leaching in the pH range of 0–7 was solubility-controlled by gypsum, while in the moderate pH range of 7–9, the leaching of Ca in bio-mineralized and raw SSPS was more likely controlled by the mineral calcium carbonate than calcium sulfate. When the mineral was in alkalinity conditions, the log activities of Ca were below the solubility line, which means other complex compounds of Ca existed.

References

Achal V., Mukherjee A. A review of microbial precipitation for sustainable construction. *Construction and Building Materials*, 2015, 93:1224–1235.

Achal V., Mukerjee A., Sudhakara Reddy M. Biogenic treatment improves the durability and remediates the cracks of concrete structures. *Construction and Building Materials*, 2013, 48:1–5.

Achal V., Mukherjee A., Sudhakara Reddy M. Microbial concrete: Way to enhance the durability of building structures. *Journal of Materials in Civil Engineering*, 2011, 23(6):730–734.

Achal V., Pan X.L. Influence of calcium sources on microbially induced calcium carbonate precipitation by bacillus sp. Cr2. *Applied Biochemistry and Biotechnology*, 2014, 173(1):307–17.

Achal V., Pan X.L., Fu Q.L., Zhang D.Y. Biomineralization based remediation of as(iii) contaminated soil by sporosarcina ginsengisoli. *Journal of Hazardous Materials*, 2012b, 201:178–184.

Achal V., Pan X., Zhang D. Bioremediation of strontium (SR) contaminated aquifer quartz sand based on carbonate precipitation induced by SR resistant Halomonas sp. *Chemosphere*, 2012a, 89(6):764–768.

Achal V., Pan X.L., Zhang D.Y., Fu Q.L. Bioremediation of pb-contaminated soil based on microbially induced calcite precipitation. *Journal of Microbiology and Biotechnology*, 2012c, 22(2):244–247.

Adolphe J.P., Loubiere J.-F., Paradas J., Soleilhavoup F. *Procédé de traitement biologique d'une surface artificielle*. European Patent 90400G97, 1990.

Agbede I.O., Manasseh J. Use of cement-sand admixture in laterite brick production for low cost housing. *Leonardo Electronic Journal of Practices Technologies*, 2008, 12(1):163–174.

Aksoy H., Unal N.E., Cokgor S., et al. A rainfall simulator for laboratory-scale assessment of rainfall-runoff-sediment transport processes over a two-dimensional flume. *Catena*, 2012, 98:63–72.

Al Qabany A., Soga K. Effect of chemical treatment used in MICP on engineering properties of cemented soils. *Geotechnique*, 2013, 63(4):331–339.

Al Qabany A., Soga K., Santamarina C. Factors affecting efficiency of microbially induced calcite precipitation. *Journal of Geotechnical and Geoenvironmental Engineering*, 2012, 138(8):992–1001.

Algaifi H.A., Bakar S.A., Sam A.R.M., et al. Insight into the role of microbial calcium carbonate and the factors involved in self-healing concrete. *Construction and Building Materials*, 2020, 254

Amidi S., Wang J. Surface treatment of concrete bricks using calcium carbonate precipitation. *Construction and Building Materials*, 2015, 80:273–278.

Anand K., Goyal S., Reddy M.S. Development of biocementitious grout using a silica fume-based bacterial agent for remediation of cracks in concrete structures. *Journal of Materials in Civil Engineering*, 2024, 36(10).

Anbu P., Kang C.H., Shin Y.J., So J.S. Formations of calcium carbonate minerals by bacteria and its multiple applications. *SpringerPlus*, 2016, 5.

Arp G., Helms G., Karlinska K., et al. Photosynthesis versus exopolymer degradation in the formation of microbialites on the atoll of kiritimati, republic of kiribati, central pacific. *Geomicrobiology Journal*, 2012, 29(1):29–65.

ASTM C 618–19. *Standard Specification for Coal Fly Ash and Raw or Calcined Natural Pozzolan for Use in Concrete*. ASTM International, West Conshohocken, PA, USA, 2019.

ASTM C 20–10. *Standard Test Methods for Apparent Porosity, Water Absorption, Apparent Gravity, and Bulk Density of Burned Refractory Brick and Shapes by Boiling Water*. ASTM International, West Conshohocken, PA, USA, 2010.

ASTM D 2166M–16. *Standard Test Method for Unconfined Compressive Strength of Cohesive Soil*. ASTM International, West Conshohocken, PA, USA, 2016a.

ASTM D 3967–16. *Standard Test Method for Splitting Tensile Strength of Intact Rock Core Specimens*. ASTM International, West Conshohocken, PA, USA, 2016b.

ASTM D 3987–06. *Standard Test Methods for Shake Extraction of Solid Waste with Water*. ASTM International, West Conshohocken, PA, USA, 2006.

ASTM D 3987–12. *Standard Practice for Shake Extraction of Solid Waste with Water*. ASTM International, West Conshohocken, PA, USA, 2012.

ASTM D 5084–03. *Standard Test Methods for Measurement of Hydraulic Conductivity of Saturated Porous Materials Using a Flexible Wall Permeameter*. ASTM International, West Conshohocken, PA, USA, 2003.

ASTM D 6272–17. *Standard Test Method for Flexural Properties of Unreinforced and Reinforced Plastics and Electrical Insulating Materials by Four-point Bending*. ASTM International, West Conshohocken, PA, USA, 2017.

Astrup T., Dijkstra J.J., Comans R.N., Van der Sloot H.A., Christensen T.H. Geochemical modeling of leaching from MSWI air-pollution-control residues. *Environmental Science and Technology*, 2006, 40(11):3551–3557.

Ataie-Ashtiani B., Beheshti A.A. Experimental investigation of clear-water local scour at pile groups. *Journal of Hydraulic Engineering*, 2006, 132(10):1100–1104.

Babu G.S., Chouksey S.K. Stress–strain response of plastic waste mixed soil. *Waste Management & Research*, 2011, 31(3):481–488.

Bahoria B., Parbat D., Nagarnaik P. Xrd analysis of natural sand, quarry dust, waste plastic (ldpe) to be used as a fine aggregate in concrete. *Materials Today: Proceedings*, 2018, 5(1):1432–1438.

Balam N.H., Mostofinejad D., Eftekhar M. Use of carbonate precipitating bacteria to reduce water absorption of aggregates. *Construction and Building Materials*, 2017, 141:565–577.

Baldermann A., Rezvani M., Proske T., et al. Effect of very high limestone content and quality on the sulfate resistance of blended cements. *Construction and Building Materials*, 2018, 188:1065–1076.

Bell F.G. Lime stabilization of clay minerals and soils. *Engineering Geology*, 1996, 42(4):223–237.

Bernardi D., DeJong J., Montoya B.M., Martinez B. Bio-bricks: Biologically cemented sandstone bricks. *Construction and Building Materials*, 2014, 55:462–469.

Braissant O., Cailleau G., Dupraz C., Verrecchia E. Bacterially induced mineralization of calcium carbonate in terrestrial environments: The role of exopolysaccharides and amino acids. *Journal of Sedimentary Research*, 2003, 73(3):485–490.

Briaud J.-L., Bernhardt M., Leclair M. The pocket erodometer test: Development and preliminary results. *Geotechnical Testing Journal*, 2012, 35(2):342–352.

Bu C., Lu X., Zhu D., et al. Soil improvement by microbially induced calcite precipitation (MICP): A review about mineralization mechanism, factors, and soil properties. *Arabian Journal of Geosciences*, 2022, 15(9):863.

Bu C., Wen K., Liu S., et al. Development of a rigid full contact mold for preparing bio-beams through microbial induced calcite precipitation *Geotechnical Testing Journal*, 2018a, 42(3):656–669.

Bu C., Wen K., Liu S., Ogbonnaya U., Li L. Development of bio-cemented constructional materials through microbial induced calcite precipitation. *Materials Structures*, 2018b, 51(1):30.

Bu C.M., Qian D., Wen K.J., Lin L. Development of innovative bio-beam using microbial induced calcite precipitation technology. In: *International Conference: Geoenvironment and Geohazard*. GSIC, 2018c, 491–498.

Bu C.M., Wen K.J., Liu S.H., et al. Development of a rigid full-contact mold for preparing biobeams through microbial-induced calcite precipitation. *Geotechnical Testing Journal*, 2019, 42(3):656–669.

Burbank M., Weaver T., Lewis R., Williams T., Williams B., Crawford R. Geotechnical tests of sands following bioinduced calcite precipitation catalyzed by indigenous bacteria. *Journal of Geotechnical Geoenvironmental Engineering*, 2013, 139(6):928–936.

Cardoso R., Pires I., Duarte S.O.D., Monteiro G.A. Effects of clay's chemical interactions on biocementation. *Applied Clay Science*, 2018, 156(3):96–103.

Carlos R.N., Concepcion J.L., Alejandro R.N., Maria T.G.M., Manuel R.G. Bacterially mediated mineralization of vaterite. *Geochim Cosmochim Acta*, 2007, 71(5):1197–1213.

Castanier S., Métayer-Levrel G.L., Perthuisot J.-P. Bacterial roles in the precipitation of carbonate minerals. In: Riding, R.E., Awramik, S.M. (eds) *Microbial Sediments*, Springer, 2000. https://doi.org/10.1007/978-3-662-04036-2_5.

Castanier S., Métayer-Levrel G.L., Perthuisot J.-P. Ca-carbonates precipitation and limestone genesis—the microbiogeologist point of view. *Sedimentary Geology*, 1999, 126(1–4):9–23.

Chauhan M.S., Mittal S., Mohanty B. Performance evaluation of silty sand subgrade reinforced with fly ash and fibre. *Geotextiles Geomembranes*, 2008, 26(5):429–435.

Chen C.M. Science mapping: A systematic review of the literature. *Journal of Data and Information Science*, 2017, 2(2):1–40.

Chen C.M., Hu Z.G., Liu S.B., Tseng H. Emerging trends in regenerative medicine: A scientometric analysis in citespace. *Expert Opinion on Biological Therapy*, 2013, 12(5):593–608.

Chen C.M., Ibekwe Sanjuan F., Hou J.H. The structure and dynamics of co-citation clusters: A multiple-perspective co-citation analysis. *Journal of the American Society for Information Science Technology*, 2014a, 61(7):1386–1409.

Chen D.M., Mi J., Chu P.F., et al. Patterns and drivers of soil microbial communities along a precipitation gradient on the mongolian plateau. *Landscape Ecology*, 2014b, 30(9):1669–1682.

Chen J., Bradshaw S., Benson C.H., Tinjum J.M., Edil T.B. Ph-dependent leaching of trace elements from recycled concrete aggregate. In: *Geocongress 2012: State of the Art and Practice in Geotechnical Engineering*. ASCE, 2012, 3729–3738.

Chen L., Shen Y., Xie A., et al. Bacteria-mediated synthesis of metal carbonate minerals with unusual morphologies and structures. *Crystal Growth & Design Journal*, 2009, 9(2):743–754.

Chen R., Miao L., Sun X., Wu L., Wang C. Comparison on different repair methods of microfracture cured by microbial. *Bulletin of the Chinese Ceramic Society*, 2019, 277(10):27–32.

Cheng L., Cord-Ruwisch R. In situ soil cementation with ureolytic bacteria by surface percolation. *Ecological Engineering*, 2012, 42(42):64–72.

Cheng L., Cord-Ruwisch R. Upscaling effects of soil improvement by microbially induced calcite precipitation by surface percolation. *Geomicrobiology Journal*, 2014, 31(5):396–406.

Cheng L., Cord-Ruwisch R., Shahin M.A. Cementation of sand soil by microbially induced calcite precipitation at various degrees of saturation. *Canadian Geotechnical Journal*, 2013, 50(1):81–90.

Cheng L., Kobayashi T., Shahin M.A. Microbially induced calcite precipitation for production of "bio-bricks" treated at partial saturation condition. *Construction and Building Materials*, 2020, 231:117095.

Cheng L., Shahin M.A., Chu J. Soil bio-cementation using a new one-phase low-pH injection method. *Acta Geotechnica*, 2018a, 14(3):615–626.

Cheng L., Shahin M.A., Mujah D. Influence of key environmental conditions on microbially induced cementation for soil stabilization. *Journal of Geotechnical and Geoenvironmental Engineering*, 2017, 143(1).

Cheng W.-C., Ni J.C., Arulrajah A., Huang H.-W. A simple approach for characterising tunnel bore conditions based upon pipe-jacking data. *Tunnelling Underground Space Technology*, 2018b, 71:494–504.

Cheng W.C., Wang L., Xue Z.F., Ni J.C., Rahman M.M., Arulrajah A. Lubrication performance of pipejacking in soft alluvial deposits. *Tunnelling Underground Space Technology*, 2019, 91:102991.

Chew S., Kamruzzaman A., Lee F. Physicochemical and engineering behavior of cement treated clays. *Journal of Geotechnical Geoenvironmental Engineering*, 2004, 130(7):696–706.

Chiet K.T.P., Kassim K.A., Chen K.B., et al. Effect of reagents concentration on biocementation of tropical residual soil. In: *Soft Soil Engineering International Conference 2015 (SEIC2015)*. IOP Publishing, 2016.

Cho Y.I., Fan C., Choi B.-G. Theory of electronic anti-fouling technology to control precipitation fouling in heat exchangers. *International Communications in Heat Mass Transfer*, 1997, 24(6):757–770.

Choi S.G., Chu J., Brown R.C., Wang K.J., Wen Z.Y. Sustainable biocement production via microbially-induced calcium carbonate precipitation: Use of limestone and acetic acid derived from pyrolysis of lignocellulosic biomass. *ACS Sustainable Chemistry*, 2017a, 5(6):5183–5190.

Choi S.G., Park S.S., Wu S.F., Chu J. Methods for calcium carbonate content measurement of biocemented soils. *Journal of Materials in Civil Engineering*, 2017b, 29.

Choi S.G., Wang K.J., Chu J. Properties of biocemented, fiber reinforced sand. *Construction and Building Materials*, 2016a, 120:623–629.

Choi S.G., Wu S.F., Chu J. Biocementation for sand using an eggshell as calcium source. *Journal of Geotechnical and Geoenvironmental Engineering*, 2016b, 142(10).

Chou C.-W., Seagren E.A., Aydilek A.H., Lai M. Biocalcification of sand through ureolysis. *Journal of Geotechnical and Geoenvironmental Engineering*, 2011, 137(12):1179–1189.

Christiansen J. The uniformity of application of water by sprinkler systems. *Agricultural Engineering*, 1941, 22:89–92.

Chu J., Ivanov V., Naeimi M., Stabnikov V., Liu H.L. Optimization of calcium-based bioclogging and biocementation of sand. *Acta Geotechnica*, 2013, 9(2):277–285.

Chuo S.C., Mohamed S.F., Setapar S.H.M., et al. Insights into the current trends in the utilization of bacteria for microbially induced calcium carbonate precipitation. *Materials*, 2020, 13(21).

Coman C., Chiriac C.M., Robeson M.S., et al. Structure, mineralogy, and microbial diversity of geothermal spring microbialites associated with a deep oil drilling in Romania. *Frontiers in Microbiology*, 2015, 6:253.

Consoli N.C., Festugato L., Heineck K.S. Strain-hardening behaviour of fibre-reinforced sand in view of filament geometry. *Geosynthetics International*, 2009a, 16(2):109–115.

Consoli N.C., Prietto P.D., Ulbrich L.A. Influence of fiber and cement addition on behavior of sandy soil. *Journal of Geotechnical Geoenvironmental Engineering*, 1998, 124(12):1211–1214.

Consoli N.C., Vendruscolo M.A., Fonini A., Dalla Rosa F. Fiber reinforcement effects on sand considering a wide cementation range. *Geotextiles Geomembranes*, 2009b, 27(3):196–203.

Consoli N.C., Zortéa F., de Souza M., Festugato L. Studies on the dosage of fiber-reinforced cemented soils. *Journal of Materials in Civil Engineering*, 2011, 23(12):1624–1632.

Cornelis G., Johnson C.A., Van Gerven T., Vandecasteele C. Leaching mechanisms of oxyanionic metalloid and metal species in alkaline solid wastes: A review. *Apply Geochemistry*, 2008, 23(5):955–976.

Cosentino D., Chenu C., Le Bissonnais Y. Aggregate stability and microbial community dynamics under drying–wetting cycles in a silt loam soil. *Soil Biology and Biochemistry*, 2006, 38(8):2053–2062.

Costa G., Polettini A., Pomi R., Stramazzo A. Leaching modelling of slurry-phase carbonated steel slag. *Journal of Hazardous Materials*, 2016, 302:415–425.

Couradeau E., Benzerara K., Gerard E., et al. An early-branching microbialite cyanobacterium forms intracellular carbonates. *Science*, 2012, 336(6080):459–62.

Cristelo N., Fernández-Jiménez A., Vieira C., Miranda T., Palomo Á. Stabilisation of construction and demolition waste with a high fines content using alkali activated fly ash. *Construction and Building Materials*, 2018, 170:26–39.

Cui M.J., Zheng J.J., Zhang R.J., Lai H.J., Zhang J. Influence of cementation level on the strength behaviour of bio-cemented sand. *Acta Geotechnica*, 2017, 12(5):971–986.

Danso H., Martinson D.B., Ali M., Williams J.B. Physical, mechanical and durability properties of soil building blocks reinforced with natural fibres. *Construction and Building Materials*, 2015, 101:797–809.

Daskalakis M.I., Rigas F., Bakolas A., et al. Vaterite bio-precipitation induced by bacillus pumilus isolated from a solutional cave in Paiania, Athens, Greece. *International Biodeterioration Biodegradation*, 2015, 99:73–84.

Dassanayake D.T., Oumeraci H. Engineering properties of geotextile sand containers and their effect on hydraulic stability and damage development of low-crested/submerged structures. *The International Journal of Ocean Climate Systems*, 2012, 3(3):135–150.

De Muynck W., Cox K., De Belie N., Verstraete W. Bacterial carbonate precipitation reduces the permeability of cementitious materials. In: *Sustainable Construction Materials and Technologies*. CRC Press, 2020, 411–416.

De Muynck W., De Belie N., Verstraete W. Microbial carbonate precipitation in construction materials: A review. *Ecological Engineering*, 2010a, 36(2):118–136.

De Muynck W., Verbeken K., De Belie N., Verstraete W. Influence of urea and calcium dosage on the effectiveness of bacterially induced carbonate precipitation on limestone. *Ecological Engineering*, 2010b, 36(2):99–111.

DeJong J.T., Fritzges M.B., Nüsslein K. Microbially induced cementation to control sand response to undrained shear. *Journal of Geotechnical Geoenvironmental Engineering*, 2006, 132(11):1381–1392.

DeJong J.T., Mortensen B., Martinez B. *Bio-Soils Interdisciplinary Science and Engineering Initiative*. Final Report on Workshop, National Science Foundation, USA, 2007, 84.

DeJong J.T., Mortensen B.M., Martinez B.C., Nelson D.C. Bio-mediated soil improvement. *Ecological Engineering*, 2010, 36(2):197–210.

DeJong J.T., Soga K., Kavazanjian E., et al. Biogeochemical processes and geotechnical applications: Progress, opportunities and challenges. *Géotechnique*, 2013, 63(4):287–301.

Deng W., Wang Y. Investigating the factors affecting the properties of coral sand treated with microbially induced calcite precipitation. *Advances in Civil Engineering*, 2018, 1–7.

Dhami N.K., Reddy M.S., Mukherjee A. Application of calcifying bacteria for remediation of stones and cultural heritages. *Frontiers in Microbiology*, 2014, 5:304.

Dhami N.K., Reddy M.S., Mukherjee A. Bionnineralization of calcium carbonates and their engineered applications: A review. *Frontiers in Microbiology*, 2013, 4.

Dhami N.K., Reddy M.S., Mukherjee A. Improvement in strength properties of ash bricks by bacterial calcite. *Ecological Engineering*, 2012, 39:31–35.

Dhami N.K., Reddy M.S., Mukherjee A. Significant indicators for biomineralisation in sand of varying grain sizes. *Construction and Building Materials*, 2016, 104:198–207.

Dilrukshi R.A.N., Nakashima K., Kawasaki S. Soil improvement using plant-derived urease-induced calcium carbonate precipitation. *Soils and Foundations*, 2018, 58(4):894–910.

Dudas M. Long-term leachability of selected elements from fly ash. *Environmental Science & Technology*, 1981, 15(7):840–843.

Dunkerley D. Flow threads in surface run-off: Implications for the assessment of flow properties and friction coefficients in soil erosion and hydraulics investigations. *Earth Surface Processes Landforms: The Journal of the British Geomorphological Research Group*, 2004, 29(8):1011–1026.

Dupraz C., Reid R.P., Braissant O., Decho A.W., Norman R.S., Visscher P.T. Processes of carbonate precipitation in modern microbial mats. *Earth-Science Reviews*, 2009, 96(3):141–162.

Dutta R.K., Khatri V.N., Panwar V. Strength characteristics of fly ash stabilized with lime and modified with phosphogypsum. *Journal of Building Engineering*, 2017, 14:32–40.

Dyer M., Viganotti M. Oligotrophic and eutrophic MICP treatment for silica and carbonate sands. *Bioinspired, Biomimetic and Nanobiomaterials*, 2016, 6(3):168–183.

El-Shora H. Properties and immobilization of urease from leaves of chenopodium album (c3). *Botanical Bulletin of Academia Sinica (Taipei)*, 2001, 42.

Engelsen C.J., Van der Sloot H.A., Wibetoe G., Justnes H., Lund W., Stoltenberg-Hansson E. Leaching characterisation and geochemical modelling of minor and trace elements released from recycled concrete aggregates. *Cement and Concrete Research*, 2010, 40(12):1639–1649.

Eyssautier-Chuine S., Marin B., Thomachot-Schneider C., et al. Simulation of acid rain weathering effect on natural and artificial carbonate stones. *Environmental Earth Sciences*, 2016, 75:1–19.

Fang Y., Gu Y., Kang Q., Wen Q., Dai P. Utilization of copper tailing for autoclaved sand–lime brick. *Construction and Building Materials*, 2011, 25(2):867–872.

Farajnia A., Shafaat A., Farajnia S., Sartipipour M., Khodadadi Tirkolaei H. The efficiency of ureolytic bacteria isolated from historical adobe structures in the production of biobricks. *Construction and Building Materials*, 2022, 317.

Fatt I. The network model of porous media. *Transactions of the AIME*, 1956, 207(01):144–181.

Fazelikia S., Abtahi S.A., Kargar M., Jafarinia M. Microbial induced calcite precipitation (MICP) potential of ureolytic bacillus sp. Isolated from the soil of eroded ecosystems for stabilizing and improving the fertility of eroded soils. *Geomicrobiology Journal*, 2023, 40(6):569–581.

Fe Rris F.G., Stehmeler L.G., Kantzas A., Mouritis F.M. Bacteriogenic mineral plugging. *The Journal of Canadian Petroleum Technology*, 1996, 35(8):56–61.

Feng K., Montoya B.M. Influence of confinement and cementation level on the behavior of microbial-induced calcite precipitated sands under monotonic drained loading. *Journal of Geotechnical and Geoenvironmental Engineering*, 2016, 142(1).

Fruchter J.S., Rai D., Zachara J.M. Identification of solubility-controlling solid phases in a large fly ash field lysimeter. *Environmental Science & Technology*, 1990, 24(8):1173–1179.

Fu J., Jiang P., Zong G., Ding J. Experimental study on compressive strength of solidified marinesilty soil mixed with microorganism. *Yangtze River*, 2021, 52(01):167–172+189.

Fujita M., Nakashima K., Achal V., Kawasaki S. Whole-cell evaluation of urease activity of pararhodobacter sp. Isolated from peripheral beachrock. *Biochemical Engineering Journal*, 2017, 124:1–5.

Fujita Y., Ferris F.G., Lawson R.D., Colwell F.S. Calcium carbonate precipitation by ureolytic subsurface bacteria. *Geomicrobiology Journal*, 2000, 17(4):305–318.

Gallagher K.L., Kading T.J., Braissant O., Dupraz C., Visscher P.T. Inside the alkalinity engine: The role of electron donors in the organomineralization potential of sulfate-reducing bacteria. *Geobiology*, 2012, 10(6):518–530.

Garcıa-Sánchez A., Alvarez-Ayuso E. Sorption of Zn, Cd and Cr on calcite. Application to purification of industrial wastewaters. *Minerals Engineering*, 2002, 15(7):539–547.

Gat D., Ronen Z., Tsesarsky M. Soil bacteria population dynamics following stimulation for ureolytic microbial-induced $CaCO_3$ precipitation. *Environmental Science & Technology*, 2016, 50(2):616–624.

Ghavami K., Toledo Filho R.D., Barbosa N.P. Behaviour of composite soil reinforced with natural fibres. *Cement Concrete Composites*, 1999, 21(1):39–48.

Gomez B.W., Dewoolkar M.M., Lens J.E., Benda C.C. Effects of fines content on hydraulic conductivity and shear strength of granular structural backfill. *Transportation Research Record*, 2014, 2462(1):1–6.

Gomez M.G., Anderson C.M., Graddy C.M.R., DeJong J.T., Nelson D.C., Ginn T.R. Large-scale comparison of bioaugmentation and biostimulation approaches for biocementation of sands. *Journal of Geotechnical and Geoenvironmental Engineering*, 2017, 143(5).

Gomez M.G., Martinez B.C., DeJong J.T., et al. Field-scale bio-cementation tests to improve sands. *Proceedings of the Institution of Civil Engineers - Ground Improvement*, 2015, 168(3):206–216.

Gowthaman S., Iki T., Nakashima K., Ebina K., Kawasaki S. Feasibility study for slope soil stabilization by microbial induced carbonate precipitation (MICP) using indigenous bacteria isolated from cold subarctic region. *SN Applied Sciences*, 2019a, 1(11).

Gowthaman S., Mitsuyama S., Nakashima K., Komatsu M., Kawasaki S. Biogeotechnical approach for slope soil stabilization using locally isolated bacteria and inexpensive low-grade chemicals: A feasibility study on Hokkaido expressway soil, Japan. *Soils and Foundations*, 2019b, 59(2):484–499.

Gowthaman S., Nakashima K., Kawasaki S. Freeze-thaw durability and shear responses of cemented slope soil treated by microbial induced carbonate precipitation. *Soils and Foundations*, 2020, 60(4):840–855.

Gyu C.S., Jian C., Kejin W., Zhiyou W. Sustainable biocement production via microbially induced calcium carbonate precipitation: Use of limestone and acetic acid derived from pyrolysis of lignocellulosic biomass. *ACS Sustainable Chemistry & Engineering*, 2017, 5(6):5183–5190.

Hamdan N., Kavazanjian Jr E., O'donnell S. Carbonate cementation via plant derived urease. In: *Proceedings of the 18th International Conference on Soil Mechanics and Geotechnical Engineering*. ISSMGE, 2013, 2–6.

Han Z.G., Zhang J.W., Bian H.L., Yue J.W., Xiao J.Z., Wei Y.Q. Study on liquefaction-resistance performance of MICP-cemented sands: Applying centrifuge shake table tests. *Journal of Geotechnical and Geoenvironmental Engineering*, 2024, 150(8).

Harkes M.P., van Paassen L.A., Booster J.L., Whiffin V.S., Van Loosdrecht M.C. Fixation and distribution of bacterial activity in sand to induce carbonate precipitation for ground reinforcement. *Ecological Engineering*, 2010, 36(2):112–117.

Hazra B., Varma A.K., Bandopadhyay A.K., et al. FTIR, XRF, XRD and SEM characteristics of Permian shales, India. *Journal of Natural Gas Science Engineering*, 2016, 32:239–255.

Heineck K.S., Coop M.R., Consoli N.C. Effect of microreinforcement of soils from very small to large shear strains. *Journal of Geotechnical Geoenvironmental Engineering*, 2005, 131(8):1024–1033.

Hejazi S.M., Sheikhzadeh M., Abtahi S.M., Zadhoush A. A simple review of soil reinforcement by using natural and synthetic fibers. *Construction and Building Materials*, 2012, 30:100–116.

Hua B., Deng B., Thornton E.C., Yang J., Amonette J.E. Incorporation of chromate into calcium carbonate structure during coprecipitation. *Water, Air, & Soil Pollution*, 2007, 179:381–390.

Hucka V., Das B. Brittleness determination of rocks by different methods. *International Journal of Rock Mechanics and Mining Sciences & Geomechanics Abstracts*, 1974, 11:389–392.

Ibrahim M.W., Hamzah A., Jamaluddin N., Ramadhansyah P., Fadzil A. Split tensile strength on self-compacting concrete containing coal bottom ash. *Procedia-Social Behavioral Sciences*, 2015, 195:2280–2289.

ISO I. *10318-1 Geosynthetics—Terms and Definitions*. International Organization for Standardization, 2015.

Izquierdo M., Querol X. Leaching behaviour of elements from coal combustion fly ash: An overview. *International Journal of Coal Geology*, 2012, 94:54–66.

Jabri E., Carr M.B., Hausinger R.P., Karplus P.A. The crystal structure of urease from klebsiella aerogenes. *Science*, 1995, 268(5213):998–1004.

Jain S. Influence of biomass and chemicals on kinetics of ureolysis-based carbonate biomineral precipitation. *Environmental Earth Sciences*, 2024, 83(2).

James R., Kamruzzaman A., Haque A., Wilkinson A. Behaviour of lime–slag-treated clay. *Proceedings of the Institution of Civil Engineers-Ground Improvement*, 2008, 161(4):207–216.

Jeefferie A., Nurul Fariha O., Mohd Warikh A., Yuhazri M., Sihombing H., J R. Preliminary study on the physical and mechanical properties of tapioca starch/sugarcane fiber cellulose composite. *Journal of Engineering Applied Sciences*, 2011, 6(4):1819–6608.

Jiang H., Cai Y., Liu J. Engineering properties of soils reinforced by short discrete polypropylene fiber. *Journal of Materials in Civil Engineering*, 2010, 22(12):1315–1322.

Jiang N.J., Liu R., Du Y.J., Bi Y.Z. Microbial induced carbonate precipitation for immobilizing pb contaminants: Toxic effects on bacterial activity and immobilization efficiency. *Science of the Total Environment*, 2019, 672:722–731.

Jiang N.J., Soga K. The applicability of microbially induced calcite precipitation (MICP) for internal erosion control in gravel-sand mixtures. *Geotechnique*, 2017a, 67(1):42–55.

Jiang N.J., Soga K., Kuo M. Microbially induced carbonate precipitation for seepage-induced internal erosion control in sand-clay mixtures. *Journal of Geotechnical and Geoenvironmental Engineering*, 2017b, 143(3).

Jiang N.-J., Yoshioka H., Yamamoto K., Soga K. Ureolytic activities of a urease-producing bacterium and purified urease enzyme in the anoxic condition: Implication for subseafloor sand production control by microbially induced carbonate precipitation (MICP). *Ecological Engineering*, 2016, 90:96–104.

Jones D. The leaching of major and trace elements from coal ash. *Environmental Aspects of Trace Elements in Coal*, 1995, 221–262.

Jongvivatsakul P., Janprasit K., Nuaklong P., Pungrasmi W., Likitlersuang S. Investigation of the crack healing performance in mortar using microbially induced calcium carbonate precipitation (MICP) method. *Construction and Building Materials*, 2019, 212:737–744.

Jonkers H.M. Bacteria-based self-healing concrete. *Heron*, 2011, 56(1–2):195.

Kahraman S. Correlation of TBM and drilling machine performances with rock brittleness. *Engineering Geology*, 2002, 65(4):269–283.

Kamennaya N., Ajo-Franklin C., Northen T., Jansson C. Cyanobacteria as biocatalysts for carbonate mineralization. *Minerals*, 2012, 2(4):338–364.

Kartal Ş., Aydın Z., Tokalıoğlu Ş. Fractionation of metals in street sediment samples by using the BCR sequential extraction procedure and multivariate statistical elucidation of the data. *Journal of Hazardous Materials*, 2006, 132(1):80–89.

Kerr P., Blevins D., Rapp B., Randall D. Soybean leaf urease: Comparison with seed urease. *Physiologia Plantarum*, 1983, 57(3):339–345.

Khan M.N.H., Amarakoon G.G.N.N., Shimazaki S., Kawasaki S. Coral sand solidification test based on microbially induced carbonate precipitation using ureolytic bacteria. *Materials Transactions*, 2015, 56(10):1725–1732.

Kim A.G., Kazonich G., Dahlberg M. Relative solubility of cations in class f fly ash. *Environmental Science & Technology*, 2003, 37(19):4507–4511.

Kim J.K., Moon Y.H., Eo S.H. Compressive strength development of concrete with different curing time and temperature. *Cement and Concrete Research*, 1998, 28(12):1761–1773.

Knorst M.T., Neubert R., Wohlrab W. Analytical methods for measuring urea in pharmaceutical formulations. *Journal of Pharmaceutical Biomedical Analysis*, 1997, 15(11):1627–1632.

Kohl R. Drop size distributions from medium-sized agricultural sprinklers. *Transaction of the ASAE*, 1974, 17(4):690–693.

Komonweeraket K., Cetin B., Aydilek A.H., Benson C.H., Edil T.B. Effects of pH on the leaching mechanisms of elements from fly ash mixed soils. *Fuel*, 2015a, 140:788–802.

Komonweeraket K., Cetin B., Benson C.H., Aydilek A.H., Edil T.B. Leaching characteristics of toxic constituents from coal fly ash mixed soils under the influence of pH. *Waste Management & Research*, 2015b, 38:174–184.

Komori T., Makishima K. Geometrical expressions of spaces in anisotropic fiber assemblies. *Textile Research Journal*, 1979, 49(9):550–555.

Kong J., hen P., Huang S., Li Q., Xu H. Experimental investigation on solidifying the desert aeolian sand by the combination of MICP and vegetation. *Journal of Zhejiang Sci-Tech University (Natural Science Edition)*, 2019, 41(5):9.

Krishnamurti K., Kate S. Changes in electrical conductivity during bacterial growth. *Nature & Insects*, 1951, 168(4265):170–170.

Kukier U., Ishak C.F., Sumner M.E., Miller W.P. Composition and element solubility of magnetic and non-magnetic fly ash fractions. *Environmental Pollution*, 2003, 123(2):255–266.

Kumar A., Walia B.S., Bajaj A. Influence of fly ash, lime, and polyester fibers on compaction and strength properties of expansive soil. *Journal of Materials in Civil Engineering*, 2007, 19(3):242–248.

Kumari D., Pan X., Lee D.J., Achal V. Immobilization of cadmium in soil by microbially induced carbonate precipitation with exiguobacterium undae at low temperature. *International Biodeterioration Biodegradation*, 2014, 94:98–102.

Kumari D., Qian X.Y., Pan X.L., Achal V., Li Q.W., Gadd G.M. Microbially-induced carbonate precipitation for immobilization of toxic metals. *Advances in Applied Microbiology*, 2016, 94:79–108.

Lagasse P.F., Zevenbergen L., Schall J.D., Clopper P. *Bridge Scour and Stream Instability Countermeasures. Experience, Selection, and Design Guidance*. United States Federal Highway Administration. Office of Bridge Technology, 2001.

Lambert S.E., Randall D.G. Manufacturing bio-bricks using microbial induced calcium carbonate precipitation and human urine. *Water Research*, 2019, 160:158–166.

Landers M.N. *Bridge Scour Data Management*. USGS Staff-Published Research, 1992, 141.

Lassu T., Seeger M., Peters P., Keesstra S.D. The Wageningen rainfall simulator: Set-up and calibration of an indoor nozzle-type rainfall simulator for soil erosion studies. *Land Degradation Development*, 2015, 26(6):604–612.

Li C., Wang Y., Zhou T., et al. Sulfate acid corrosion mechanism of biogeomaterial based on MICP technology. *Journal of Materials in Civil Engineering*, 2019, 31(7).

Li C., Yao D., Liu S., et al. Improvement of geomechanical properties of bio-remediated aeolian sand. *Geomicrobiology Journal*, 2017, 35(2):132–140.

Li G., Yang M., Meng Y., et al. The assessment of correlation between rock drillability and mechanical properties in the laboratory and in the field under different pressure conditions. *Journal of Natural Gas Science Engineering*, 2016a, 30:405–413.

Li J., Chen C.M. *Citespace: Text Mining and Visualization in Scientific Literature*. Capital University of Economics and Business Press, 2016.

Li L., Amini F., Zhao Q., et al. Development of a flexible mold for bio-mediated soil materials. In: *IFCEE 2015*. ASCE, 2015a, 2339–2348.

Li L., Shao W., Li Y., Cetin B. Effects of climatic factors on mechanical properties of cement and fiber reinforced clays. *Geotechnical and Geological Engineering*, 2015b, 33:537–548.

Li M.D., Li L., Ogbonnaya U., Wen K.J., Tian A., Amini F. Influence of fiber addition on mechanical properties of MICP-treated sand. *Journal of Materials in Civil Engineering*, 2015c, 28(4):04015166.

Li M.D., Wen K., Li Y., Zhu L. Impact of oxygen availability on microbially induced calcite precipitation (MICP) treatment. *Geomicrobiology Journal*, 2018a, 35(1):15–22.

Li M.M., Fang C., Kawasaki S., Achal V. Fly ash incorporated with biocement to improve strength of expansive soil. *Scientific Reports*, 2018b, 8(1):2565.

Li M.M., Fu Q.L., Zhang Q.Z., Achal V., Kawasaki S. Bio-grout based on microbially induced sand solidification by means of asparaginase activity. *Scientific Reports*, 2015d, 5:16128.

Li W., Chen W.S., Zhou P.P., Zhu S.L., Yu L.J. Influence of initial calcium ion concentration on the precipitation and crystal morphology of calcium carbonate induced by bacterial carbonic anhydrase. *Chemical Engineering Journal*, 2013, 218:65–72.

Li X.W., Tao J.L., van Paassen L.A. Reactive transport modeling of microbial-induced calcite precipitation treatment through shallow underwater injection. *Computers and Geotechnics*, 2024, 174.

Li Y., Richardson J.B., Niu X., Jackson O.J., Laster J.D., Walker A.K. Dynamic leaching test of personal computer components. *Journal of Hazardous Materials*, 2009, 171(1–3):1058–1065.

Li Y., Wen K.J., Li L., Huang W., Bu C.M., Amini F. Experimental investigation on compression resistance of bio-bricks. *Construction and Building Materials*, 2020, 265.

Li Y.F., Lu X.Y., Liu S.H., et al. The progress and trend of microbially induced carbonate precipitation (MICP) research: A bibliometric analysis. *Environmental Earth Sciences*, 2023, 82(23).

Lin H., Suleiman M.T., Brown D.G., Kavazanjian E. Mechanical behavior of sands treated by microbially induced carbonate precipitation. *Journal of Geotechnical and Geoenvironmental Engineering*, 2016, 142(2).

Ling H., Qian C.X. Effects of self-healing cracks in bacterial concrete on the transmission of chloride during electromigration. *Construction and Building Materials*, 2017, 144:406–411.

Liu L., Shen Y., Liu H., Chu J. Application of bio-cement in erosion control of levees. *Geotechnical Mechanics*, 2016, 37(12).

Liu S., Du K., Huang W., Wen K., Amini F., Li L. Improvement of erosion-resistance of bio-bricks through fiber and multiple MICP treatments. *Construction and Building Materials*, 2021, 271.

Liu S., Du K., Wen K., Armwood-Gordon C., Li L., Tang Q. Influence of rainfall-induced erosion on the stability of sandy slopes treated by MICP. *Advances in Civil Engineering*, 2022, 2022:1–10.

Liu S., Wen K., Armwood C., et al. Enhancement of MICP-treated sandy soils against environmental deterioration. *Journal of Materials in Civil Engineering*, 2019, 31(12): 04019294.1–04019294.13.

Loste E., Wilson R.M., Seshadri R., Meldrum F.C. The role of magnesium in stabilising amorphous calcium carbonate and controlling calcite morphologies. *Journal of Crystal Growth*, 2003, 254(1–2):206–218.

Luis A., Deng L., Shao L., Li H.A. Triaxial behaviour and image analysis of Edmonton clay treated with cement and fly ash. *Construction and Building Materials*, 2019, 197:208–219.

Luo M., Bai J., Jing K., Ding Z., Yang D., Qian C. Self-healing of early-age cracks in cement mortars with artificial functional aggregates. *Construction and Building Materials*, 2021, 272:139–152.

Mahawish A., Bouazza A., Gates W.P. Effect of particle size distribution on the bio-cementation of coarse aggregates. *Acta Geotechnica*, 2017, 13(4):1019–1025.

Mahedi M., Cetin B., Dayioglu A.Y. Leaching behavior of aluminum, copper, iron and zinc from cement activated fly ash and slag stabilized soils. *Waste Management & Research*, 2019, 95:334–355.

Maria C.V., Samucle D.B., Carmine F., Massimo V. Usage of infinitesimals in the Menger's Sponge model of porosity. *Applied Mathematics and Computation*, 2012, 218(16):8187–8195. https://doi.org/10.1016/j.amc.2011.06.013.

Martinez B.C., DeJong J.T., Ginn T.R., et al. Experimental optimization of microbial-induced carbonate precipitation for soil improvement. *Journal of Geotechnical and Geoenvironmental Engineering*, 2013, 139(4):587–598.

McLean J.E., Bledsoe B.E., States. U. *Environmental Protection Agency. Technology Innovation Office Superfund Technology Support Center for Ground Water*. Technology Innovation Office, Office of Solid Waste and Emergency Response, 1992.

Métayer-Levrel G.L., Castanier S., Orial G., Loubie're J.F., Perthuisot J.P. Applications of bacterial carbonatogenesis to the protection and regeneration of limestones in buildings and historic patrimony. *Sedimentary Geology*, 1999, 126(1–4):25–34.

Millennium C., Innovation O., Engineering C., Resources B., Studies D., Council N. *Geological and Geotechnical Engineering in the New Millennium: Opportunities for Research and Technological Innovation*. 2006.

Mississippi Department of Environmental Quality (MDEQ). *Regulations for the Beneficial Use of Nonhazardous Solid Wastes*. Jackson, Mississippi, 2005.

Mitchell J.K., Santamarina J.C. Biological considerations in geotechnical engineering. *Journal of Geotechnical and Geoenvironmental Engineering*, 2005, 131(10):1222–1233.

Moazed H., Bavi A., Boroomand-Nasab S., Naseri A., Albaji M. Effects of climatic and hydraulic parameters on water uniformity coefficient in solid set systems. *Journal of Applied Sciences*, 2010, 10(16):1792–1796.

Mombelli D., Mapelli C., Barella S., Di Cecca C., Le Saout G., Garcia-Diaz E. The effect of chemical composition on the leaching behaviour of electric arc furnace (EAF) carbon steel slag during a standard leaching test. *Journal of Environmental Chemical Engineering*, 2016, 4(1):1050–1060.

Montoya B.M., DeJong J.T. Stress-strain behavior of sands cemented by microbially induced calcite precipitation. *Journal of Geotechnical and Geoenvironmental Engineering*, 2015, 141(6).

Morales L., Garzón E., Romero E., Sánchez-Soto P.J. Microbiological induced carbonate ($CaCO_3$) precipitation using clay phyllites to replace chemical stabilizers (cement or lime). *Applied Clay Science*, 2019, 174:15–28.

Moreno N., Querol X., Andrés J., et al. Physico-chemical characteristics of European pulverized coal combustion fly ashes. *Fuel*, 2005, 84(11):1351–1363.

Mori D., Uday K.V. A review on qualitative interaction among the parameters affecting ureolytic microbial-induced calcite precipitation. *Environmental Earth Sciences*, 2021, 80(8).

Mörsdorf G., Kaltwasser H. Ammonium assimilation in proteus vulgaris, Bacillus pasteurii, and Sporosarcina ureae. *Archives of Microbiology*, 1989, 152:125–131.

Mortensen B., Haber M., DeJong J., Caslake L., Nelson D. Effects of environmental factors on microbial induced calcium carbonate precipitation. *Journal of Applied Microbiology*, 2011, 111(2):338–349.

Mugwar A.J., Harbottle M.J. Toxicity effects on metal sequestration by microbially-induced carbonate precipitation. *Journal of Hazardous Materials*, 2016, 314:237–248.

Muthukumaran A., Ilamparuthi K. Laboratory studies on geotextile filters as used in geotextile tube dewatering. *Geotextiles Geomembranes*, 2006, 24(4):210–219.

Mwandira W., Mavroulidou M., Joshi S., Gunn M.J. Fruit and vegetable waste used as bacterial growth media for the biocementation of two geomaterials. *Science of the Total Environment*, 2024, 947.

Mwandira W., Nakashima K., Kawasaki S. Bioremediation of lead-contaminated mine waste by pararhodobacter sp. Based on the microbially induced calcium carbonate precipitation technique and its effects on strength of coarse and fine grained sand. *Ecological Engineering*, 2017, 109:57–64.

Nagy B., Kustermann A. Rehabilitation of porous building components and masonry by MICP injection method. *Buildings*, 2023, 13(5).

Namikawa T., Koseki J. Evaluation of tensile strength of cement-treated sand based on several types of laboratory tests. *Soils Foundations*, 2007, 47(4):657–674.

Naveed M., Duan J., Uddin S., Suleman M., Hui Y., Li H. Application of microbially induced calcium carbonate precipitation with urea hydrolysis to improve the mechanical properties of soil. *Ecological Engineering*, 2020, 153.

Nemati M., Greene E., Voordouw G. Permeability profile modification using bacterially formed calcium carbonate: Comparison with enzymic option. *Process Biochemistry*, 2005, 40(2):925–933.

Okwadha G.D., Li J. Optimum conditions for microbial carbonate precipitation. *Chemosphere*, 2010, 81(9):1143–1148.

Oliveira P.J.V., Freitas L.D., Carmona J.P.S.F. Effect of soil type on the enzymatic calcium carbonate precipitation process used for soil improvement. *Journal of Materials in Civil Engineering*, 2016, 29(4):1–7.

Oliveira R.C., Hammer P., Guibal E., Taulemesse J.-M., Garcia Jr O. Characterization of metal–biomass interactions in the lanthanum (iii) biosorption on sargassum sp. Using SEM/EDX, FTIR, and XPS: Preliminary studies. *Chemical Engineering Journal*, 2014, 239:381–391.

Omoregie A.I., Palombo E.A., Nissom P.M. Bioprecipitation of calcium carbonate mediated by ureolysis: A review. *Environmental Engineering Research*, 2021, 26(6).

Omoregie A.I., Palombo E.A., Ong D.E., Nissom P.M. Biocementation of sand by Sporosarcina pasteurii strain and technical-grade cementation reagents through surface percolation treatment method. *Construction and Building Materials*, 2019, 228:116828.

Patrick Jr W.H., Mahapatra I.C. Transformation and availability to rice of nitrogen and phosphorus in waterlogged soils. *Advances in Agronomy*, 1968, 20:323–359.

Phillips A.J., Gerlach R., Lauchnor E., Mitchell A.C., Cunningham A.B., Spangler L. Engineered applications of ureolytic biomineralization: A review. *Biofouling*, 2013a, 29(6):715–33.

Phillips A.J., Lauchnor E., Eldring J., et al. Potential CO2 leakage reduction through biofilm-induced calcium carbonate precipitation. *Environmental Science & Technology*, 2013b, 47(1):142–149.

Porter H., Blake J., Dhami N.K., Mukherjee A. Rammed earth blocks with improved multifunctional performance. *Cement and Concrete Composites*, 2018, 92:36–46.

Qian C.X., Yu X.N., Wang X. A study on the cementation interface of bio-cement. *Materials Characterization*, 2018, 136:122–127.

Qian C.X., Zheng T.W., Rui Y.F. Living concrete with self-healing function on cracks attributed to inclusion of microorganisms: Theory, technology and engineering applications-a review. *Science China-Technological Sciences*, 2021, 64(10):2067–2083.

Qian X., Fang C., Huang M., Achal V. Characterization of fungal-mediated carbonate precipitation in the biomineralization of chromate and lead from an aqueous solution and soil. *Journal of Cleaner Production*, 2017, 164:198–208.

Qiu J.S., Tng D.Q.S., Yang E.H. Surface treatment of recycled concrete aggregates through microbial carbonate precipitation. *Construction and Building Materials*, 2014, 57:144–150.

Querol X., Umana J., Alastuey A., Ayora C., Lopez-Soler A., Plana F. Extraction of soluble major and trace elements from fly ash in open and closed leaching systems. *Fuel*, 2001, 80(6):801–813.

Rahardjo H. *Effect of Rainfall on the Slope Stability of Residual Soil in Singapore*. Nanyang Technological University, 1999.

Rahim S.-S., Seyed M.A.Z., Ali N., Brendan C.O.K. Improving sand with microbial-induced carbonate precipitation. *Proceedings of the Institution of Civil Engineers - Ground Improvement*, 2015, 168(3):217–230.

Rahman M., Rashid M., Hossain M.A., Hasan M., Hasan M. Performance evaluation of bamboo reinforced concrete beam. *International Journal of Engineering Technology*, 2011, 11(4):142–146.

Rahman M.M., Hora R.N., Ahenkorah I., Beecham S., Karim M.R., Iqbal A. State-of-the-art review of microbial-induced calcite precipitation and its sustainability in engineering applications. *Sustainability*, 2020, 12(15).

Rebata-Landa V. *Microbial Activity in Sediments: Effects on Soil Behavior*. PhD Dissertation, Georgia Institute of Technology, 2007.

Riding R.E., Awramik S.M. *Microbial Sediments*. Springer Berlin Heidelberg, 2000.

Rowshanbakht K., Khamehchiyan M., Sajedi R.H., Nikudel M.R. Effect of injected bacterial suspension volume and relative density on carbonate precipitation resulting from microbial treatment. *Ecological Engineering*, 2016, 89:49–55.

Salifu E., MacLachlan E., Iyer K.R., Knapp C.W., Tarantino A. Application of microbially induced calcite precipitation in erosion mitigation and stabilisation of sandy soil foreshore slopes: A preliminary investigation. *Engineering Geology*, 2016, 201:96–105.

Samonin V., Elikova E. A study of the adsorption of bacterial cells on porous materials. *Microbiology (Beijing)*, 2004, 73(6).

Sarda D., Choonia H.S., Sarode D., Lele S. Biocalcification by Bacillus pasteurii urease: A novel application. *Journal of Industrial Microbiology Biotechnology*, 2009, 36(8):1111–1115.

Shao W., Cetin B., Li Y., Li J., Li L. Experimental investigation of mechanical properties of sands reinforced with discrete randomly distributed fiber. *Geotechnical and Geological Engineering*, 2014, 32:901–910.

Sharma M., Satyam N., Reddy K.R. State of the art review of emerging and biogeotechnical methods for liquefaction mitigation in sands. *Journal of Hazardous Toxic and Radioactive Waste*, 2021, 25(1).

Shields A. *Anwendung der aehnlichkeitsmechanik und der turbulenzforschung auf die geschiebebewegung*. PhD Thesis, Technical University Berlin, 1936.

Shiraishi F. Chemical conditions favoring photosynthesis-induced $CaCO_3$ precipitation and implications for microbial carbonate formation in the ancient ocean. *Geochimica et Cosmochimica Acta*, 2012, 77:157–174.

Silva R., Negri R., de Mattos Vidal D. A new image-based technique for measuring pore size distribution of nonwoven geotextiles. *Geosynthetics International*, 2019, 26(3): 261–272.

Small H. Co-citation in the scientific literature: A new measure of the relationship between two documents. *Journal of the American Society for Information Science and Technology*, 1973, 24(4):265–269.

Sondi I., Salopek-Sondi B. Influence of the primary structure of enzymes on the formation of $CaCO_3$ polymorphs: A comparison of plant (canavalia e nsiformis) and bacterial (bacillus p asteurii) ureases. *Langmuir*, 2005, 21(19):8876–8882.

Soon N.W., Lee L.M., Khun T.C., Ling H.S. Factors affecting improvement in engineering properties of residual soil through microbial-induced calcite precipitation. *Journal of Geotechnical and Geoenvironmental Engineering*, 2014, 140(5):04014006.

Sridhar S., Bhatt N., Suraishkumar G.K. Mechanistic insights into ureolysis mediated calcite precipitation. *Biochemical Engineering Journal*, 2021, 176.

Steindl F.R., Baldermann A., Galan I., et al. Chemical resistance of eco-concrete–experimental approach on ca-leaching and sulphate attack. *Construction and Building Materials*, 2019, 223:55–68.

Stocks Fischer S., Galinat J.K., Bang S.S. Microbiological precipitation of CaCO3. *Soil Biology and Biochemistry*, 1999, 31(11):1563–1571.

Stuckrath C., Serpell R., Valenzuela L.M., Lopez M. Quantification of chemical and biological calcium carbonate precipitation: Performance of self-healing in reinforced mortar containing chemical admixtures. *Cement and Concrete Composites*, 2014, 50:10–15.

Su P., Zhang J., Li Y. Investigation of chemical associations and leaching behavior of heavy metals in sodium sulfide hydrate stabilized stainless steel pickling sludge. *Process Safety Environmental Protection*, 2019, 123:79–86.

Su Y.L., Zheng T.W., Qian C.X. Application potential of bacillus megaterium encapsulated by low alkaline sulphoaluminate cement in self-healing concrete. *Construction and Building Materials*, 2021, 273.

Sun X.H., Miao L.C., Chen R.F., Wang H.X., Xia J.X. Surface rainfall erosion resistance and freeze-thaw durability of bio-cemented and polymer-modified loess slopes. *Journal of Environmental Management*, 2022, 301.

Sun X.H., Miao L.C., Tong T.Z., Wang C.C. Study of the effect of temperature on microbially induced carbonate precipitation. *Acta Geotechnica*, 2018, 14(3):627–638.

Tang C., Shi B., Gao W., Chen F., Cai Y. Strength and mechanical behavior of short polypropylene fiber reinforced and cement stabilized clayey soil. *Geotextiles Geomembranes*, 2007, 25(3):194–202.

Tariq K., Maki T. Mechanical behaviour of cement-treated sand. *Construction and Building Materials*, 2014, 58:54–63.

Terzis D., Bernier-Latmani R., Laloui L. Fabric characteristics and mechanical response of bio-improved sand to various treatment conditions. *Géotechnique Letters*, 2016, 6(1):50–57.

Terzis D., Laloui L. A decade of progress and turning points in the understanding of bio-improved soils: A review. *Geomechanics for Energy and the Environment*, 2019, 19.

Tiano P., Cantisani E., Sutherland I., Paget J.M. Biomediated reinforcement of weathered calcareous stones. *Journal of Cultural Heritage*, 2006, 7(1):49–55.

Tong H., Ma W., Wang L., Wan P., Hu J., Cao L. Control over the crystal phase, shape, size and aggregation of calcium carbonate via a l-aspartic acid inducing process. *Biomaterials*, 2004, 25(17):3923–3929.

U.S. Department of Transportation. *User Guidelines for Waste and Byproduct Materials in Pavement Construction*. Federal Highway Administration Research and Technology, 2016, FHWA-RD-97–148.

U.S. Environmental Protection Agency (USEPA). *Method 9045D: Soil and Waste pH*. SW 846, Office of Solid Waste and Emergency Response. USEPA, 1986.

USEPA. *Hazardous Waste Management System; Identification and Listing of Hazardous Waste; Toxicity Characteristic Revisions*. USEPA, 1990.

USEPA Standard Method 1311. *Toxicity Leaching Characteristic Leaching Procedure*. USEPA, 1992a.

USEPA Standard Method 1312. *Synthetic Precipitation Leaching Procedure*. USEPA, 1992b.

USEPA Standard Method 1312. *Synthetic Precipitation Leaching Procedure*. USEPA, 1994.

USEPA Standard Method 1313. *Liquid-Solid Partitioning as a Function of Extract pH Using a Parallel Batch Extraction Procedure*. USEPA, 2012.

USEPA Standard Method 3050B. *Acid Digestion of Sediments, Sludges, and Soils*. USEPA, 1996.

Vafai K. *Handbook of Porous Media*. CRC Press, 2005.

Van Dijk A., Bruijnzeel L., Rosewell C. Rainfall intensity–kinetic energy relationships: A critical literature appraisal. *Journal of Hydrology and Hydromechanics*, 2002, 261(1–4):1–23.

van Paassen L.A. *Biogrout, Ground Improvement by Microbial Induced Carbonate Precipitation*. Delft University of Technology, 2009.

van Paassen L.A., Daza C.M., Staal M., Sorokin D.Y., van der Zon W., van Loosdrecht M.C.M. Potential soil reinforcement by biological denitrification. *Ecological Engineering*, 2010a, 36(2):168–175.

van Paassen L.A., Ghose R., van der Linden T.J.M., van der Star W.R.L., van Loosdrecht M.C.M. Quantifying biomediated ground improvement by ureolysis: Large-scale biogrout experiment. *Journal of Geotechnical and Geoenvironmental Engineering*, 2010b, 136(12):1721–1728.

Vita M.C., De Bartolo S., Fallico C., Veltri M. Usage of infinitesimals in the Menger's sponge model of porosity. *Applied Mathematics and Computation*, 2012, 218(16):8187–8195.

Walker P. *The Australian Earth Building Handbook: Hb 195, Standard Australia International Ltd*. New South Wales Department of Agriculture Science Bulletin, 2002.

Wang G., Xu H., Ding X., Gao Y., Chen P., Hu X. Microbial induced solidification of municipal solid waste incineration fly ash. In: *Proceedings of the 8th International Congress on Environmental Geotechnics Volume 3: Towards a Sustainable Geoenvironment*. Springer, 2019, 363–368.

Wang J.Y., Soens H., Verstraete W., De Belie N. Self-healing concrete by use of microencapsulated bacterial spores. *Cement and Concrete Research*, 2014, 56:139–152.

Wang S., Wang X.B., Han X.G., Deng Y. Higher precipitation strengthens the microbial interactions in semi-arid grassland soils. *Global Ecology and Biogeography*, 2018a, 27(5):570–580.

Wang Y., Ding W., Zhang J., et al. Comparative study of MICP repair fluid and traditional concrete crack repair materials. *Hina Concrete and Cement Products*, 2018b, (5):10–14.

Warren L.A., Maurice P.A., Parmar N., Ferris F.G. Microbially mediated calcium carbonate precipitation: Implications for interpreting calcite precipitation and for solid-phase capture of inorganic contaminants. *Geomicrobiology Journal*, 2001, 18(1):93–115.

Wen K., Li Y., Amini F., Li L. Impact of bacteria and urease concentration on precipitation kinetics and crystal morphology of calcium carbonate. *Acta Geotechnica*, 2019, 15(1):17–27.

Wen K.J., Bu C.M., Liu S.H., Li Y., Li L. Experimental investigation of flexure resistance performance of bio-beams reinforced with discrete randomly distributed fiber and bamboo. *Construction and Building Materials*, 2018a, 176(10):241–249.

Wen K.J., Li Y., Liu S.H., Bu C.M., Li L. Development of an improved immersing method to enhance microbial induced calcite precipitation treated sandy soil through multiple treatments in low cementation media concentration. *Geotechnical and Geological Engineering*, 2018b, 37(2):1015–1027.

Westoby M.J., Brasington J., Glasser N.F., Hambrey M.J., Reynolds J.M. 'Structure-from-motion' photogrammetry: A low-cost, effective tool for geoscience applications. *Geomorphology*, 2012, 179:300–314.

Whiffin V.S. *Microbial* $CaCO_3$ *Precipitation for the Production of Biocement*. Murdoch University, 2004.

Whiffin V.S., Paassen L.A.V., Harkes M.P. Microbial carbonate precipitation as a soil improvement technique. *Geomicrobiology Journal*, 2007, 24(5):417–423.

White C., Wilkinson S.C., Gadd G.M. The role of microorganisms in biosorption of toxic metals and radionuclides. *International Biodeterioration Biodegradation*, 1995, 35(1–3):17–40.

Wiktor V., Jonkers H.M. Quantification of crack-healing in novel bacteria-based self-healing concrete. *Cement and Concrete Composites*, 2011, 33(7):763–770.

Wu C. Towards linear-time incremental structure from motion. In: *2013 International Conference on 3D Vision-3DV 2013*. IEEE, 2013, 127–134.

Wu C., Agarwal S., Curless B., Seitz S.M. Multicore bundle adjustment. In: *CVPR 2011*. IEEE, 2011, 3057–3064.

Xiao P., Liu H.L., Xiao Y., Stuedlein A.W., Evans T.M. Liquefaction resistance of bio-cemented calcareous sand. *Soil Dynamics and Earthquake Engineering*, 2018, 107:9–19.

Xiao Y., He X., Wu W., et al. Kinetic biomineralization through microfluidic chip tests. *Acta Geotechnica*, 2021a, 16(10):3229–3237.

Xiao Y., Stuedlein A.W., Ran J.Y., et al. Effect of particle shape on strength and stiffness of biocemented glass beads. *Journal of Geotechnical and Geoenvironmental Engineering*, 2019, 145(11).

Xiao Y., Wang Y., Wang S., et al. Homogeneity and mechanical behaviors of sands improved by a temperature-controlled one-phase MICP method. *Acta Geotechnica*, 2021b, 16(5):1417–1427.

Xiao Y., Zhang Z.C., Stuedlein A.W., Evans T.M. Liquefaction modeling for biocemented calcareous sand. *Journal of Geotechnical and Geoenvironmental Engineering*, 2021c, 147(12).

Xie S., Qi L., Zhou D. Investigation of the effects of acid rain on the deterioration of cement concrete using accelerated tests established in laboratory. *Atmospheric Environment*, 2004, 38(27):4457–4466.

Xie Y.H., Tang C.S., YIin L.Y., et al. Water stability improvement of clayey soil based on microbial induced calcite precipitation. *Journal of Zhejiang University (Engineering Science)*, 2019, 53(8):10.

Xu J., Wang X., Wang B. Biochemical process of ureolysis-based microbial CaCO3 precipitation and its application in self-healing concrete. *Applied Microbiology and Biotechnology*, 2018, 102(7):3121–3132.

Xu J.M., Chen Z.T., Cheng F., Liu Z.Q., Zheng Y.G. Exploring a cellulose-immobilized bacteria for self-healing concrete via microbe-induced calcium carbonate precipitation. *Journal of Building Engineering*, 2024, 95.

Xu X., Han J.T., Cho K. Formation of amorphous calcium carbonate thin films and their role in biomineralization. *Chemistry of Materials*, 2004, 16(9):1740–1746.

Xu X., Li Q., Lai Y., Pang W., Zhang R. Effect of moisture content on mechanical and damage behavior of frozen loess under triaxial condition along with different confining pressures. *Cold Regions Science and Technology*, 2019, 157:110–118.

Yang W.S., Xu L., Wang Z., et al. Synchronous removal of ammonia nitrogen, phosphate, and calcium by heterotrophic nitrifying strain pseudomonas sp. Y1 based on microbial induced calcium precipitation. *Bioresource Technology*, 2022, 363.

Yang Y., Chu J., Cao B., Liu H., Cheng L. Biocementation of soil using non-sterile enriched urease-producing bacteria from activated sludge. *Journal of Cleaner Production*, 2020, 262:121315.

Yasuhara H., Hayashi K., Okamura M. Evolution in mechanical and hydraulic properties of calcite-cemented sand mediated by biocatalyst. *Geo-frontiers 2011: Advances in Geotechnical Engineering*, 2011, 3984–3992.

Yasuhara H., Neupane D., Hayashi K., Okamura M. Experiments and predictions of physical properties of sand cemented by enzymatically-induced carbonate precipitation. *Soils Foundations*, 2012, 52(3):539–549.

Yetimoglu T., Salbas O. A study on shear strength of sands reinforced with randomly distributed discrete fibers. *Geotextiles Geomembranes*, 2003, 21(2):103–110.

Yu X.N., Qian C.X., Xue B. Loose sand particles cemented by different bio-phosphate and carbonate composite cement. *Construction and Building Materials*, 2016, 113:571–578.

Yun H.-D. Flexural behavior and crack-damage mitigation of plain concrete beam with a strain-hardening cement composite (SHCC) layer at tensile region. *Composites Part B: Engineering*, 2013, 45(1):377–387.

Zaimoglu A.S., Yetimoglu T. Strength behavior of fine grained soil reinforced with randomly distributed polypropylene fibers. *Geotechnical and Geological Engineering*, 2012, 30:197–203.

Zhang J., Su P., Li L. Bioremediation of stainless steel pickling sludge through microbially induced carbonate precipitation. *Chemosphere*, 2022, 298:134213.

Zhang J., Su P., Li L. Microbial induced carbonate precipitation modified steel slag: Mechanical improvement and erosion resistance to sulfate attack. *Journal of Cleaner Production*, 2023, 405.

Zhang J., Su P., Li Y., Li L. Environmental investigation of bio-modification of steel slag through microbially induced carbonate precipitation. *Journal of Environmental Sciences (China)*, 2021a, 101:282–292.

Zhang J., Su P., Wen K., Li Y., Li L. Mechanical performance and environmental effect of coal fly ash on MICP-induced soil improvement. *KSCE Journal of Civil Engineering*, 2020, 24(11):3189–3201.

Zhang J., Wen K., Li L. Bio-modification of coal fly ash using urease-producing bacteria. *Fuel*, 2021b, 286.

Zhang Y., Cetin B., Likos W., Edil T. Impacts of pH on leaching potential of elements from MSW incineration fly ash. *Fuel*, 2016, 184:815–825.

Zhao Q., Li L., Li C., Li M., Amini F., Zhang H. Factors affecting improvement of engineering properties of MICP-treated soil catalyzed by bacteria and urease. *Journal of Materials in Civil Engineering*, 2014a, 26(12):04014094.

Zhao Q., Li L., Li C., Zhang H., Amini F. A full contact flexible mold for preparing samples based on microbial-induced calcite precipitation technology. *Geotechnical Testing Journal*, 2014b, 37(5).

Zheng T.W., Su Y.L., Qian C.X., Zhou H.Y. Low alkali sulpho-aluminate cement encapsulated microbial spores for self-healing cement-based materials. *Biochemical Engineering Journal*, 2020a, 163.

Zheng T.W., Su Y.L., Zhang X., Zhou H.Y., Qian C.X. Effect and mechanism of encapsulation-based spores on self-healing concrete at different curing ages. *ACS Applied Materials & Interfaces*, 2020b, 12(47):52415–52432.

Zhu T., Dittrich M. Carbonate precipitation through microbial activities in natural environment, and their potential in biotechnology: A review. *Frontiers in Bioengineering and Biotechnology*, 2016, 4:4.

Zhu Y.J., Kim M.C., Chen C.M. An investigation of the intellectual structure of opinion mining research. *Information Research-An International Electronic Journal*, 2017, 22(1).

Index